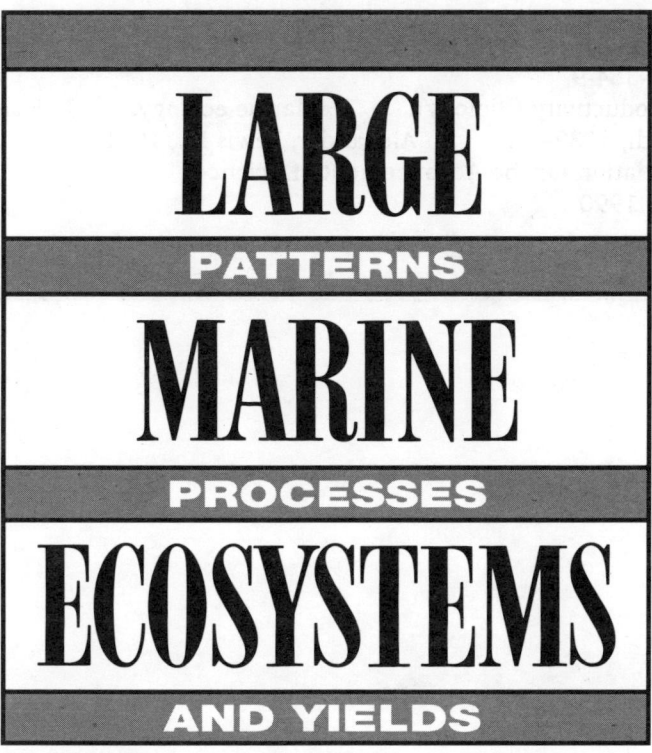

LARGE MARINE ECOSYSTEMS

PATTERNS, PROCESSES AND YIELDS

Edited by *Kenneth Sherman,*
Lewis M. Alexander and
Barry D. Gold

AMERICAN ASSOCIATION FOR THE ADVANCEMENT OF SCIENCE

Library of Congress Cataloging-in-Publication Data

Large marine ecosystems: Patterns, processes, and yields / edited by Kenneth Sherman, Lewis M. Alexander, and Barry D. Gold
    p.   cm.
    Includes index.
    ISBN 0-87168-384-9
    1. Primary productivity (Biology).   2. Marine ecology.   3. Fishery resources. I. Sherman, Kenneth, 1932– .   II. Alexander, Lewis M., 1921– .   III. Gold, Barry D. IV. American Association for the Advancement of Science.
QH541.5.S3P38   1990
574.5'—dc20   90-38263
                                                                                                                       CIP

Publication No. 90-30S

© 1990 by the American Association for the Advancement of Science
1333 H Street, N.W., Washington, D.C. 20005

Contents

Preface ... v

Perspective: Large Marine Ecosystems ... vii
    Kenneth Sherman and Barry D. Gold

Contributors ... xiii

Part One: Perturbations and Yields of Large Marine Ecosystems

Introduction ... 3
    Kenneth Sherman

1. The Weddell Sea: A High Polar Ecosystem ... 5
    Gotthilf Hempel

2. Environmental Influence on Recruitment and Biomass Yields in the Norwegian Sea Ecosystem ... 19
    Bjornar Ellertsen, Petter Fossum, Per Solemdal, Svein Sundby, Snorre Tilseth

3. Fluctuation in the Cod Biomass of the West Greenland Sea Ecosystem in Relation to Climate ... 36
    Holger Hovgård and Erik Buch

4. The Caribbean Sea: A Large Marine Ecosystem in Crisis ... 44
    William J. Richards and James A. Bohnsack

5. Productivity and Fisheries Potential of the Banda Sea Ecosystem ... 54
    Jenne J. Zijlstra and Martien A. Baars

Part Two: Biodynamics of Large Marine Ecosystems

Introduction ... 69
    Brian J. Rothschild

6. Biodynamics of the Sea: Preliminary Observations on High Dimensionality and the Effects of Physics on Predator-Prey Interrelationships ... 71
    Brian J. Rothschild and Thomas R. Osborn

7. Physical-Optical-Biological Scales Relevant to Recruitment in Large Marine Ecosystems ... 82
    Thomas D. Dickey

8. Direct Simulation of the Effect of Turbulence on Planktonic Contact Rates ... 99
    Thomas Osborn, Hidekatsu Yamazaki, Kyle Squires

9. Application of Molecular Techniques to the Study of Marine
   Recruitment Problems   104
   Dennis A. Powers, Fred W. Allendorf, Thomas T. Chen

10. Application of Image Analysis in Demographic Studies of
    Marine Zooplankton in Large Marine Ecosystems   122
    Mark S. Berman

11. Growth, Survival, and Recruitment in Large Marine Ecosystems 132
    Geoffrey C. Laurence

12. Perspectives on Larval Fish Ecology and Recruitment Processes:
    Probing the Scales of Relationships   151
    Christopher T. Taggart and Kenneth T. Frank

Part Three: Theory and Management of Large Marine Ecosystems

Introduction   167
   Lewis M. Alexander and Barry D. Gold

13. Scaling Pattern and Process in Marine Ecosystems   169
    Robert E. Ricklefs

14. Physical and Biological Scales and the Modelling of Predator-Prey
    Interactions in Large Marine Ecosystems   179
    Simon A. Levin

15. Biomass Potential of Large Marine Ecosystems: A Systems Approach 188
    Nicholas J. Bax and Taivo Laevastu

16. Productivity, Perturbations, and Options for Biomass Yields in
    Large Marine Ecosystems   206
    Kenneth Sherman

17. Geographic Perspectives in the Management of Large Marine
    Ecosystems   220
    Lewis M. Alexander

18. Interrelationships of Law in the Management of Large Marine
    Ecosystems   224
    Martin H. Belsky

Index   235

Preface

This volume is arranged in three parts. The first examines, through case studies, the effects of perturbations on biomass yields in five large marine ecosystems (LMEs) over several decades. The second part includes studies on the basic biodynamic processes shaping the survival, genetics, and recruitment of marine populations. In the third part of the volume, the contributors focus on the theory underlying patterns and processes in LMEs as well as the dynamics affecting biomass yields of the systems. The section concludes with thoughtful discussions of the interrelationships of science, geography, and law in the management of LMEs.

As in the previous two AAAS volumes on LMEs,[1] the results of the studies have been brought together to increase the awareness among scientists, managers, and students that LMEs are scientifically, technically, and legally tractable regimes for the conservation and management of living resources.

The editors are grateful to the wide cross-section of professionals who were willing to take time out of busy schedules to prepare and present their findings in this volume. We should like to acknowledge the financial assistance for the symposium and preparation of the volume provided by the National Science Foundation, the Marine Mammal Commission, the Center for Ocean Management Studies of the University of Rhode Island, the National Oceanic and Atmospheric Administration's National Marine Fisheries Service, and the Office of Oceanography and Marine Assessment, National Ocean Service. We are especially indebted to Jennie Dunnington who, once again, has demonstrated her outstanding skill in overseeing the editorial production of the manuscript.

—*K. Sherman and L. M. Alexander*
Narragansett, RI

—*B. D. Gold*
Washington, DC
May 1990

---

1 K. Sherman and L. M. Alexander (Editors). 1986. Variability and management of large marine ecosystems. AAAS Selected Symposium 99. Westview Press, Boulder, CO; K. Sherman and L. M. Alexander (Editors). 1989. Biomass yields and geography of large marine ecosystems. AAAS Selected Symposium 111. Westview Press, Boulder, CO.

Perspective

Large Marine Ecosystems

Kenneth Sherman and Barry D. Gold

Studies on the patterns, processes, and probable causes of perturbations in biomass yields of large marine ecosystems (LMEs) presented in this volume have been stimulated, in part, from discussions at two earlier symposia. The first was held during the 1984 AAAS annual meeting in New York, and the second at the 1987 annual meeting in Chicago. The discussion of cause-and-effect relationships influencing changing levels of biomass yields was focused principally on the fish components of LMEs located within the Exclusive Economic Zones (EEZs) of several countries in the northern and southern hemispheres. The LMEs have been defined as relatively large regions of the global EEZs, generally on the order of > 200,000 km^2, characterized by unique bathymetry, hydrography, and productivity, within which marine populations have adapted reproductive, growth, and feeding strategies (Sherman and Alexander, 1986).

On a global basis, nearly 95 percent of the biomass yields from the oceans are produced within the 200-mile baselines of the Exclusive Economic Zones of coastal nations. It is within these zones that the studies of marine population dynamics and management practices have been most highly developed. These are areas that are becoming increasingly stressed from pollution, natural environmental perturbation, and overexploitation of living resources. It is against this background and the potential global problems associated with atmospheric increases in levels of greenhouse gases and decreases in ozone that scientists and resource managers have focused attention on LMEs as regional units for the implementation of management actions leading to sustained and predictable yields of living resources.

From the syntheses of the biological, physical, chemical, and biomass yield information on the large marine ecosystems reviewed at the previous two symposia, several generalities emerge regarding the factors driving biomass fluctuations. Of the 23 LMEs reviewed, perturbations in biomass yields in six ecosystems (Kuroshio Current, Oyashio Current, Humboldt Current, California Current, Iberian Coastal, and Benguela Current) were attributed principally to large-scale natural environmental changes, including circulation, temperature, and productivity. In contrast, excessive fishing mortality has been attributed as the principal driving force for long-term changes in the biomass yields of the Gulf of Thailand ecosystem, Yellow Sea ecosystem, and the Northeast U.S. Continental Shelf ecosystem. Major changes to the ecology of the Great Barrier Reef ecosystem have been attributed to direct impact of starfish predation. Changes in productivity and composition of the fish community observed in the Baltic ecosystem were attributed primarily to pollution effects and only secondarily to fishing effects.

Among the ecosystems reviewed in the present volume — which stems from a third set of LME symposia held at the 1988 AAAS annual meeting in Boston — natural environmental perturbations appear to control productivity in the Antarctic ecosystem, although it has been clearly demonstrated that biomass yields from the resident fish communities inhabiting island regions have been declining as a result of overfishing of the parent stocks (CCAMLR, 1988). The boreal Norwegian Sea and West Greenland Sea ecosystems appear to be primarily driven by natural environmental changes in temperature and circulation, and secondarily by excessive fishing mortality in the case of the West Greenland cod biomass. The advantages of a holistic research approach to improve understanding of the biological and physical forces affecting mortality, recruitment, and biomass yields of Arcto-Norwegian cod are clearly evident in the "life-table" approach successfully adopted by Dr. Ellertsen and his research team investigating the Norwegian Sea ecosystem (Chapter 2).

It appears that disease vectors (epizootics) and overfishing are imposing increasing stress in the Caribbean Sea, one of two tropical ecosystems reviewed. The other, the Banda Sea ecosystem, is in a pristine state wherein biomass productivity and potential yield are under the influence of the monsoons and man-the-predator has had no detectable impact on potential yields. For the remaining LMEs reviewed, the available information for designating principal, secondary, or tertiary driving forces affecting variability in biomass yields was inconclusive.

There is a growing awareness of the utility of the LME approach to resource management among marine scientists, geographers, economists, government representatives, and lawyers. Concerns regarding the socioeconomic and political difficulties in management across national boundaries — as in the case of the Sea of Japan ecosystem, for example, or the North Sea ecosystem, or the 38 nations sharing the resources of the Caribbean Sea ecosystem — remain. In at least one LME, the Antarctic, the international community of 20 nations — including the United States, Soviet Union, Japan, Norway, United Kingdom, and other countries with significant fisheries interests — has progressed to a management regime based on an ecosystem perspective in the adoption and implementation of the Commission for the Conservation of Antarctic Marine Living Resources (CCAMLR, 1988). Also, within the EEZ of the United States, efforts are underway to implement ecosystem management. An atlas depicting the LMEs of the United States has been published by NOAA (1988), and recently, a consortium of federal and state scientists and managers developed a plan for the management of the Northern California Current ecosystem (Bottom et al., 1989). The Northeast Coast of Australia is presently under an ecosystems-oriented management regime (Sainsbury, 1988).

Global change in the form of ozone depletion, warming, and the greenhouse effect may become a source of stress on the biomass production of the oceans. The rather dramatic decadal fluctuations in marine biomass yields documented in the previous LME symposia, when considered in light of the growing concerns over global change, should serve to accelerate the movement toward adoption of LMEs as regional units for the conservation and management of living marine resources under existing maritime law. In the present volume (Chapter 18) and in earlier papers on the topic, Dean Belsky argues that based on customary law, nation-states can now apply an ecosystem approach to management of large marine ecosystems.

Adoption of LME management regimes will need to be accompanied by vigorously pursued research programs. Fisheries yields of the biomass produced within LMEs have not been effectively optimized on a sustained basis. Stocks continue to collapse under the stress of high levels of fishing, increasing pollution, and natural environmental perturbations, or some combination of the three. Important questions remain to be answered in relation to the utility of food chain dynamics as an input to yield models and management strategies for improving decision making on resource allocations. Other topics for examination include the increasing coastal-shelf enrichment trends in relation to increasing frequency of plankton blooms and their potential impact on the structure of food chains.

In an effort to better define the role of LMEs as global management units, considerable attention is being directed to the need for a better understanding of LMEs in relation to biodynamics (Rothschild and Osborn, Chapter 6). At the interface of small-scale interactions between physics and biology, Dickey (Chapter 7) and Osborn, Yamazaki, and Squires (Chapter 8) have addressed the biophysics of LMEs in relation to theoretical and empirical studies focused on the effects of turbulence on planktonic stages of marine populations. Accompanying the biodynamics theory are the pragmatic applications of recent advances in molecular biology and biotechnology developed by Powers, Allendorf, and Chen (Chapter 9) and Berman (Chapter 10) and empirical observations made of the recruitment process by Laurence (Chapter 11) and Taggart and Frank (Chapter 12). Collectively, these authors are concerned with improving the understanding and the predictions of the effects of secondary production on the sustained productivity of LMEs.

The boundaries of LMEs are not firmly fixed. However, a significant advance has been made in the identification of LMEs in the Atlantic rim by Prescott (1989) and in the Pacific rim by Morgan (1989). The way forward is outlined from a geographer's perspective in this volume by Alexander.

Application of ecosystem models and theory in support of LME management are given in the reports by Levin (Chapter 14) and Bax and Laevastu (Chapter 15). Levin argues that LMEs are useful testing areas for ecological theory; from his perspective, the fundamental theoretical problem in relation to biomass variability in LMEs is to understand the processes controlling the annual recruitment cycle of fish and other species; as with the contributions on biodynamics, attention is given to the coupling of physics and biology. In addition, however, predator-prey interactions are also emphasized. Levin suggests a sequential approach for moving ahead in the integration of theoretical and empirical strategies for elucidating the recruitment process. Included is a hierarchical movement from statistical observations of distributional patterns of physics and biology to an examination of patchiness, testing of competing models, and integration of validated component models as part of an iterative intellectual, experimental, and field observational process.

A pragmatic approach to ecosystem management with an emphasis on fisheries is given by Bax and Laevastu. They present a method for estimating the abundance of all species of interest within a defined ecosystem in relation to top-down predator-prey interactions. They examine the energy budgets of several LMEs. Based on empirical evidence they argue that fish-fish predation is the largest source of annual mortality with each system usually dominated by one keystone predator species. The role of ecosystem model simulation is underscored with plausible examples of probable species interactions based on predator-prey information that can lead to management strategies based on selected removal or enhancement of predators. The authors conclude their presentation with an argument for moving forward with the development of management policy at the ecosystem level, based on information already on hand, and the best estimates of the results of various interventions on the system. Other important ecosystem populations require more study when one considers a top-down approach to the management of LMEs, including the interactions among fish, marine mammals, and sea birds, a significant topic that has been the focus of several important studies (Croxall, 1987; Burger, 1988).

With regard to major environmental issues, the potential impacts of climate and global change on resource populations will need to be considered. Also of concern is the growing frequency of coastal eutrophication events, the dumping of urban wastes, toxic chemicals, and the continuing introduction and impact of petrogenic hydrocarbons on sustained ecosystem productivity (ICES 1977, 1980, 1986; Smayda, in prep.).

The contributions to the LME symposia bring into focus the issue of scale in applied marine studies conducted in support of resource management. Large-scale changes in biomass yields have been documented in the contributions to the symposia. The variability in biomass levels has been described in the context of region-wide patterns in the LMEs examined. As the trend for the management of living resources moves from single-species to multispecies assemblages, it becomes increasingly important to encompass entire ecosystems as management units. This approach will ensure that management measures designed to optimize the productivity of target species assemblages will also include consideration for related competitor/predator populations and their environments. By matching the sampling effort to the time and space scales of the processes that are of most direct influence to growth and survival of living marine resource populations, forecasts of biomass yield trends among the species can be improved for LMEs. Studies of changes in abundance and population renewal of resource species in general and fish stocks on a large marine ecosystem scale is in agreement with the proposition by Ricklefs (1987) that ecologists should begin to address community processes on a regional basis.

Although the designation of LMEs as global management units is an evolving scientific and geopolitical process, sufficient progress has been made to allow for useful comparisons of the processes influencing changes in biomass yields in different LMEs. Effective management strategies will in part be contingent on the identification of the major driving forces resulting in large-scale changes in biomass yields. Management of species responding to strong environmental signals will be enhanced by improving the understanding of the physical factors forcing biological changes. Whereas in other LMEs where the prime driving force is predation, options can be explored for implementing adaptive management strategies aimed at reducing the losses of long-term biomass yield through controlled manipulations of predatory stocks when it can be quantitatively demonstrated that this is the best decision among plausible options (Sissenwine, in prep.; Collie, in prep.). Remediation actions are required to ensure that "pollution" of coastal waters is reduced in the Baltic and does not become a principal driving force in any other LMEs. By comparing the results of research and management regimes among the LMEs, it should be possible in the future to accelerate the understanding of how the nonsteady-state ecosystems respond to and recover from stress.

References

Bottom, D. L., K. K. Jones, J. D. Rodgers, and R. F. Brown. 1989. Management of living resources: A research plan for the Washington and Oregon continental margin. Oregon Dept. Fish. Wildl. Publ. No. NCRI-T-89-004. 80 pp.

Burger, J. (Editor). 1988. Seabirds & other marine vertebrates: Competition, predation and other interactions. Columbia Univ. Press, New York. 339 pp.

Collie, J. S. In preparation. Adaptive strategies for management of fisheries resources in large marine ecosystems. *In* Food chains, yields, models, and management of large marine ecosystems, K. Sherman and B. D. Gold, Eds.

Commission for the Conservation of Antarctic Marine Living Resources [CCAMLR]. 1988.

Croxall, J. P. (Editor). 1987. Seabirds feeding ecology and role in marine ecosystems. Cambridge Univ. Press, Cambridge, UK. 408 pp.

International Council for the Exploration of the Sea (ICES). 1977. Petroleum hydrocarbons in the marine environment. Proceedings from ICES workshop held in Aberdeen, UK, 9–12 September 1975. Rapp. P.-v. Cons. int. Explor. Mer 171. 230 pp.

ICES. 1980. Biological effects of marine pollution and the problems of monitoring. Proceedings from ICES workshop held in Beaufort, NC, 26 February–2 March 1979. Rapp. P.-v. Cons. int. Explor. Mer 179. 346 pp.

ICES. 1986. Contaminant fluxes through the coastal zone. A symposium held in Nantes, France, 14–16 May 1984. Rapp. P.-v. Cons. int. Explor. Mer 186. 485 pp.

Morgan, J. 1989. Large marine ecosystems in the Pacific Ocean. *In* Biomass and geography of large marine ecosystems. pp. 377–394. Ed. by K. Sherman and L. M. Alexander. AAAS

Selected Symposium 111. Westview Press, Boulder, CO. 493 pp.

National Oceanic and Atmospheric Administration [NOAA]. 1988. Fishery resource programs. Folio Map No. 7, A national atlas: Health and use of coastal waters, United States of America. U.S. Dept. of Commerce, NOAA, National Ocean Service, Office of Oceanography and Marine Assessment, Washington, DC.

Prescott, J. R. V. 1989. The political division of large marine ecosystems in the Atlantic Ocean and some associated seas. *In* Biomass and geography of large marine ecosystems. pp. 395–442. Ed. by K. Sherman and L. M. Alexander. AAAS Selected Symposium 111, Westview Press, Boulder, CO. 493 pp.

Report of the seventh meeting of the Scientific Committee. Hobart, Australia. SC-CAMLR-VII. 211 pp.

Ricklefs, R. E. 1987. Community diversity: Relative roles of local and regional processes. Science 235:167–171.

Sainsbury, K. J. 1988. The ecological basis of multispecies fisheries, and management of a demersal fishery in tropical Australia. *In* Fish population dynamics, 2nd edition. Ed. by J. A. Gulland. John Wiley & Sons, New York.

Sherman, K., and L. M. Alexander (Editors). 1986. Variability and management of large marine ecosystems. AAAS Selected Symposium 99. Westview Press, Boulder, CO. 319 pp.

Sissenwine, M. P., V. C. Anthony, and E. B. Cohen. In preparation. Resource productivity and fisheries management, Northeast Shelf Ecosystem. *In* Food chains, yields, models, and management of large marine ecosystems, K. Sherman and B. D. Gold, Eds.

Smayda, T. J. In preparation. Global epidemic of noxious phytoplankton blooms and food chain consequences. *In* Food chains, yields, models, and management of large marine ecosystems, K. Sherman, L. M. Alexander, B. D. Gold, Eds.

Contributors

Lewis M. Alexander, *Center for Ocean Management Studies, University of Rhode Island, Kingston, RI*

Fred W. Allendorf, *Department of Zoology, University of Montana, Missoula, MT*

Martien A. Baars, *Netherlands Institute for Sea Research, Texel, The Netherlands*

Nicholas J. Bax, *NOAA/National Marine Fisheries Service Northwest and Alaska Fisheries Center, Seattle, WA*

Martin H. Belsky, *Albany Law School, Albany, NY*

Mark S. Berman, *NOAA/National Marine Fisheries Service Northeast Fisheries Center, Narragansett Laboratory, Narragansett, RI*

James A. Bohnsack, *NOAA/National Marine Fisheries Service Southeast Fisheries Center, Miami Laboratory, Miami, FL*

Erik Buch, *Greenland Fisheries Institute, Copenhagen, Denmark*

Thomas T. Chen, *Department of Biology, Johns Hopkins University, Baltimore, MD*

Thomas D. Dickey, *Ocean Physics Group, University of Southern California, Los Angeles, CA*

Bjornar Ellertsen, *Department of Aquaculture, Institute of Marine Research, Bergen-Nordnes, Norway*

Petter Fossum, *Department of Aquaculture, Institute of Marine Research, Bergen-Nordnes, Norway*

Barry D. Gold, *National Academy of Sciences, Washington, DC*

Gotthilf Hempel, *Alfred Wegner Institute for Polar Research, Federal Republic of Germany*

Holger Hovgård, *Greenland Fisheries Institute, Copenhagen N., Denmark*

Taivo Laevastu, *NOAA/National Marine Fisheries Service Northwest and Alaska Fisheries Center, Seattle, WA*

Geoffrey C. Laurence, *NOAA/National Marine Fisheries Service Northeast Fisheries Center, Narragansett Laboratory, Narragansett, RI*

Simon A. Levin, *Ecosystem Research Center, Division of Biological Sciences, Cornell University, Ithaca, NY*

Thomas R. Osborn, *Chesapeake Bay Institute, Baltimore, MD*

Dennis A. Powers, *Department of Biology, Johns Hopkins University, Baltimore, MD*

William J. Richards, *NOAA/National Marine Fisheries Center Southeast Fisheries Center, Miami Laboratory, Miami, FL*

Robert E. Ricklefs, *Department of Biology, University of Pennsylvania, Philadelphia, PA*

Brian J. Rothschild, *Chesapeake Biological Laboratory, University of Maryland, Solomons, MD*

Kenneth Sherman, *NOAA/National Marine Fisheries Service Northeast Fisheries Center, Narragansett Laboratory, Narragansett, RI*

Per Solemdal, *Department of Aquaculture, Institute of Marine Research, Bergen-Nordnes, Norway*

Kyle Squires, *Department of Mechanical Engineering, Stanford University, Stanford, CA*

Svein Sundby, *Department of Aquaculture, Institute of Marine Research, Bergen-Nordnes, Norway*

Christopher T. Taggart, *Department of Oceanography, Dalhousie University, Halifax, Nova Scotia, Canada*

Snorre Tilseth, *Department of Aquaculture, Institute of Marine Research, Bergen-Nordnes, Norway*

Hidekatsu Yamazaki, *Chesapeake Bay Institute, Johns Hopkins University, Baltimore, MD*

Jenne J. Zijlstra, *Netherlands Institute for Sea Research, Texel, The Netherlands*

Part One:
Perturbations and Yields of Large Marine Ecosystems

Part One:

Introduction

In a conference held in 1985 on the status and future of ecosystem science, one of the points generally agreed to by the participants was that "experimental manipulation of entire ecosystems is a powerful approach for elucidating ecosystem function and assessing anthropogenic stress."[1] Although "controlled" manipulations of large marine ecosystems (LMEs) have not been achieved, comparative studies of the effects of changes on a regional scale among LMEs can be instructive. There is a growing recognition among ecologists that the events and processes that shape feeding migrations and reproductive strategies are occurring on the regional or LME scale. Several of the LMEs described in this section have been perturbed in some manner that does provide insight as to cause and effect of stress. Others are presently in a relatively pristine state. Among the more pristine LMEs described is the Weddell Sea. Gotthilf Hempel's review of this "high polar" ecosystem is preceded with an instructive comparison of the differences in hydrography, bathymetry, and faunal assemblages of northern and southern polar seas. Chapter 1 is based on the extensive studies conducted from the RV *Polarstern* of the Alfred Wegner Institute for Polar and Marine Research in Bremerhaven during five summer cruises and one winter cruise between 1983 and 1987.

Hempel's chapter represents a comprehensive synthesis of the contemporary status of the Weddell Sea ecosystem from primary production and phytoplankton up the food chain to zooplankton, krill, benthos, fish, birds, and mammals. The basic and applied scientific contributions of the studies conducted by the Wegner Institute are pertinent to the conservation and management of Antarctic living marine resources. Much of the fisheries assessment activity of the Convention for the Conservation of Antarctic Marine Living Resources (CCAMLR) is directed to the Antarctic Peninsula area and the islands of South Georgia, the South Orkneys, and the South Shetlands. The discovery and quantitative assessment of low biomass of the demersal fish community and contrasting higher biomass of the pelagic fish, *Pleuragramma antarcticum*, provides a highly significant contribution to the CCAMLR information base. In addition, the study provides quantitative information of predator-prey interactions among the seal, whale, fish, penguin, and krill populations. As CCAMLR is the first international organization to manage resources from an ecosystem perspective, these observations should be most welcome. Chapter 2, authored by a Norwegian team led by Bjornar Ellertsen, on the influence of the Norwegian Sea ecosystem on the growth and survival of the biomass of Arcto-Norwegian cod, represents an initial effort to carry forward a systematic laboratory, mesocosm, and field study of mortality throughout the first year of life of a marine fish species. The study has historical significance as this is the cod stock that stimulated Johan Hjort to expound the "critical phase" hypothesis in 1914 as a mechanism for explaining variability in year-class strength and fish population size. Unlike the pioneer Hjort, however, the contemporary Norwegian research team was able to bring to the problem modern technology in the form of ships, new acoustic instrumentation, and an extensive time series of physical and biological measurements relating to early life history and predator-prey dynamics of larval and juvenile cod.

The Arcto-Norwegian cod appears to respond favorably to warm conditions, which

---

1 G. E. Likens, *et al.*, 1987. Status and Future of Ecosystem Science. Cary Conference, May 1985, Occasional Publication of The Institute of Ecosystem Studies. The New York Botanical Garden, Millbrook, NY. No. 3, February 1987.

also favor the breeding and swarming of their copepod prey. The Norwegian team concludes from their study that high biomass of cod reflects adaptation of the species to produce good-to-excellent survival during years of warmer-than-normal temperatures in the Lofoten spawning areas of the Norwegian Sea ecosystem, followed by high biomass yields several years later when the cod are recruited to the fishery. The favorable survival response of cod to the warm temperature, followed by high levels of prey abundance, is indicative of a biological response to an environmental signal as the primary driving force affecting the natural production cycle of Arcto-Norwegian cod biomass.

In the western Atlantic, the long-term climatic signal appears to have been the dominant feature affecting the biomass yield of cod in the West Greenland Sea ecosystem. In Chapter 3, principal investigators Holger Hovgård and Erik Buch of the Greenland Fisheries Research Institute have assembled one of the longest time series of temperature anomalies used in contemporary marine ecosystem analyses dating from 600 A.D. to the late 1980s. Although a stand-alone direct correlation to temperature is not convincing, the circumstantial evidence supported by the population analyses relating to fishing mortality and cod demographics argues effectively that the stock was reduced not only by cooling temperatures, but also by failure of good recruitment at low stock sizes and in the variable advected recruitment from the Icelandic spawning stock upstream in the Irminger Sea.

In contrast to the influence of environmental conditions as a principal driving force in biomass perturbations of boreal ecosystems, the Caribbean Sea ecosystem is presently under stress from the extensive, unexplained mortalities of corals and an epizootic disease that has led to catastrophic losses of the sea urchin, *Diadema*, an important grazer on the coral reef ecosystem. William Richards and James Bohnsack, the authors of Chapter 4, estimate fish biomass yields from the Caribbean at an annual level of approximately 500,000 metric tons. Their chapter serves to raise the level of concern for the Caribbean as an LME which, at present, is not the subject of any international management regime but is, however, under stress from epizootic events and human interventions. The authors focus on the ecosystem-wide problems associated with fisheries management, mortality of corals, and expanding tourism, and argue cogently for the establishment of appropriate institutional arrangements that would allow the 38 nations of the region to manage the resources of the Caribbean from an ecosystem perspective.

For another tropical ecosystem, Jenne Zijlstra and Martien Baars present in Chapter 5 new information for the Banda Sea ecosystem. The observations are based on a time series of surveys that measured chlorophyll, primary production, zooplankton, and fish biomass. The study corroborated earlier postulates that the ecosystem would be under the principal influence of upwellings associated with southeast and northwest monsoons. Based on extensive physical, chemical, and biological oceanographic measurements, the authors estimate that the Banda Sea ecosystem has the potential for supporting an annual biomass of pelagic fish of approximately 600,000–900,000 metric tons (mt), compared to an average annual yield of about 30,000 mt. The estimated annual average primary production estimate of 400–500 of $C/m^2/yr$ is surprisingly similar to the values estimated for parts of the North Sea and the Northeast Continental Shelf of the United States. The comprehensive synthesis of available productivity information by the authors will be of great assistance to any future development of fisheries for the Banda Sea ecosystem.

—*Kenneth Sherman*

Chapter 1

The Weddell Sea
A High Polar Ecosystem

Gotthilf Hempel

Abstract

The ecological information presented is based on the results of studies during one winter and five summer cruises of the RV *Polarstern* during 1983–1987. The heavy ice cover of the Weddell Sea ecosystem contributes to the overall low productivity of the pelagic component of the system. The pelagic fish *Pleuragramma antarcticum* replaces krill and the keystone species for the warm blooded top predators, including crabeater seals, Weddell seals, and emperor penguins. Substantial pressure on key prey species is exerted on *Pleuragramma* and benthopelagic fish species by emperor penguins and Weddell seals in the vicinity of breeding sites. One of the largest shelf areas around the Antarctic continent is found in the southern part of the Weddell Sea. The northeast part of the shelf is rich in suspension feeders (e.g., sponges, bryozoans). Bryozoans are the principal component of the southern shelf community, and holothurians dominate the southern trench community. The relatively high abundance of epibenthic biomass is attributed to the high sedimentation rate of ice algae and phytoplankton that are not grazed down, but sink as aggregates in sedimentary bursts recorded in sediment traps reaching an annual peak in December. The permanent ice cover of the Weddell Sea ecosystem precludes the potential for any large-scale fisheries, including whaling.

On Polar Seas in General

The Weddell Sea ecosystem is one of the most "polar" parts of the Southern Ocean. General summary descriptions of the Southern Ocean ecosystems in comparison to Northern Polar waters have been published inter alia by Knox and Lowry (1977), Nemoto and Harrison (1981), and Hempel (1985, 1988). The following characteristic features of the polar seas may be pointed out.

Common to most polar seas are seasonal or permanent ice covers, year-round low temperatures, and intense seasonality in irradiance. With respect to those factors, the Arctic Mediterranean is more polar than most of the Antarctic circumpolar current system. On the other hand, the Southern Ocean is older and has been cold for a longer geological period.

The Arctic is characterized by a meridional current system, the Antarctic by latitudinal circular currents; both systems also possess large gyres. The Arctic seas have very broad shelf areas. They are well stratified with seasonal nutrient exhaustion of the surface layer. The Southern Ocean is mostly a well-mixed, deep-water upwelling system. Its shelves are narrow and deep.

The Arctic bottom fauna consist of a small number of species of all major taxa. These species are mostly eurythermal and may have successfully invaded the North Polar waters from the boreal Atlantic or Pacific. In the Antarctic benthos, only a few taxonomic groups (e.g., Porifera, Echinodermata, Isopoda, Amphipoda, Polychaeta) have evolved into a large number of stenothermal species, while the remainder of the high taxa are poorly represented in the Antarctic.

North/south differences in the plankton are less striking; euphausiids, however, are virtually missing in the Arctic proper, while they are key elements of the Antarctic system. The waters adjacent to the Arctic Basin are rich in pelagic fish, quite in contrast to the Antarctic.

Polar food webs show the following characteristics: (i) annual primary production is poor and highly seasonal; (ii) the benthos is largely decoupled from the pelagic zone, except in shallow Arctic shelf areas; (iii) birds and mammals are the dominant top predators; (iv) giant species occur in several families, but species of "normal" body size are also frequent; (v) the classic short Antarctic food chain of diatoms-krill-whales is just one important line in a complex system. The unusually large trophic steps of about eight orders of magnitude in weight are only possible because of locally high food concentrations in phytoplankton blooms, carpets of ice algae, and in krill swarms. In turn, the condensed biomass of these swarms and of each individual whale makes them attractive to commercial exploitation; (vi) the majority of polar animals grow slowly and reproduce late; (vii) metabolic rate is low in the benthic organisms, but high in very mobile necton, particularly krill, penguins, and whales; (viii) long periods of seasonal starvation further lower the overall ecological efficiency in polar animals; (ix) most polar waters are poor fishing areas because of low net production; and (x) in the Arctic, only the shallow outlets of the Barents and Bering Seas permit a substantial exploitation of finfish.

Antarctic finfish stocks are small. Most whale stocks have been severely decimated, leaving krill as the only large fishing resource, possibly with an MSY of the order of 50 million tons, compared with whale MSY of 2 million tons and finfish of far less than 1 million tons.

Total biomass of krill is of the order of hundreds of millions of tons, with a maximum life span of more than five years. Interannual fluctuations in distribution and possible changes in year-class strength add to the uncertainties of any krill fishery. Studies of population dynamics of krill are hampered by the lack of methods for accurately determining age and tracing migrations. Any krill fishery would compete with baleen whales, seals, and penguins.

Zonation of plankton coincides with the fluctuating ice cover, particularly in the Antarctic circumpolar current system. In both hemispheres, the ice-free zone adjacent to the Antarctic Convergence and the Arctic Front respectively is largely occupied by an oceanic copepod community. The seasonal pack-ice zone between the spring and autumn boundary of maximum and minimum pack-ice extent is a broad belt around the Antarctic continent. This zone is termed Seasonal Ice Zone (SIZ). It sustains the proverbial krill system (Figure 1.1).

The Antarctic coastal belt comprising the permanent ice zone (PIZ) — broadest in the Weddell, Bellingshausen, and Amundsen Seas — resembles the central Arctic Ocean. Until very recently, because of their year-round ice cover, these high polar waters in the Antarctic and in the Arctic were little studied except for some observations by drifting vessels or ice islands.

The Weddell Sea

Within the PIZ, the Weddell Sea (Figure 1.2) is a well-defined, semi-enclosed system bordered by ice shelves on three sides and occupied by a large cyclonic gyre of the East Wind Drift in the center and a coastal current along the shelf. In the south and east, the westward-flowing coastal current follows the contours of the continental shelf. It is deflected to the north by the Antarctic Peninsula and turns east again in the Weddell-Scotia Confluence. Part of the coastal current diverges from the mainstream near Halley Bay and forms a cyclonic gyre centered over the southern shelf. The cold shelf water with temperatures below $-1.8°C$ transported in the coastal current is separated from the warmer water of the central Weddell Sea by a steep slope front. The southern half of the Weddell Sea has one of the largest shelf areas of the Antarctic continent. Along the eastern coast, the sea ice

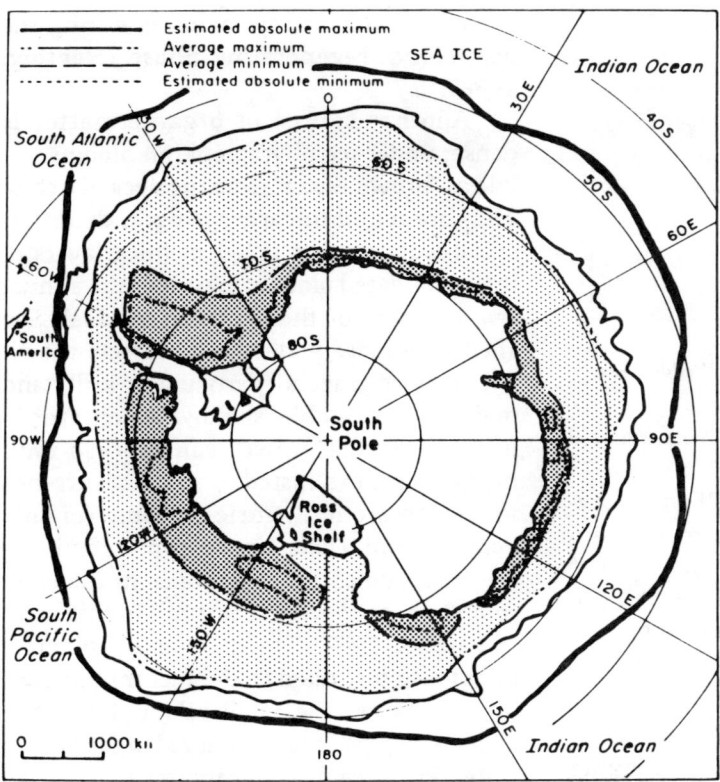

Figure 1.1. Seasonal pack-ice distribution in the Antarctic Ocean (modified after Ackley, 1981). The dark shaded area marks the permanent ice zone (PIZ), the light shaded area represents the seasonal ice zone (SIZ).

breaks up frequently, resulting in large polynyas. But also in other areas, the pack ice is under permanent strain, converging and diverging to form ridges and leads.

Most of the ecological information presented in this chapter was collected in the years 1983–1987 during five summer cruises and one winter cruise of RV *Polarstern*.

Phytoplankton

During most of the year, the surface mixed layer is deep (~100 m) and not favorable for phytoplankton development. In winter and early spring, record Secchi-disc depth of 70–80 m (Gieskes et al., 1987) coincided with exceptionally sparse phytoplankton stocks (0.01 mg chlorophyll $\cdot$ m^{-3}) under pack ice and in the coastal polynya. A few live algal cells in good physiological conditions were found down to 500 m. Primary production levels in the polynya and under thin sea ice were correspondingly low (< 10 mg C $\cdot$ m^{-2} $\cdot$ d^{-1}) even in mid-November in spite of 24 hours of daylight. In late November, melting started and a thin surface layer was formed. Then phytoplankton developed rapidly, probably inoculated from cells in the water column and from the sea ice community. By January/February, phytoplankton stocks along the southeastern coast increased up to 1.5 mg Chl-a $\cdot$ m^{-3} and production levels were high on the shelf (750 mg C $\cdot$ m^{-2} $\cdot$ d^{-1}) but much lower further offshore (v. Bodungen et al., 1988). Further south, up to 1,700 mg C $\cdot$ m^{-2} $\cdot$ d^{-1} were recorded off the Filchner/Ronne ice shelf (v. Broeckel, 1985). A southward increase in production levels was also reported by El-Sayed and Taguchi (1981). Nothing is known about phytoplankton production in the inner Weddell Sea in late autumn and winter (March–September). With the disappearance of summer stratification, thickening of ice cover, and decreasing daily irradiance, phytoplankton is assumed to decrease quickly.

Estimates for the annual primary pro-

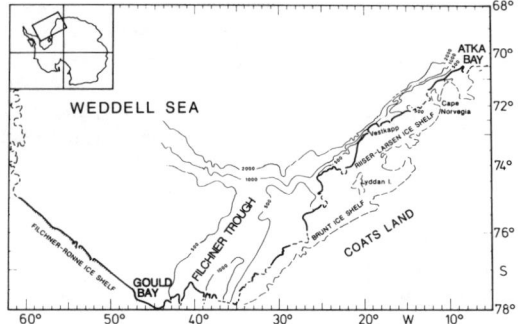

Figure 1.2. Geographical map of the Weddell Sea.

duction for the Southern Ocean in general range between 16 and 25–130 g C · m^{-2} · yr^{-1} for deep and inshore/shelf waters, respectively (Sakshaug and Holm-Hansen, 1984). Those estimates are based on C^{14} measurements, mainly in summer. The annual production rates in the Weddell Sea are presumably at the very low end of the scale of the Southern Ocean. However, more recent production estimates based on depletion of nitrate between winter and summer indicate much higher productivity. Thus, Jennings et al. (1984) estimated primary production for a 90-day period (ending in January) to be about 30 g C · m^{-2} in the northern Weddell Sea and v. Bodungen et al. (1988) calculated a range of 46–58 g C · m^{-2} until February in the southeastern Weddell Sea. These values represent only the "new" nitrate-based production. If recycled or ammonia-based production is included, the figures are likely to be at least two or three times higher.

Phytoplankton species composition in the Weddell Sea varies considerably in the mesoscale range, but the dominant species do not differ significantly from those in the Southern Ocean in general. Thus, many centric and pennate diatoms, the gelatinous *Phaeocystis*, and the silicoflagellate *Distephanus*, together with nanoflagellates, comprise the bulk of the phytoplankton (Noethig, 1988). Here again, except for bloom situations where large centric diatoms dominate (El-Sayed, 1971), the smaller (< 10 μm) fraction contributes most to production (v. Broeckel, 1985). Protozooplank-

ton also attain a large biomass during the post-bloom regenerating phase (Noethig, 1988).

Another source of organic matter is dense, under-ice phytoplankton blooms visible as "brown water" from the deck of a ship. They are found in early spring in the interstitial water within 0.5–1 m thick layers of unconsolidated platelet ice under the annual sea ice cover of the southern Weddell Sea shelf. The loose platelet ice protects the algae against grazers (particularly krill) and maintains the stable, shallow water layer in position so that growth can proceed until nutrients are exhausted — a rare occurrence in the Antarctic. Centric diatoms (mainly *Porosira* and *Thalassiosira* spp.) dominate the crop, and chlorophyll concentrations up to 200 μg · l^{-1} were measured in this layer (Smetacek, 1987). The high diatom biomass might also be linked to the fact that no herbivores, except the small calanoid copepod *Stephus longipes*, were observed in the brown water. Presumably they cannot swim and filter in the brash ice platelets and needles. Similar brown water under ice was observed by El-Sayed (1971) further to the west in summer.

Meadows of ice algae growing from the underside of sea ice are ubiquitous in the Weddell Sea but have a patchy distribution. Under-ice observations with a remotely operated vehicle (ROV) revealed patch diameters of a few meters and visibly high concentrations of ice algae in crevasses caused by ridging. Direct studies on primary production of ice algae in the Weddell Sea are missing. Sullivan et al. (1985) estimated an annual production of approximately 5 g C · m^{-2} · y^{-1} for McMurdo Sound, while Kottmeier and Sullivan (1987) calculated that winter production in Antarctic sea ice (August–October) could range from 6 to 18 × 10^{12} g C. This is about 10% of the estimate of annual primary production by phytoplankton in the Southern Ocean (El-Sayed, 1971). Evidence that sea ice microalgae begin to grow even earlier in the season was obtained during July–August 1986 when pack ice was found to contain mean chlorophyll concentrations of 1.9 $\mu g \cdot l^{-1}$ and thus up to several orders of magnitude higher than in the water column at that period (Dieckmann, 1987). Little is known about the fate of Weddell Sea phytoplankton and ice algae in terms of microbial decomposition and grazing. Heterotrophic planktonic microorganisms — as well as herbivorous euphausiids, copepods, and salps — are of little importance in the water column, at least in winter.

Zooplankton and krill

Zooplankton has been sampled and analyzed for the eastern Weddell Sea in summer, winter, and spring. For the cyclopoid copepods, Fransz (1988) described different modes of overwintering and reproduction. In late winter, surprisingly large numbers of small copepods with eggs and nauplii were found. Fransz concluded that in late winter/early spring, small copepods occur in two different developmental stages: as nauplii which survive on their yolk reserves and as resting copepodites living on the reserves of the previous summer. Early in the season (October/November) when food is scarce in the Weddell Sea, some females produce eggs in small numbers. The same species reproduce more in summer with food in their stomachs.

The early start of reproduction can be interpreted as an adaptation to the occurrence of phytoplankton being rather unpredictable in space and time in the PIZ.

In late winter, small cyclopoid copepods (less than 1 mm in length) outnumber the large calanoid copepods in the upper 200 m of the Weddell Sea by two to three orders of magnitude and even their total biomass is higher. When phytoplankton become available, the consumption rate of the small species exceeds that of the large species by about one order of magnitude (Fransz, 1988). These small carnivorous or omnivorous cyclopoid copepods are the main food of fish larvae (Kellermann, 1987). In the southern Weddell Sea, unlike other regions of the Antarctic Ocean, fish larvae constitute a major part of zooplankton.

Lipids play a particularly important role in the existence of polar marine animals. The formation of their energy reserves compensates for the high seasonality of phytoplankton abundance. Much of the reserves are used for reproduction. Wax esters, the main reserve in most copepod and *Euphausia crystallorophias*, are particularly useful for long seasons of starvation between phytoplankton blooms. Krill *(Euphausia superba)* lack wax esters. Triglycerides help to overcome shorter starvation periods in krill and other opportunistic species that switch from one food source to the other (Hagen, 1988).

Boysen-Ennen and Piatkowski (1988) have analysed two size fractions of macrozooplankton by RMT 1+8 (330 resp. 4500 μm mesh size) in the upper 300 m. They described three distinct communities of epipelagic meso- and macro-zooplankton of the eastern Weddell Sea in summer revealed by cluster analysis (Figure 1.3).

The Southern Shelf community is poor in numbers but relatively rich in mostly neritic species that are mainly herbivorous (e.g., *Euphausia crystallorophias*) or omnivorous (e.g., the copepod *Metridia gerlachei*) (Figure 1.4). This community is similar to the macroplankton community described from shelf waters at the Antarctic Peninsula (Piatkowski, 1987).

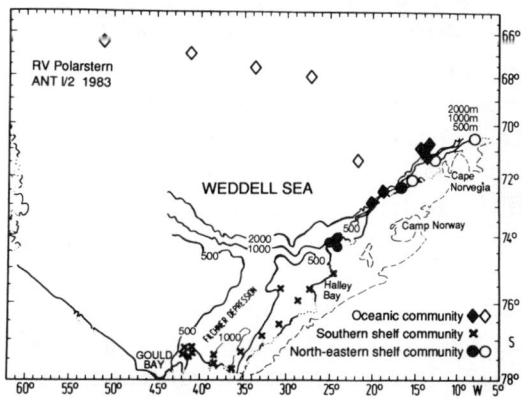

Figure 1.3. Zooplankton communities of the Weddell Sea. Geographical extension as revealed by cluster analysis (Canberra-metric). (From Boysen-Ennen and Piatkowski, 1988.)

Figure 1.4. Percentages of the main species and the feeding types in the three macrozooplankton communities of the Weddell Sea. (From Boysen-Ennen and Piatkowski, 1988.) fff = fine filter feeders, gff = gross filter feeders, c = carnivore, o = omnivore.

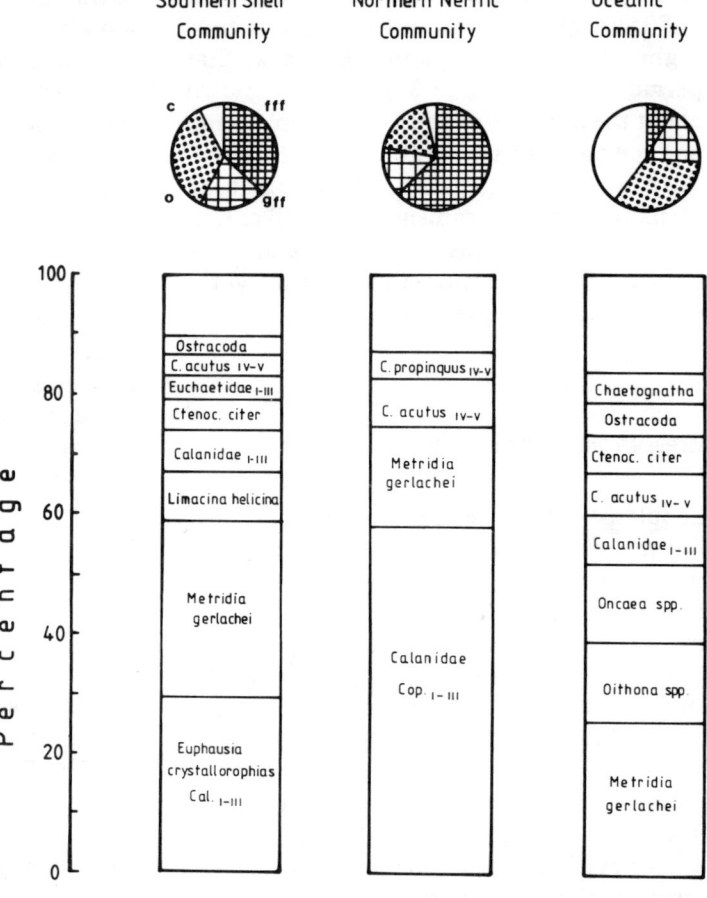

The Northeastern Shelf community consists of neritic and oceanic species and is confined to the narrow shelf slope of the northeastern Weddell Sea, where a convergence zone separates the cold shelf waters from the warmer water masses of the Eastwind Drift. It has a high total abundance, with copepodites of *Calanus propinquus* and *Calanoides acutus* dominating. Species diversity is low, the majority of the species are filter feeders, and the macrozooplankton consists mainly of euphausiids and carnivorous chaetognaths. Macrozooplankton biomass in terms of wet volume can attain about 280 ml · 1000 m^{-2} (Figure 1.5; Piatkowski, 1987). This is in the range of mean values at the Antarctic Peninsula and can be attributed to the temporary high concentrations of euphausiids in the Northeastern Shelf community.

In the oceanic community, no single species dominates. Gelatinous species like the siphonophores *Diphyes antarctica* and *Dimophyes arctica* (and the salp *Salpa thompsoni*) are frequent. Carnivores and omnivores make up about two-thirds of the individuals caught. The species composition is similar to other offshore regions in the Southern Ocean. Zooplankton biomasses are the lowest except in areas where single species are distributed in patches (e.g., salps).

More detailed studies on the zooplankton communities in the southeastern Weddell Sea were carried out in January/February 1985 and October/November 1986 (Hubold et al., 1988; Hubold and Hempel, 1987; Hempel and Hempel in prep.). Plankton tows by bongo net (300 µm mesh size) covered the upper 200 m (sometimes 500 m). In addition, a multiple opening/

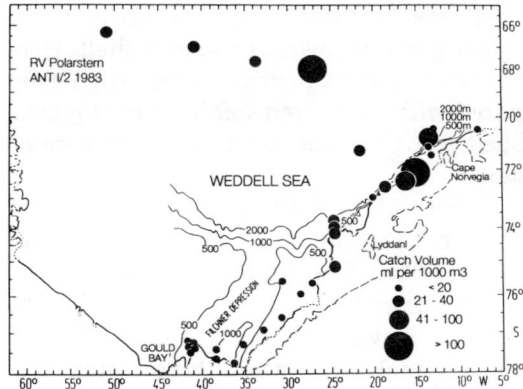

Figure 1.5. Catch volume (in ml · 1000m^{-3}) of macrozooplankton in the Weddell Sea. (From Piatkowski, 1987.)

closing net (200 μm mesh size) was used in the study of vertical distribution of zooplankton in the upper 1,000 m.

A station grid off Vestkapp extending 50 miles offshore revealed the presence of the Southern Shelf community on the shelf and upper slope and the oceanic community further off shore. In the Southern Shelf community off Vestkapp, typical taxa of the oceanic community like *Thysanoessa sp.* and *Rhincalanus gigas* are concentrated in the warm deep water in winter. In summer they rise into the surface layer and penetrate onto the shelf.

In the course of the coastal current from Vestkapp in Filchner Trough, zooplankton decreases in abundance. The shifts in species composition of the zooplankton were a matter of relative abundance of certain taxa rather than their complete absence in the South. At first glance, winter plankton off Dawson-Lambton Glacier was extremely scarce and summer plankton off Filchner was dominated by gastropods, polychaetes, larval *Pleuragramma antarcticum*, and *Euphausia crystallorophias* rather than by large copepods (Hempel et al., 1983).

Cluster analysis demonstrated regional similarities between the Vestkapp samples in January and Filchner in February. There was a major shift in zooplankton composition off Vestkapp, possibly caused by coastal transport, from January to February when total zooplankton abundance dropped by half. Copepods like *Metridia gerlachei* and *Calanoides acutus* decreased in number near the surface, but became more abundant in deeper layers and turned into a resting state (Schnack, pers. commun.). In February a massive invasion of early larvae of *Euphausia superba* was observed over the shelf and slope (average 1000 indiv. · 100 m^{-3}). In late winter the abundance of surface zooplankton near Vestkapp was less than 10% of the summer values. Larval fish and euphausiids were virtually absent. Only copepods, ostracods, and chaetognaths occurred in numbers above 10 individuals · 100 m^{-3}. Those low winter figures for the upper 200 m can be attributed only partially to seasonal downward migration.

Published information on the life of krill in winter and in the pack ice was scarce until recently (Boyd et al., 1984). Meanwhile, underwater observations by divers and ROVs (Garrison et al., 1986; O'Brien, 1988; Marschall, 1988; Daly and Macaulay, 1988) have demonstrated that krill occur in great numbers in the pack ice where they thrive on ice algae by grazing on the underside of the sea ice. Obviously this is an additional source of food, at least sufficient for overwintering. In shallow water, krill may also feed on detritus on the sea bed (Kawaguchi et al., 1986). De Laca (pers. commun.) saw krill swarms stirring up and filtering benthic detritus and diatoms. In winter, distribution of adult krill extends much further in the south than in summer, when krill was virtually missing in all plankton samples south of Halley (Piatkowski, 1987). Wherever RV *Polarstern* broke sea ice in October/November 1986, krill were washed onto the ice floes (Marschall, 1988), while there was virtually no krill in plankton samples taken at the same site. That is not surprising in view of the absence of phytoplankton. Also, copepods as another food source for omnivorous krill were too scarce to sustain the energy demands of swimming krill (Price et al., 1988). Some adult krill may stay year round in the pack ice. This was indicated by RV *Polarstern* observations in March 1983 in the heavy pack of the northwestern Weddell Sea. On the other hand, larvae and juvenile *Euphausia superba* have not been observed in the pack ice proper; they seem to be confined to open water and the marginal ice zone.

The winter feeding of krill and its very wide distribution in the pack ice add to the picture of krill as being highly adaptable. For most krill, the swarm life in summer is the period of rapid growth and reproduction in the phytoplankton blooms in the water column outside the pack ice. During the rest of the year, krill find food and shelter in the rafted and ridged pack ice. Marschall (1988) interpreted several morphological and behavioral features of krill as adaptations to a pseudobenthic life on the underside of the sea ice. All this is in contrast to the traditional view of krill (e.g., Rosenberg et al., 1986) as a sole filter feeder being confined to the open waters and the ice margins of the Seasonal Ice Zone (SIZ), and fasting except for the summer phytoplankton blooms.

Additional heterotrophic consumers using the abundant sea ice algae as a food source include Protozoa (e.g., foraminifers, ciliates, flagellates) and Metazoa (e.g., nematodes, copepods) which live within the ice in abundance that exceeds concentrations in the water columns by several orders of magnitude (Horner, 1985; Spindler and Dieckmann, 1986).

Benthos

The southern shelf of the Weddell Sea, with a depth of 200–600 m, has a width of up to 500 km in front of the Filchner/Ronne Ice Shelf. At the east coast it reaches 100 km and along the peninsula the shelf extends up to 225 km. Because of the ice shelves there is no littoral zone along the 2,500-km coastline. The sediments are mostly of glacial origin. Sand, gravel, and stones are mixed. They are particularly coarse in the near-shore areas of swift bottom currents. Soft bottom is only found on the deeper parts of the shelf. Spiculae of sponges and debris of Bryozoa cover the sea floor over wide areas of the upper parts of the northeastern shelf. Here the total benthos biomass is very high compared with the barren soft bottom of glacial shelf troughs such as the Filchner depression.

Since 1983, RV *Polarstern* has sampled the benthos of the shelf and slope of the eastern Weddell Sea almost every year. Voss (1988) described 430 taxa mainly taken by Agassiz Trawl between 200 and 1,200 m. Cluster analysis revealed three major communities (Figure 1.6). The northeastern shelf community is rich in suspension feeders, mainly sponges and bryozoans, and has a high number of species and high diversity. The large sponges of the northeastern shelf form a complex system of microhabitats for a large number of species. This rich benthic fauna is similar to that of the Ross Sea (Bullivant, 1967; Dayton et al., 1970) and the coast of the Indian sector (Gruzov et al., 1968; Ushakov, 1963) of the Southern Ocean, suggesting a rather uniform circumpolar high-Antarctic macro-benthos that is different from the fauna of the Scotia subregion.

The Southern Shelf community, with Bryozoans as a major component, occupies sandy and silty bottom. The number of species and their diversity are lower than further to the northeast.

The Southern Trough community was found at the deeper stations of the two shelf troughs in front of the eastern and western

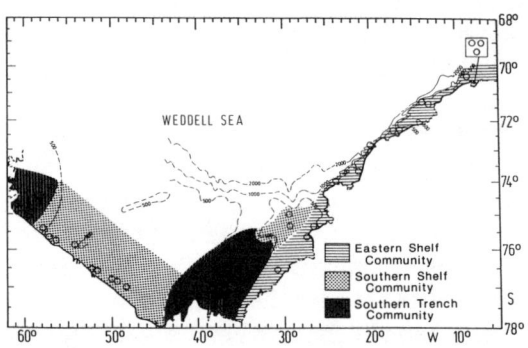

Figure 1.6. Geographical extension of macrozoobenthos communities in the Weddell Sea as revealed by cluster analysis. (From Voss, 1988.)

corner of the Filchner/Ronne ice shelf. Here species number and diversity are again lower. Suspension feeders are virtually missing while grazers, particularly holothurians, are abundant. Obviously the coastal current on the shelf results in a high suspension of material, as was directly observed by ROV. Lower current velocities in the trough and transformation of the coastal current into a cyclonic gyre centered over the depression result in more sedimentation and less resuspension and hence better feeding conditions for mobile grazers.

A special photographic study of the holothurians of the same region demonstrated a high variety in the "lebensformen" (modes of life) from endobenthic to sessile and mobile epibenthic and up to epizoic forms living on top of large sponges (Gutt, 1988).

The high abundance of epibenthos is unexpected in view of the low biomass of primary producers and the great water depth of the shelf. It might be explained, however, by relatively high sedimentation after phytoplankton blooms, which are not much subjected to grazing and decomposition in the water column and quickly sink down as aggregates. These blooms are presumably of short duration, hence rarely or only locally recorded (e.g., El-Sayed, 1971). They are possibly seeded by algae that grow rapidly in melting sea ice. The organic fallout from melting sea ice may also contribute to this rain. Green "fluff" on the sea bottom has been observed on many photographs taken along the southeastern shelf of the Weddell Sea (Gutt, pers. commun.). Sedimentation in short bursts has been recorded in sediment traps moored year round in Bransfield Strait where most of the year's sedimentation occurred in December (Wefer et al., 1988). The suspension feeders are extremely sluggish and presumably very slow growing. They are not cold adapted in terms of relatively high metabolism at low temperatures but are adjusted to a life of "sit and wait" with low energy demand as demonstrated by Clarke (1983) and others for the Antarctic shallow-water fauna.

For comparison, the same trawls and camera used in the Weddell Sea were also applied on the Arctic Belgica Bank on the East Greenland shelf. In contrast to the highly diverse Weddell Sea benthos, which consists mainly of suspension-feeding sponges and bryozoans, these two groups were largely missing on Belgica Bank where less-specialized asteroids and ophiurids dominated (Piepenburg, 1988). Their numbers were also three times lower than in the Weddell Sea.

Fish

The fish of the Weddell Sea are similar to those of the Ross Sea and other high Antarctic regions in species composition and diversity. Apart from nototheniids, only a few small liparids, zoarcids, and rajids are to be found. Sixty fish species belonging to eight families were caught by bottom trawl in depths of 200–1,200 m (Ekau, 1988; Schwarzbach, 1988). Nototheniids and channichthyids dominated the fish on the shelves by biomass. Bathydraconids and artedidraconids are represented mostly by smaller individuals. The fish community of the Weddell Sea is distinctly different from that of the Scotia Sea and the Antarctic Peninsula. There, for example, the genus *Notothenia* is dominant, whereas *Trematomus* is the most important one on the high Antarctic shelves of the Weddell Sea.

Food and feeding of Weddell Sea fish has recently been studied in some detail (Kock et al., 1984; Hubold, 1985a; Schwarzbach, 1988; Ekau, 1988). Many fish species depend on benthos for food despite the low energy value of the bulk of the epibenthos. Some groups of demersal nototheniids tend to use pelagic food resources as well. This is reflected in their food composition, where euphausiids, pteropods, and copepods dominate (Schwarzbach, 1988), and in their morphological adaptations (Ekau, 1988). Bathydraconids feed mainly benthopelagically, while artedidraconids selectively pick polychaetes or amphipods. Channichthyids show the strictest specialization, feeding pelagically on fish and euphausiids. Interspecific food competition is somewhat reduced by vertical segregation of the feeding zones of the various fish species.

Demersal fish biomass on the eastern shelf and slope rarely exceeds $1 \text{ t} \cdot \text{km}^{-2}$, decreasing from northeast to southwest (Ekau, 1988). This is at least one order of magnitude less than originally found off Elephant Island and South Georgia (Kock, 1986) (Figure 1.7) and reported from most other demersal fishing grounds of the world. In comparison to nototheniids from the Antarctic Peninsula, species from the Weddell Sea are small and grow slowly (Ekau, 1988). Because of their low biomass, demersal fish stocks are not an important year-round food resource for top predators. Demersal fish are mainly taken as supplementary food in winter and spring, when pelagic resources (e.g., krill, Pleuragramma antarcticum) are scarce.

In contrast to the demersal stocks, pelagic fish become increasingly important towards the south, where a mean biomass of $1.3 \text{ t} \cdot \text{km}^{-2}$ was estimated. Hence, the southern Weddell Sea houses one of the few major pelagic fish populations of the Antarctic. More than 90% both in numbers and weight of this population is contributed by the fully pelagic nototheniid Pleuragramma antarcticum (Hubold and Ekau, 1987), while myctophids and mesopelagic oceanic fish are absent from the southern Weddell Sea. Spawning of Pleuragramma occurs along the coast, and larvae and juveniles are an abundant component of the surface zooplankton (Figure 1.8). Middle size classes seem to migrate northward, where they thrive on copepods and Euphausia superba. Adults live bentho-pelagically on the southernmost parts of the shelf and depend on E. crystallorophias (Hubold, 1984, 1985a, 1985b). Growth rate of P. antarcticum is at the lower end of all known Antarctic fish. Hence pelagic fish production must be considered low in the Weddell Sea (Hubold and Tomo, 1989).

Although abundance and production of Pleuragramma antarctitum cannot be compared with, for example, capelin in the Subarctic, the species plays a similar role in the

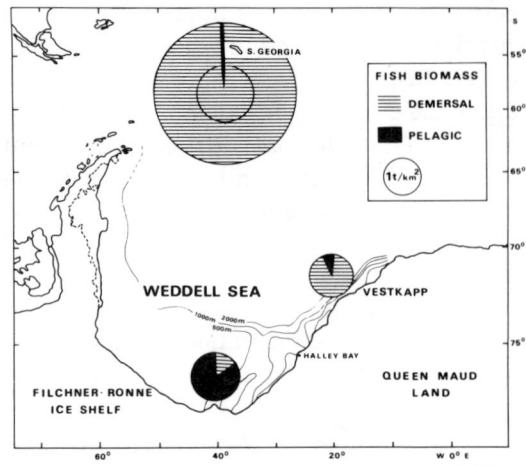

Figure 1.7. Biomass of pelagic and demersal fish in the Weddell Sea (from Hubold, pers. commun.; and Ekau, 1988) compared to South Georgia before onset of fisheries (outer circle $14 \cdot \text{km}^{-2}$, from Everson, 1977) and after heavy exploitation (inner circle, from Kock, 1986).

pelagic system of the inner Weddell Sea, forming an important link between the secondary producers and top predators. The presence of some killer and bottle-nose whales in the Weddell Sea even in late winter may be linked to Pleuragramma antarcticum which is found in the stomachs of Weddell and crabeater seals, and of emperor penguins (Ploetz, 1986; Klages, 1989).

Birds and mammals

During summer the bird and mammal fauna of the southern and central Weddell Sea consist mainly of Weddell and crabeater seals, Emperor and adelie penguins, and of minke and killer whales. Stomachs of crabeater seals taken in the southern Weddell Sea in summer contained mainly Euphausia crystallorophias and Pleuragramma antarcticum. In winter only Weddell seals and emperor penguins remain in their coastal habitats and reproduce on fast ice under the inlet cliffs of the ice shelves. In November 1986 Stonehouse and Hempel (in prep.) carried out an overall census along the coast of the eastern

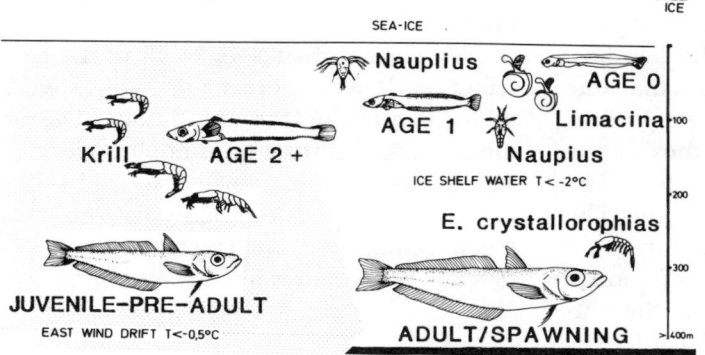

Figure 1.8. Schematic life cycle of *Pleuragramma antarcticum* in the Weddell Sea. (From Hubold, 1985b.)

Weddell Sea and found eight major breeding sites, normally shared by both species (Figure 1.9). Emperors were more abundant south of 75°S while seals concentrated more to the North. The total number of 150,000 adult emperors and 20,000 adult Weddell seals means substantial feeding pressure in the vicinity of the breeding sites. Weddell seals in summer feed mainly on the pelagic silverfish *Pleuragramma antarcticum* (Ploetz, 1986) and in winter on benthopelagic ice fish (Ploetz et al., 1987) while emperors have a mixed diet of krill, squid, and *P. antarcticum*, at least in winter (Klages, 1989).

The abundance of squid in the Weddell Sea and in the Southern Ocean in general is still unknown because of the lack of adequate sampling methods. The very small number of squid obtained so far by conventional catching methods is in contrast to the substantial number of squid beaks found in the stomachs of seals and penguins.

Conclusion

Most of the Weddell Sea is covered by sea ice year round. The ice cover breaks up irregularly, forming ephemeral leads, but also polynyas, which have a longer life, particularly along the eastern and southern ice shelves. In summer the coastal polynya houses an important phytoplankton bloom. In winter and early spring the polynya phytoplankton is particularly poor in the immediate vicinity of the ice shelves compared with production at the underside of the annual ice. Overall primary production in the Weddell Sea is low and mainly limited to the very short summer period. On the other hand, the patches of ice algae attached to the underside and crevasses of packice are feeding grounds for *Euphausia superba*, particularly in spring. Generally speaking, the annual sea ice of the Weddell Sea is of an advantage to the food web, as it does not preclude primary production and provides feeding ground and shelter for many organisms. This, together with the high nutrient supply to the euphotic zone, gives the Weddell Sea an advantage over the Arctic Basin with its less dynamic cover of multiyear ice and its permanent haline stratification and nutrient exhaustion.

Compared with the productivity of the

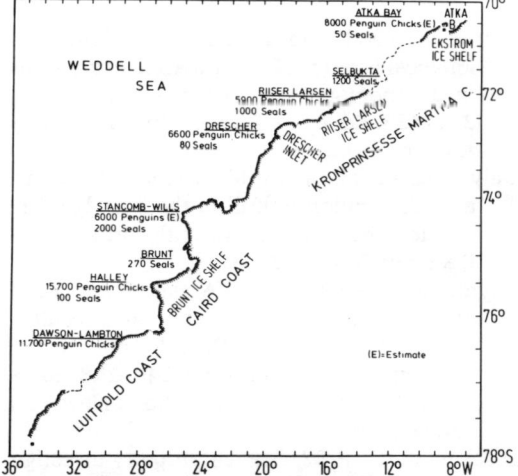

Figure 1.9. Distribution of emperor penguins and Weddell seals along the eastern Weddell Sea coast. (From Hempel and Stonehouse, 1987.)

Seasonal Ice Zone, the pelagic communities are poor in the Weddell Sea and other parts of the Antarctic coastal belt. The benthos, however, is much richer in biomass than one would expect from the low pelagic production. The P/B ratio in the benthos of the Weddell Sea is presumably very low. The same holds true for the fish fauna. The pelagic fish *Pleuragramma antarcticum* seems to be ecologically more important in the Weddell Sea than elsewhere in the Antarctic. It replaces krill as a food basis for the warm-blooded top predators. Abundance and year-round occurrence of Weddell seals and emperor penguins might have significant effects on their prey stocks, at least locally.

The old fauna of the Weddell Sea seems to be adapted to low food supply and near-freezing temperatures by low rates of growth and reproduction. This means any recovery from human impacts would take much longer than elsewhere.

In contrast to the Seasonal Ice Zone with its krill system, the Permanent Ice Zone of the Weddell Sea does not hold any promise for fisheries, including whaling.

Acknowledgments

The ecological description of the Weddell Sea as given in this chapter is largely based on the Ph.D. theses and postdoctoral studies of the following authors: Elisabeth Boysen-Ennen, Gerhard Dieckmann, Werner Ekau, Julian Gutt, Wilhelm Hagen, Gerd Hubold, Adolf Kellermann, Silke Kuehl-Kellermann, Peter Marschall, Eva-Maria Noethig, Uwe Piatkowski, Dieter Piepenburg, Jochen Ploetz, Renate Scharek-Mutlu, Wiebke Schwarzbach, and Joachim Voss.

Most of the studies were carried out at the Alfred-Wegener-Institute for Polar and Marine Research and the Institute for Polar Ecology of the University of Kiel. Further comments were provided by Irmtraut Hempel (zooplankton), Victor Smetacek (primary production), and Michael Spindler (ice fauna).

My sincere thanks go to all of the contributors. It was a great pleasure to work with them during the expeditions. The tedious work of sorting the samples, identifying species — often poorly described in the literature — and of compiling the results is still not completed, but a picture of the Weddell Sea ecosystem is already emerging.

References

Ackley, S.F. 1981. A review of sea-ice weather relationships in the Southern Hemisphere. *In* Sea level, ice, and climate change. Proceedings of the Canberra Symposium, Dec. 1979. pp. 127–159. IAHS Publ. No. 131.

Bodungen, B. von, Noethig, E. M., and Sui, Q. 1988. New production of phytoplankton and sedimentation during summer 1985 in the southeastern Weddell Sea. J. Compar. Biochem. Physiol. 90B:475–487.

Boyd, C. M., Heyraud, M., and Boyd, C. N. 1984. Feeding of the antarctic krill *Euphausia superba*. J. Crust. Biol. 4(Spec. No. 1):123–141.

Boysen-Ennen, E., and Piatkowski, U. 1989. Meso- and macrozooplankton communities in the Weddell Sea. Polar Biol. 9:17–35.

Broeckel, K. von. 1985. Primary production data from the southeastern Weddell Sea. Polar Biol. 4:75–80.

Bullivant, J. S. 1967. The fauna of the Ross Sea. Part 5. Ecology of the Ross Sea benthos. N. Z. Dept. Scient. Industr. Res., Bull. 176, pp. 49–75.

Clarke, A. 1983. Life in cold water: The physiological ecology of polar marine ectotherms. *In* Oceanography and marine biology 21. pp. 341–453. Ed. by M. Barnes. University Press, Aberdeen, UK.

Daly, K. L., and Macaulay, M. C. 1988. Abundance and distribution of krill in the ice edge zone of the Weddell Sea, austral spring 1983. Deep-Sea Res. 35:21–41.

Dayton, P. K., Robilliard, G. A., and Paine, R. T., 1970. Benthic faunal zonation as a result of anchor ice at McMurdo Sound, Antarctica. *In* Antarctic ecology, Vol. 1. pp. 244–258. Ed. by M. W. Holdgate. Academic Press, New York and London.

Dieckmann, G. 1987. Sea ice property investigations. *In* The winter-expedition of RV *Polar-*

stern to the Antarctic (ANT V/1-3). pp. 96–99. Ed. by S. Schnack-Schiel. Ber. Polarforsch. 39.

Ekau, W. 1988. Oekomorphologie nototheniider Fische aus dem Weddellmeer, Antarktis. Ber. Polarforsch. 51. 140 pp.

El-Sayed, S. Z. 1971. Observations on a phytoplankton bloom in the Weddell Sea. In Biology of the Antarctic Seas IV. pp. 301–312. Ed. by G. A. Llano and U. E. Wallen. Ant. Res. Ser. 17.

El-Sayed, S. Z., and Taguchi, S. 1981. Primary production and standing crop along the ice edge in the Weddell Sea. Deep-Sea Res. 28:1017–1032.

Everson, I. 1977. The living resources of the Southern Ocean. FAO GLO/SO/77/1, Rome. 156 pp.

Fransz, G. 1988. Vernal abundance, structure and development of epipelagic copepod populations under the sea ice of the eastern Weddell Sea. Polar Biol. 9:107–114.

Garrison, D. L., Sullivan, C. W., and Ackley, S. F. 1986. Sea ice microbial communities in Antarctica. BioScience 36:243–250.

Gieskes, W. W. C., Veth, C., Woehrmann, A., and Graefe, M. 1987. Secchi disc visibility world record shattered. EOS 68:123.

Gruzov, E. N., Propp, M. V., and Pushkin, A. F. 1968. Biological communities of coastal areas of the Davis Sea (based on observations of divers). Soviet Antarctic Expedition Information Bulletin 6(6):523–533.

Gutt, J. 1988. On the distribution and ecology of Sea Cucumbers (Holothuroidea, Echinodermata) in the Weddell Sea (Antarctica). Ber. Polarforsch. 41. 87 pp.

Hagen, W. 1988. Zur Bedeutung der Lipide im antarktischen Zooplankton. Ber. Polarforsch. 49. 129 pp.

Hempel, G. 1985. On the biology of polar seas, particularly the Southern Ocean. In Marine biology of polar regions and effects of stress on marine organisms. pp. 3–33. Ed. by J. S. Gray and M. E. Christiansen. John Wiley, London.

Hempel, G., 1988. Antarctic marine research in winter: The winter Weddell Sea project 1986. Polar Record 24:43–48.

Hempel, I., and Hempel, G. In preparation. Macrozooplankton in the eastern Weddell Sea in winter. Polar Biol.

Hempel, G., and Stonehouse, B. 1987. Aerial counts of penguins, seals and whales in the eastern Weddell Sea. In The winter-expedition of RV Polarstern to the Antarctic (ANT V/1-3). pp. 227–230. Ed. by S. Schnack-Schiel. Ber. Polarforsch. 39.

Hempel, I., Hubold, G., Kaczmaruk, B., Keller, R., and Weigmann-Haass, R. 1983. Distribution of some groups of zooplankton in the inner Weddell Sea in summer 1979/80. Ber. Polarforsch. 9. 36 pp.

Horner, R. A. 1985. Sea ice biota. CRC Press, Inc., Boca Raton, FL. 215 pp.

Hubold, G. 1984. Spatial distribution of Pleuragramma antarcticum (Pisces: Nototheniidae) near the Filchner and Larsen Ice Shelves (Weddell Sea, Antarctica). Polar Biol. 3:231–236.

Hubold, G. 1985a. Stomach contents of the Antarctic silverfish Pleuragramma antarcticum from the southern and eastern Weddell Sea (Antarctica). Polar Biol. 5:43–48.

Hubold, G. 1985b. On the early life history of the high Antarctic silverfish Pleuragramma antarcticum. In Antarctic nutrient cycles and food webs. Proceedings of the 4th SCAR Symposium on Antarctic Biology. pp. 445–451. Ed. by W. R. Siegfried, P. R. Condy, and R. M. Laws. Springer, Berlin, Heidelberg, New York.

Hubold, G., and Ekau., W. 1987. Midwater fish fauna of the Weddell Sea, Antarctica. Proc. V Congr. europ. Ichthyol. Stockholm, 1985. pp. 391–396.

Hubold, G., and Hempel, I. 1987. Seasonal variability of zooplankton in the southern Weddell Sea. Meeresforschung 31(3–4):185–192.

Hubold, G., Hempel, I., and Meyer, M. 1988. Zooplankton communities in the southern Weddell Sea (Antarctica). Polar Biol. 8:225–233.

Hubold, G., and Tomo, A. 1989. Age and growth of Antarctic silverfish Pleuragramma antarcticum Boulenger, 1902 from southern Weddell Sea and Antarctic Peninsula. Polar Biol. 9:205–212.

Jennings, J. C., Gordon, L. J., and Nelson, D. M. 1984. Nutrient depletion indicates high primary productivity in the Weddell Sea. Nature 309:51–53.

Kawaguchi, K., Ishikawa, S., and Matsuda, O. 1986. The overwintering strategy of Antarctic krill (Euphausia superba Dana) under the coastal fast ice off the Ongul Islands in Luetzow-Holm Bay, Antarctica. Mem. Natl. Inst. Polar Res., Spec. Issue 44:67–85.

Kellermann, A. 1987. Food and feeding of postlarval and juvenile Pleuragramma antarcticum (Pisces: Notothenioidei) in the seasonal pack-ice zone off the Antarctic Peninsula. Polar Biol. 7:307–315.

Klages, N. 1989. Food and feeding ecology of emperor penguins in the eastern Weddell Sea during the chick-rearing period. Polar. Biol. 9:385-390.

Knox, G. A., and Lowry, J. K. 1977. A comparison between the benthos of the Southern Ocean and the North Polar Ocean with special reference to the Amphipoda and Polychaeta. In Polar oceans. pp. 423–462. Ed. by M. J. Dunbar. Arctic Institute of North America, Calgary, Canada.

Kock, K-H. 1986. The state of exploited Antarctic fish stocks in the Scotia Arc Region during SIBEX (1983–1985). Arch. FischWiss. 37(Beih 1):129–186.

Kock, K-H., Schneppenheim, R., and Siegel, V. 1984. A contribution to the fish fauna of the Weddell Sea. Arch. FischWiss. 34(2/3):103–120.

Kottmeier, S. T., and Sullivan, C. W. 1987. Late winter primary production and bacterial production in sea ice and seawater west of the Antarctic Peninsula. Mar. Ecol. Prog. Ser. 35:175–186.

Marschall, H-P. 1988. The overwintering strategy of Antarctic krill under pack ice. Polar Biol. 9:129–135.

Nemoto, T., and Harrison. G. 1981. High latitude ecosystems. In Analysis of marine ecosystems. pp. 95–126. Ed. by A. R. Longhurst. Academic Press, London.

Noethig, E-M. 1988. Untersuchungen zur Oekologie des Phytoplanktons in der suedoestlichen Weddell See im australen Sommer 1985 (Januar/Februar, ANT III/3). Ber Polarforsch. 53. 118 pp.

O'Brien, D. P. 1988. Direct observations of the behavior of *Euphausia superba* and *Euphausia crystallorophias* (Crustacea: Euphausiacea) under pack ice during the Antarctic spring of 1985. J. Crust. Biol. 7:437–448.

Piatkowski, U. 1987. Zoogeographische Untersuchungen und Gemeinschaftsanalysen antarktischem Makroplankton. Ber. Polarforsch. 34. 150 pp.

Piepenburg, D. 1988. Zur Zusammensetzung der Bodenfauna in der westlichen Framstrasse. Ber. Polarforsch. 52. 118 pp.

Ploetz, J. 1986. Summer diet of Weddell seals (*Leptonychotes weddellii*) in the eastern and southern Weddell Sea, Antarctica. Polar Biol. 6:97–102.

Ploetz, J., et al. 1987. Weddell seals and emperor penguins in Drescher Inlet. pp. 222–227. Ed. by S. Schnack-Schiel. Ber. Polarforsch. 39.

Price, H. J., Boyd, K. R., and Boyd, C. M. 1988. Omnivorous feeding behavior of the Antarctic krill *Euphausia superba*. Mar. Biol. 97:67–77.

Rosenberg, A. A., Beddington, J. R., and Besson, M. 1986. Growth and longevity of krill during the first decade of pelagic whaling. Nature 324:152–154.

Sakshaug, E., and Holm-Hansen, O. 1984. Factors governing pelagic production in polar oceans. In Marine phytoplankton and productivity. Lecture notes on coastal and esturine studies 8. pp. 1–18. Ed. by O. Holm-Hansen, L. Bolis, and R. Gilles. Springer-Verlag.

Schwarzbach, W. 1988. Die Fischfauna der oestlichen und suedlichen Weddell-See: Geographische Verbreitung, Nahrung und trophische Stellung der Fischarten. Ber. Polarforsch. 54. 94 pp.

Smetacek, V. 1987. The under-ice water layer. In The winter-expedition of RV *Polarstern* to the Antarctic. pp. 182–190. Ed. by S. Schnack-Schiel. Ber. Polarforsch. 39.

Spindler, M., and Dieckmann, G. S. 1986. Distribution and abundance of the planktonic foraminifer *Neogloboquadrina pachyderma* in sea ice of the Weddell Sea (Antarctica). Polar Biol. 5:185–191.

Stonehouse, B., and Hempel, G. In preparation. Census of emperor penguins (*Aptenodytes forsteri*) and Weddell seals (*Leptonychotes weddellii*) on the ice shelves of the eastern Weddell Sea. Polar Biol.

Sullivan, C. W. S., Palmisano, A. C., Kottmeier, S., and McGrath-Grossi, S. 1985. The influence of light on growth and development of sea-ice microbial communities in McMurdo Sound. In Antarctic nutrient cycles and food webs. pp. 78–83. Ed. by R. Siegfried, P. R. Condy, and R. M. Laws. Springer-Verlag, Berlin.

Ushakov, P. V. 1963. Some characteristics of the distribution of bottom fauna off the coast of East Antarctica. Soviet Antarctic Expedition No 40, 5–13. Inform. Bull. Soviet Antarctic Expetition 4:287–292.

Voss, J. 1988. Zoogeography and community analysis of macrozoobenthos of the Weddell Sea (Antarctica). Ber. Polarforsch. 45. 145 pp.

Wefer, G., Fischer, G., Fuetterer, D., and Gersonde, R. 1988. Seasonal particle flux in the Bransfield Strait (Antarctica). Deep-Sea Res. 35:891–898.

Chapter 2

Environmental Influence on Recruitment and Biomass Yields in the Norwegian Sea Ecosystem

*Bjornar Ellertsen, Petter Fossum, Per Solemdal,
Svein Sundby, Snorre Tilseth*

Abstract

Results from egg and larval surveys of the Arcto-Norwegian cod stock are used together with long time series of data on cod peak spawning, temperature, and zooplankton and compared with abundance indices of Arcto-Norwegian cod. The effect of environmental factors at the spawning ground are discussed in relation to recruitment variability.

Mortality estimates of egg and larval stages compared with postlarval and O-group abundance indices have shown that the year-class strength is established during the first two months after peak spawning. These findings indicate that the critical mortality mechanism in the egg and larval stage point toward environmentally controlled mortality as regulatory factors in year-class variability.

Introduction

The variation in fish population size caused by variations in year-class strength was first documented by Hjort (1914). He also suggested an explanation for these variations:

It may well be imagined that a certain — though possibly brief — lapse of time might occur between the period when the young larvae first require extraneous nourishment and the period when such nourishment is first available. If so, it is highly probable that an enormous mortality would result. It would then also be easy to understand that even the richest spawning might yield but a poor amount of fish, while poorer spawning, taking place at a time more favorable in respect to the future nourishment of the young larvae, might often produce the richest year classes.

This idea, the critical period or match-mismatch concept, has been the subject of several investigations trying to document the mass mortality of early larvae, as reviewed by Marr (1956), May (1974), and Dahlberg (1979). Due to overwhelming sampling problems, no decisive answer on the effect of starvation was documented and other factors, like predation on the larval stage, were proposed to be more important for the annual variation in year-class strength (Anonymous, 1975).

Against this background it is interesting to note that two of the more comprehensive recruitment studies on cod, starting in the mid 1970s at the Northeast Fisheries Center (United States), and the Institute of Marine Research (Norway), have both focused on the effect of starvation of early larvae.

In 1984 a new concept was put forward by the Americans focusing on predation on postlarvae and O-group as the main regulator of year-class strength (Anonymous, 1984). In 1987 a change to a broader concept of the recruitment problem was again put forward (Anonymous, 1987).

Consequently we collectively feel that we need to change the research strategy to a

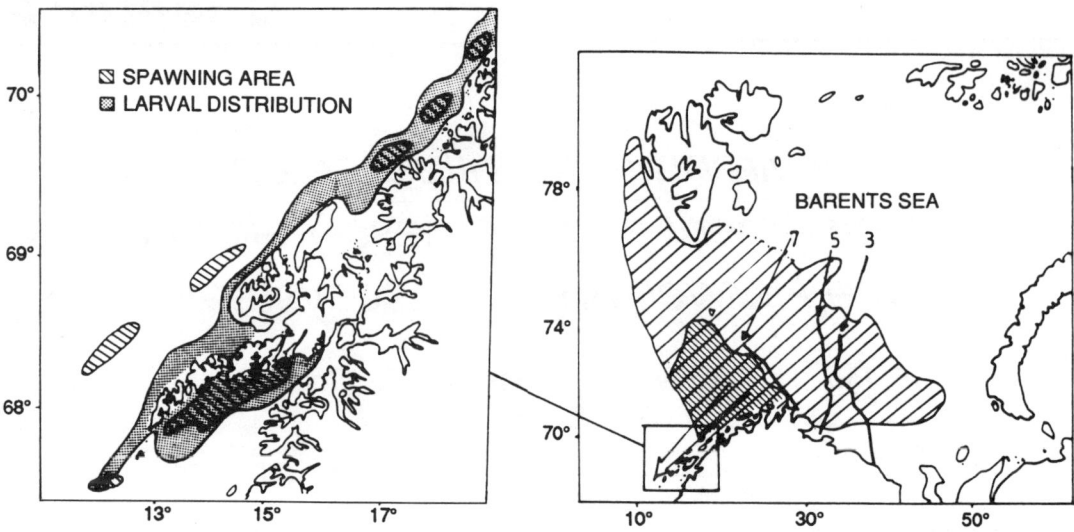

Figure 2.1. The distribution of the Arcto-Norwegian cod, ▧ Postlarvae June–July, ▨ O-group August–Sept., "3" = eastern boundary of the 3-year old cod, ⟹ spawning migration.

broader evaluation of the problem, de-emphasizing the mechanistic approach for the time being and more toward providing information on mortality rates within a life table context before we can resume specific hypothesis testing.

The studies were carried out on the Arcto-Norwegian (also named Northeast Arctic) cod (*Gadus morhua* L.), a species distributed in the Atlantic water masses of the Barents Sea and the west coast of Spitsbergen. The adult fish experience temperatures between 0° and 10°C, while the main part of the stock is found at temperatures between 2° and 6°C (Loeng and Sundby, 1986). Spawning occurs outside the feeding area along the coast of Norway. Distribution of Arcto-Norwegian cod is given in Figure 2.1.

Three different activities were initiated at the start of the project: (i) basic studies of larvae in the laboratory; (ii) enclosure experiments which developed into the mass production of O-group cod (Kvenseth and Øiestad, 1984); (iii) field investigations, mainly in the Lofoten area, including investigations on the vertical distribution and migration of cod larvae and nauplii, the effect of wind, transport and spreading processes, and the patchiness of nauplii.

Spawning, Development, and Distribution of Eggs

The spawning areas of Arcto-Norwegian cod are patchy, located in the Norwegian coastal current off mid- and north-Norway along a 1,200-km coastline. The spawning areas are the same sites every year, although the magnitude of the spawning at the sites may vary. Most of the eggs are spawned along a limited part of the coastline, in Lofoten and Vesterålen, between 67°30'N and 68°30'N. From 1983 to 1985, 60–70% of the total egg number were spawned here, when the production varied between 3.1×10^{13} and 4.7×10^{13} eggs (Sundby and Bratland, 1987). Especially in Lofoten where the spawning schools are very dense, the peak egg concentration may exceed 10,000 eggs $\cdot$ m^{-2} surface (Wiborg, 1950; Sundby, 1980). Such high concentrations are caused not only by dense spawning, but are also due to higher retention at the Lofoten spawning grounds than at the other spawning sites (Sundby and Bratland, 1987). A characteristic feature of all the larger and most of the medium-sized spawning fields is that they are found either in bays close to the coast (the Lofoten fields), or on offshore banks where the bottom topography induces a clockwise circulation (Sundby, 1984). Both in the bays and on the banks where spawning takes place, the residence time of the water masses is prolonged, and consequently this will also be the case for eggs that develop in these areas, an analog to retention areas (Iles and Sinclair, 1982). However, not all of the major spawning grounds are found in areas of high residence time of the water masses. Especially at some of the Vesterålen offshore spawning grounds, cod eggs are subjected to rapid transport and dispersion (Sundby and Bratland, 1987). Altogether, about 20–30% of the eggs are spawned at sites where there is short residence time of the water masses. Most of these eggs are spawned close to the offshore front between the coastal water and the Atlantic water. In this region both water masses move rapidly northwards along the shelf edge.

The influence of the changing winds has been thoroughly described in Lofoten (Ellertsen et al., 1981a,b; Furnes and Sundby, 1981; Sundby, 1981). Northeasterly winds in Vestfjorden increase the egg transport out of the fjord, while southwesterly winds increase the spreading within the fjord. Calm wind conditions give a low rate of both transport and spreading.

The spawning occurs in March and April. In Lofoten, where the largest spawning sites are found, the spawning intensity has been monitored each year since 1975 (Solemdal, 1982a; Ellertsen et al., 1984). There is a remarkable constancy in the spawning procedure each year; starting during the first days of March, it reaches maximum intensity during the first week of April and gradually comes to an end within the first half of May (Figure 2.2). The spawning intensity has also been monitored at several other major spawning fields (Anonymous, 1983), revealing a tendency towards later spawning at the grounds farther north. The spawning period is delayed by about two weeks at the northernmost spawning field relative to the Lofoten area (Sundby and Bratland, 1987).

Pedersen (1984, 1985) examined the

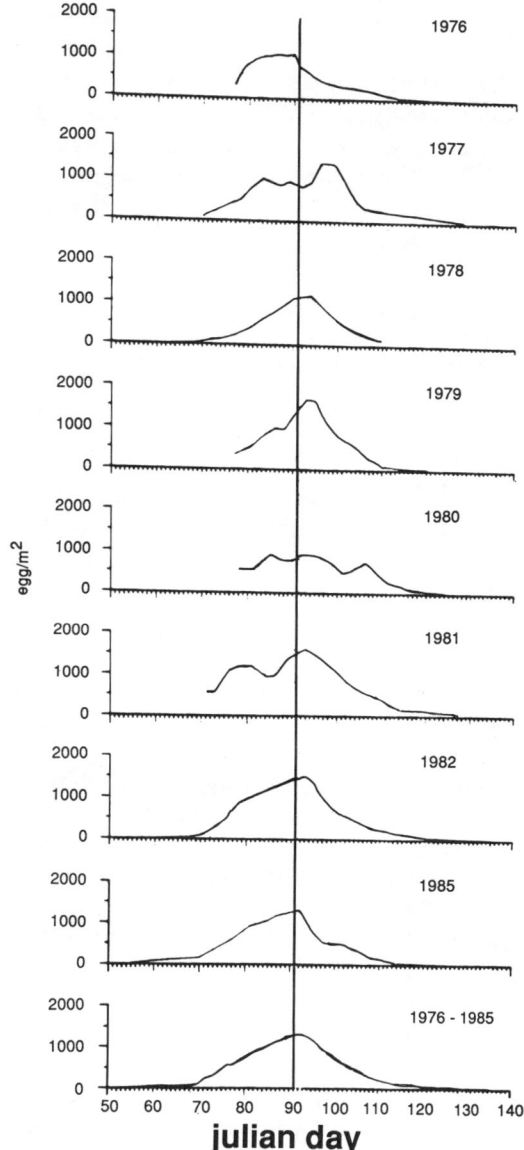

Figure 2.2. Spawning intensity curves for the years 1976–1983 and the mean spawning curve. Vertical bar represents mean date of 50% spawning for the period 1976–1983. (From Pedersen, 1984.)

long-term changes of the maximum spawning intensity in Lofoten based on cod roe investigations from fish catch statistics since 1929. He found a tendency towards later spawning, approximately 7–15 days, during the period from 1930 to 1960.

The Arcto-Norwegian cod spawn in the thermocline between the cold coastal water and warmer Atlantic water, at temperatures between 4° and 6°C, both pelagically and where these layers intersect the bottom (Gade, 1894; Sund, 1925). Distribution of 3 to 5-day-old eggs are shown in Figures 2.3 and 2.4. The depth of the thermocline may vary considerably from year to year (Eggvin, 1932), but short-term variations caused by wind effects may also be of great importance for the depth of the thermocline, causing upwelling or downwelling at the spawning fields (Furnes and Sundby, 1981; Ellertsen et al., 1981a).

The eggs are buoyant compared to the natural environment, and hence tend to rise towards the surface layers (Solemdal, 1970; Solemdal and Sundby, 1981). The neutral buoyancy and the egg diameter for the entire egg population are Gaussian distributed, and range from 29.5 to 33.0 p.s.u., and from 1.2 to 1.6 mm, respectively. From these values the ascending speed of the eggs was calculated to range from 0.2 to 1.8 mm $\cdot$ s^{-1} (Solemdal and Sundby, 1981; Sundby, 1983). Solemdal and Sundby (1981) showed that neutral buoyancy is correlated to the weight of the eggshell, whose thickness may vary between 5 and 9 μm (Davenport et al., 1981). The specific weight of the eggs increases through the development (Sundnes

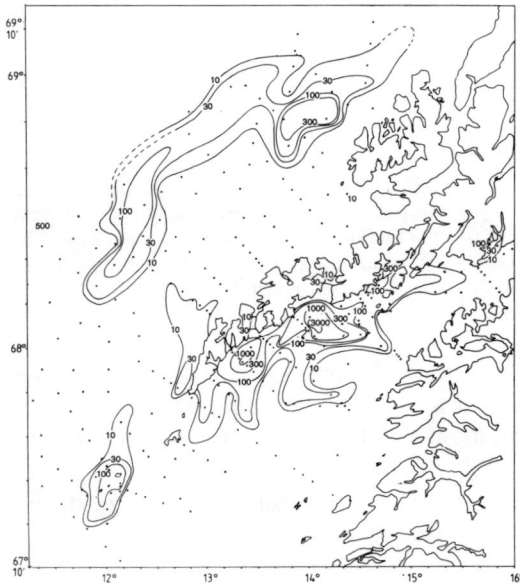

Figure 2.3. Distribution of 3-to-5-day-old eggs, 3–10 April 1985. (From Sundby and Bratland, 1987.)

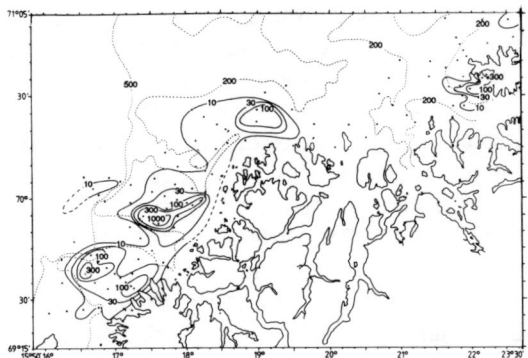

Figure 2.4. Distribution of 3-to-5-day-old eggs, 10–14 April 1985. (From Sundby and Bratland, 1987.)

et al., 1965). The temperatures 2.5° and 5°C used in Strømme's (1977) experiments placed the time from spawning to hatching at 30 and 18 days, respectively. It has been shown that environmental temperature at the spawning fields may go below 1°C in extreme years, as in 1966 and 1981 (Ellertsen et al., 1987), during which the time from spawning to hatching is considerably longer. In 1981 the hatching time for the earliest spawned eggs was about 35 days (Ellertsen et al., 1987; Solemdal, 1984).

Knutsen and Tilseth (1985) found that large eggs produce large larvae. Kjesbu (1989) investigated the first of 10–15 batches of eggs spawned during eight weeks and found that large female spawners produce large eggs. During spawning a significant reduction in egg size and dry weight was recorded. Solemdal and Sundby (1981) also found that size of eggs in the field decreases as the spawning season progresses.

The eggs are found in the entire upper mixed layer, but the concentration increases towards the surface. During calm conditions high concentrations are found at the surface, and most of the eggs are found above 20 m depth. During rough weather the eggs are mixed downward due to the influence of wave action and turbulence (Figure 2.5). When the wind speed exceeds 15 m $\cdot$ s^{-1}, the vertical concentration profile of cod eggs shows only a very slight increase towards the surface and may be found in relatively large concentrations down to the bottom of the

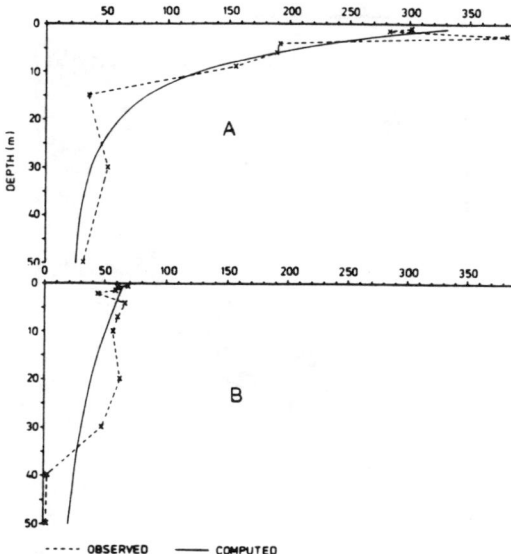

Figure 2.5. The vertical distribution of cod eggs during calm conditions (**A**) and during rough weather conditions (**B**). (From Sundby, 1983.)

mixed layer. Models for how the vertical distribution of pelagic and bathypelagic eggs in general are influenced by buoyancy and turbulence are described by Sundby (1983), Sundby (1989), and Westgård (1988). The vertical spreading of the pelagic Arcto-Norwegian cod eggs is mainly determined by the wind-induced turbulence, and only to a very small extent by the buoyancy distribution.

Egg mortality estimates for the years 1983 and 1984 have been made in Lofoten where the retention is high (Ellertsen et al., 1987; Fossum, 1988), based on population fecundity estimates (Sundby and Solemdal, 1984; Sundby and Bratland, 1987) and subsequent measurements of late egg stages. In both years the mortality from spawning to hatching was 90% (Fossum, 1988) (Figure 2.10). The causes of mortality are not well known, but there is evidence that predation by herring may be considerable (Melle and Ellertsen, 1984; Melle, 1985). The stock fecundity estimates are assumed to have a high precision, since they were in very good agreement with acoustic measurements of the biomass of spawning fish (Gødo et al., 1984; Gødo et al., 1985).

Distribution and Feeding of Early Larvae

The cod larvae in the Lofoten area are, like the eggs, trapped in the mixed water masses of the cold coastal current, and consequently their distribution is determined by the dynamics of the coastal current. The maximum recorded concentration of larvae is approximately 200 per m^2 surface, which is about 5% of the maximum recorded concentration of newly spawned eggs. The number of larvae estimated during one survey varied between years from 2×10^9 to 1.2×10^{11} (Ellertsen et al., 1987). The horizontal distribution of larvae has been recorded since 1979. The horizontal distribution of larvae for the two years 1981 and 1983 is shown in Figure 2.6.

1981 and 1983 represent two extreme years with respect to temperature during incubation, with mean March/April temperatures in the upper 30 m of the Vestfjord 1.6° and 3.6°C, respectively. The peak of hatching varied by approximately two weeks between 1981 and 1983 (Solemdal, 1984; Ellertsen et al., 1987). The annual temperature variations of the coastal current will influence the time and location of first feeding. Spawning curves and first feeding curves for the two extreme years 1981 and 1983 are shown in Figure 2.7.

Investigations on the vertical distribution of cod larvae, carried out with large pumps (Solemdal, 1982b; Solemdal and Ellertsen, 1984), showed that the majority of first feeding larvae in the Lofoten waters are distributed in the upper 30 m of the water column (Ellertsen et al., 1979; Wiborg, 1948). Diurnal vertical migration depends on the ability of the larvae to move as well as vertical turbulence. During extremely calm conditions larvae "control" their vertical position in the water column and show diurnal migration (Ellertsen et al., 1984) (Figure 2.8).

Like most marine fish larvae, cod larvae are visual feeders (Ellertsen et al., 1980) and select their prey on the basis of size (Ellertsen et al., 1979). Larval gut content analysis

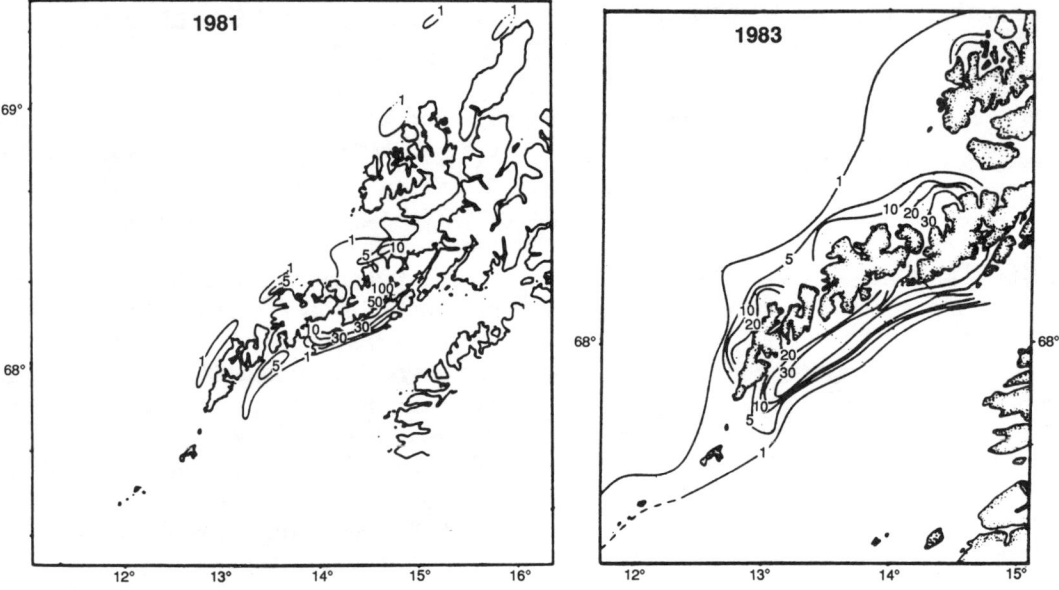

Figure 2.6. Distribution and abundance of cod larvae in 1981 and 1983. (From Ellertsen *et al.*, 1987.)

from the Lofoten area shows that nauplii of *Calanus finmarchicus* are the most dominant prey item (Wiborg, 1948; Ellertsen *et al.*, 1984). A larval bioassay technique (Lasker, 1975) tested in Lofoten showed that the cod larvae were able to capture nauplii at rather low concentrations (Ellertsen *et al.*, 1979), and that larvae were feeding on copepod fecal pellets. Fish larvae have been frequently reported with green food remains in the gut (Lebour, 1919). Wiborg (1948) reports the same findings in cod larvae from the Lofoten area. Nordeng and Bratland (1971) have identified the phytoplankters *Peridinium pellucidum* and *Coscinodiscus* sp. in the gut of cod larvae from the same area. Analysis of polyunsaturated fatty acids in the cod larvae from Lofoten shows influence of specific fatty acids from phytoplankton (Tilseth *et al.*, 1987). The significance of this observation from a nutritional point of view is not yet fully understood.

Ellertsen *et al.* (1984) examined cod larval distribution and prey availability in relation to physical factors in Lofoten during larval first feeding. Feeding conditions were influenced by prey density. Increasing wind broke down patches of high prey concentration, and consequently the feeding ability was reduced.

Ellertsen *et al.* (1987) examined the prey density and larval distribution in different regions of the Lofoten area and found higher prey densities, higher larval feeding incidence, and higher number of larvae with full gut in the areas of Vestfjorden where the retention is large. In the surrounding areas these parameters were much more variable. Comparison of gut fullness of stage 7 larvae (Fossum, 1986) caught in the upper 40 m

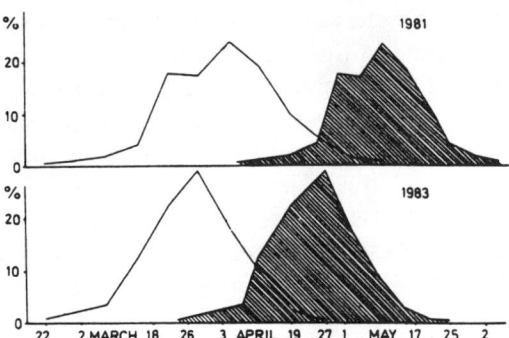

Figure 2.7. Spawning curves (**open**) and first feeding curves (**hatched**) from the years 1981 and 1983. (From Ellertsen *et al.*, 1987.)

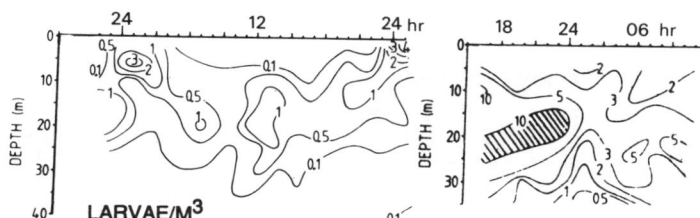

 Figure 2.8. Under extremely calm wind conditions (**left**), the larvae are able to show vertical migration. Under less calm conditions, the larvae are more evenly distributed (**right**). (From Ellertsen et al., 1984.)

during daytime and the integrated nauplii concentration in the water column showed that maximum gut fullness was reached when the nauplii concentration exceeded 10 per liter. The empirical data indicate a critical prey density of 5–10 nauplii per liter (Figure 2.9).

Ellertsen et al. (1987) and Fossum (1988) examined survival of cod eggs and larvae within the Lofoten spawning area in 1983 and 1984, when strong year classes were produced. Both years, only 10% of the eggs reached hatching, and 2–3% reached first feeding (Figure 2.10). Larval length/dry weight results for the first feeding periods between 1982–1985 indicate that the first feeding larval condition is correlated to the year-class strength.

In experiments designed to study the effect of plankton species, size, and light on cod larval feeding, the larvae were fed the flagellates *Dunaliella* sp. (7–9 μm) and *Peridinium trochoidum* (50–80 μm). *Dunaliella* sp. was not purposely ingested but concentrated on visceral arches. When the visceral arches were clogged, the flagellates were swallowed.

The larvae were able to spit out the particles when the jaw became functional, which was clearly demonstrated by the substantial reduction of larvae with "green guts" from the fourth to the fifth day after hatching concomitantly with the development of a functional jaw (Ellertsen et al., 1980).

The cod larvae proved to be active feeders on bigger phytoplankters. When given *P. trochoidum* as the only food, the feeding incidence of the larvae increased from the yolk sac stage 5 to yolk absorption and reached 90% on day 7 at 1,000 lux, and some feeding also occurred in complete darkness, probably due to the high particle density in the aquaria. Even at moderate densities the prey was consumed (Ellertsen et al., 1976), although the larvae would prefer particles bigger than 100 μm if those were present at the same time as *P. trochoidum* (Ellertsen et al., 1979).

Most marine fish larvae are visual feeders, and active feeding occurs only above a certain light intensity (Blaxter, 1966). Cod larvae fed *Artemia salina* were able to capture *Artemia* nauplii at 0.4 lux, but not at 0.1 lux, and highest feeding incidence was observed at 1.4 lux (Ellertsen et al., 1976, 1980). Providing that the light intensity threshold for visual feeding of cod larvae on nauplii is close to 0.1 lux, there are 22–24 hours available for feeding per diurnal period in May in the Lofoten area (Blaxter, 1966).

Tilseth and Ellertsen (1984b) studied larval cod feeding strategy, consumption, and digestion rates in laboratory experiments. Their results showed that feeding strategy varied with food availability, gut content volume, and state of digestion. First feeding cod larvae showed maximum gut fullness at average 3.5 nauplii (220 μm carapax length). The digestion rate was less

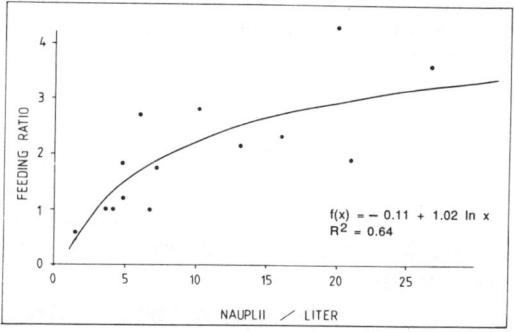

Figure 2.9. Feeding ratio in cod larvae in relation to food density. (From Ellertsen et al., 1987.)

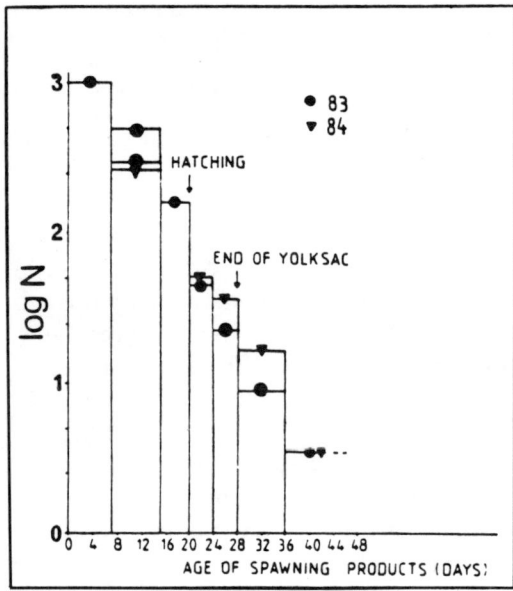

Figure 2.10. Survival of the egg and larvae, from spawning to larval age 20 days. (From Ellertsen et al., 1987.)

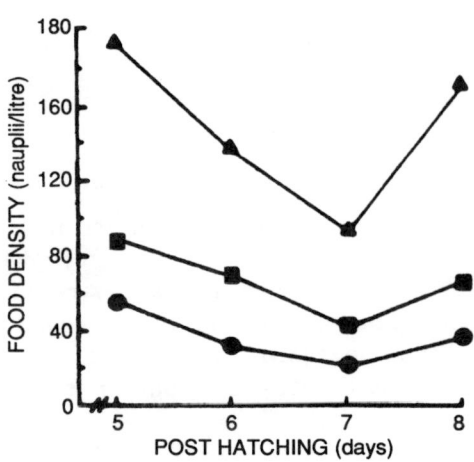

Figure 2.11. Food density requirements in first feeding larvae of cod. The curves are based on maximum (▲), mean (■), and minimum (●) values of swimming speed, feeding success, and catabolic energy demand. (From Solberg and Tilseth, 1984.)

than 30 minutes, depending on the gut volume. The evacuation rate was dependent on the food availability. Solberg and Tilseth (1984) examined cod larval metabolic requirements and feeding ability and related this to prey density requirements during first feeding. They found that the larvae were consuming more energy than supplied from the yolk at the end of the yolk sac stage. They pointed out that an average larva should need about 50 nauplii per liter to meet its metabolic requirements. They also pointed out that their simplified model was sensitive to larval search volume and feeding success (Figure 2.11).

Laboratory experiments showed that the cod larvae developed a functional mouth at 5°C, four days after hatching. The neutral buoyancy at hatching increased from 30 to 35 p.s.u. at the end of the yolk sac stage. Through the starvation period, the neutral buoyancy is first reduced and finally increased at the moribund stage (Tilseth and Strømme, 1976).

A study of the feeding, growth, and mortality of a cod larval cohort with prey production and temperature (Moksness, 1978) agreed well with the growth and mortality based on a mathematical model (Ellertsen et al., 1981c) (Figure 2.12). The mesocosm work developed into the mass production of O-group cod (Kvenseth and Øiestad, 1984).

Abundance of Nauplii

In investigating the vertical distribution of copepod nauplii different plankton pumps and particle rate meters have been used (Ellertsen et al., 1976, 1979; Mohus, 1981; Tilseth and Ellertsen, 1984a).

The highest concentrations of nauplii usually occur at 5–15 m depth, occasionally at the very surface (Ellertsen et al., 1979; Tilseth and Ellertsen, 1984a), and the general impression is that very high concentrations of nauplii at the surface usually occur in the evening during very calm weather (Ellertsen et al., 1979).

Almost all *C. finmarchicus* nauplii in the Lofoten occur within the upper 50 meters due to spawning close to the surface, the sinking velocity of eggs, and the incubation

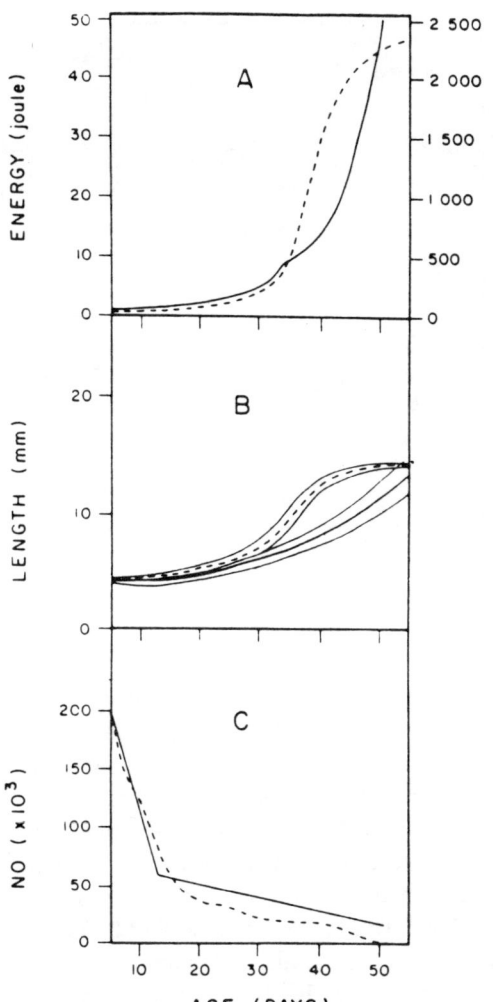

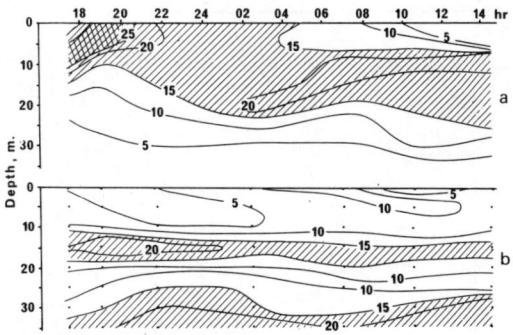

Figure 2.13. Vertical distribution of copepod nauplii, 10–11 May 1977. Relative abundance among eight sampling depths. (a) copepod nauplii > 120 μm, (b) copepod nauplii < 120 μm.

Figure 2.12. (A) Estimated mean dry weight and energy content of the cod larvae from samples in the basin (solid line) and computed mean energy content (dashed line) versus age. (B) Estimated mean standard length in mm of the larval population ± 1 SD (solid line) and computed length ± 1 SD (dashed line) versus age. (C) Estimated number of larvae in the population (solid line) and computed number (dashed line) versus age. (From Ellertsen et al., 1981c.)

time (Harding et al., 1951; Marshall and Orr, 1955; Salzen, 1956). However, the vertical distribution of copepod nauplii is not constant throughout a 24-hour period. Numerous 24-hour stations between 1975–1987 have revealed an increasing concentration in the upper 0–5 m in the evening versus a daytime maximum at 5–15 m (Ellertsen et al., 1979; Tilseth and Ellertsen, 1984a; and unpublished data). Based on 1977 data, we can conclude that while the medium- and larger-sized nauplii show a diurnal vertical migration, the smallest ones (< 120 μm carapax length) do not seem to migrate (Figure 2.13).

The first feeding cod larvae only migrate vertically during extremely calm weather conditions. Usually, their maximum concentration is at 10–20 m depth during the 24-hour period (Ellertsen et al., 1979; Tilseth and Ellertsen, 1984a). Cod larvae show a reduced feeding intensity at night (Ellertsen et al., 1979; Tilseth and Ellertsen, 1984b), coinciding in general with a period of reduced nauplii concentrations at 10–20 m.

The copepod nauplii are unevenly distributed horizontally in the Vestfjord and outside the Lofoten archipelago. Distributions for 1982 and 1984 are shown in Figure 2.14. Concentrations are generally highest in sheltered areas like the Austnesfjord (10–20 n · l^{-1}), followed by Vesterålsfjord (1–20 n · l^{-1}), the Lofoten east side (1–10 n · l^{-1}), and the west side of the Lofoten islands, usually 1–5 n · l^{-1} (Ellertsen et al., 1984, 1987).

Later investigations (1987) covering the whole Vestfjord south to 67°N, in February–May, reveal that the main production of *C. finmarchicus* nauplii starts in the southernmost area, reaching the central and northern parts of the Vestfjord about 1–2 weeks after the onset in the south (unpub-

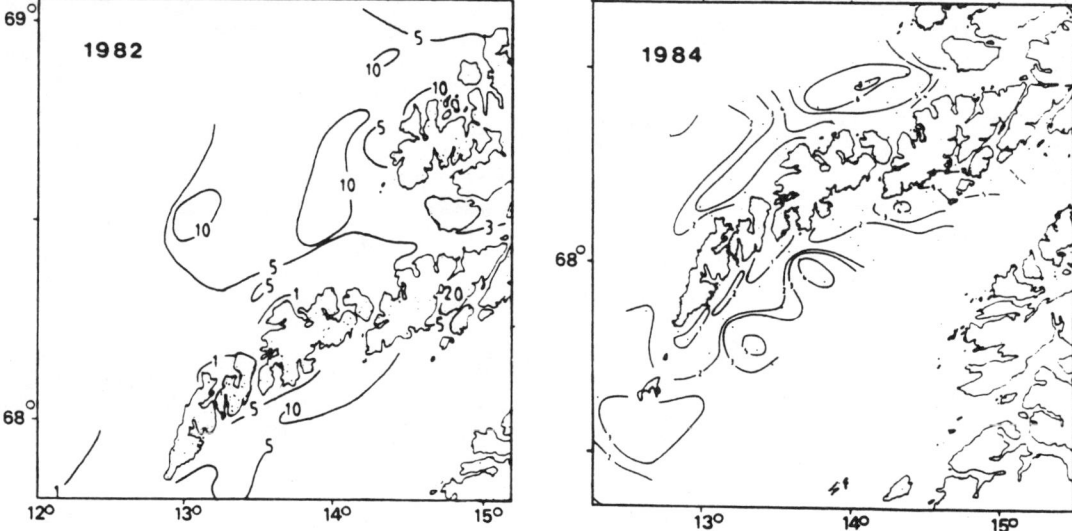

Figure 2.14. Distribution and abundance of copepod nauplii 1982 and 1984, number · l^{-1}. (From Ellertsen et al., 1987.)

lished data). The observed concentrations of nauplii in a north-south direction are therefore dependent upon the phase in the production cycle at the time of observation.

The occurrence of *C. finmarchicus* at a fixed station in Lofoten was analyzed for 1960–1984 (some years omitted due to insufficient sampling). In general the *C. finmarchicus* samples consist of 60–90% adult females at the end of March, but only 4–5% females a month later when the copepodite stage I (C I) and II (C II) dominate. A linear regression between the date, D, for maximum occurrence of *C. finmarchicus* C I and the mean sea temperature, t, of the upper 30 m in April (Figure 2.15), gives a correlation coefficient, r^2, of 0.72. The linear regression becomes

$$D = 176.6 - 17.55t$$

Nauplii are not sampled adequately from this fixed station due to coarse mesh size (180 μm) in the net.

As seen from Figure 2.15, the time of spawning of *C. finmarchicus* is directly affected by the temperature. In the extremely warm year 1960, when the mean monthly temperatures of the upper 30 m in March and April were 3.5° and 4.4°C respectively, the time of maximum occurrence of C I was about 1 April. This implies that spawning was most intense in late February/early March. In the extremely cold year 1981, when the March and April temperatures were 0.7° and 1.9°C, respectively, maximum occurrence of C I was observed in late May, indicating maximum spawning intensity in mid-April. In 1960 Baranenkova (1965) also reported an unusually early spawning of *C. finmarchicus* in the Norwegian Sea. That year the nauplii production was finished

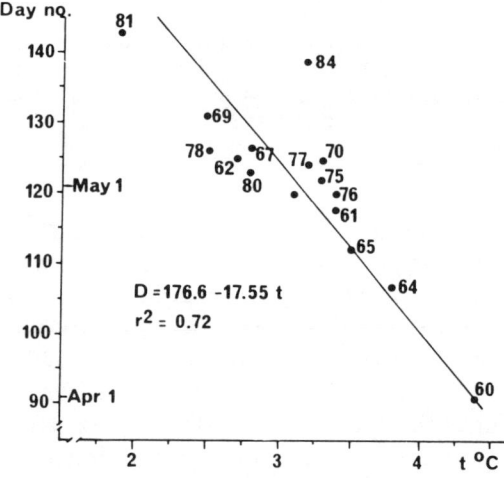

Figure 2.15. Time of maximum occurrence of *Calanus finmarchicus* copepodite stage I versus temperature.

when the first feeding cod larvae occurred in the area, perhaps explaining the below-average year class.

The match/mismatch between nauplii and cod larvae occurrence in the other years is the subject of the present investigation. Preliminary analyses seem to indicate that a higher degree of overlapping in time of larvae and nauplii takes place in years with medium temperatures.

Postlarval, O-Group, and Young Cod

Abundance indices of 2-to-3-month-old cod have been estimated for the years 1979–1981, and 1983–1987 (Bjørke and Sundby, 1984, 1986, 1987; Bjørke et al., 1988), based on midwater trawl surveys in June–July. Quantitative estimates were made from 1979. In 1982 the sampling was insufficient due to bad weather, and a reliable index could not be estimated.

The length of the larvae at this stage varies from about 20 mm to 45 mm depending on geographical area and year. Most years they are distributed in the southwestern part of the Barents Sea and in the eastern part of the Norwegian Sea. The postlarval cod is mainly found in coastal water, with horizontal distribution largely determined by the mesoscale bottom topography. Especially Tromsøflaket (bank) and Ingøydjupet (trough) have a strong influence on the circulation (Loeng and Sundby, 1986) (Figure 2.16). Some years there is an easterly and some years there is a westerly distribution of the larvae, but every year a large fraction (40–90%) of the year class is found at the bank Tromsøflaket (Bjørke and Sundby, 1987). In 1983, 90% of the year class at the postlarval stage was concentrated within the clockwise vortex at Tromsøflaket. This year the strongest year class since 1970 was produced, but we have no reason to believe that this was caused by the limited distribution of postlarvae. The highest temperatures are found in the western part of the area of distribution, coinciding with the highest fre-

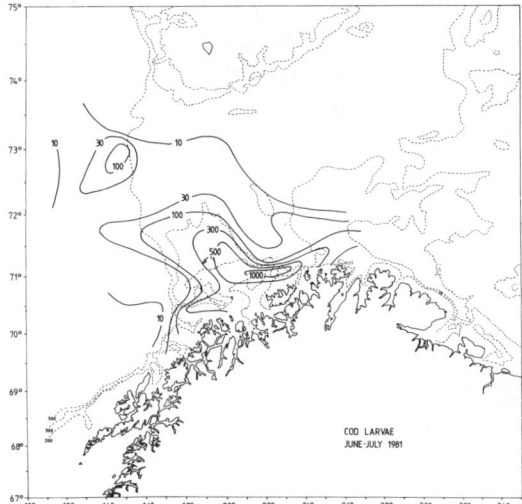

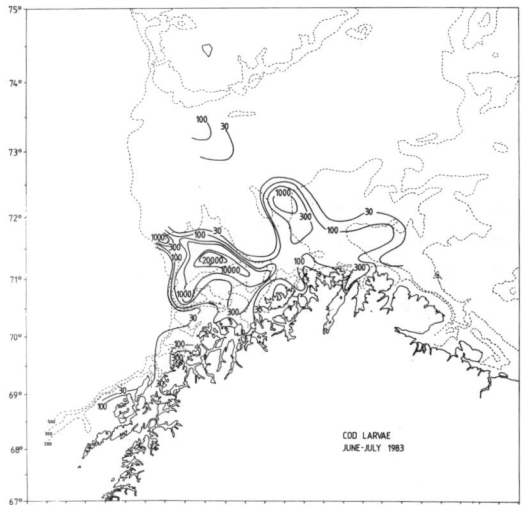

Figure 2.16. The distribution of cod postlarvae in June–July in 1981 and 1983. (From Bjørke and Sundby, 1987.)

quency of large larvae.

O-group surveys have been carried out on 4-to-5-month-old fish in the Barents Sea each year in August–September since 1965 (Anonymous, 1985) and results reported to the ICES (Figure 2.17). The surveys are based on acoustic measurements and trawl sampling. Although the O-group cod is more widely dispersed and advected farther into the Barents Sea than the concentrated postlarval cod, the same characteristic features between years appear with respect to westerly or easterly distribution of both post-larvae and O-group (Bjørke and Sundby, 1987). Randa (1984) analysed the O-group indices of cod and compared them to the abundance of 3-year-old fish, estimated by virtual population analysis. The correlation coefficient was 0.92, indicating that the O-group indices may be used as a reasonable measure for the year-class strength. Bjørke and Sundby (1987) compared the postlarval indices to the O-group indices and found a correlation coefficient of 0.82, significant at the 5% level.

Saetersdal and Loeng (1987) compared the year-class strength of cod at the young fish stage with the temperature conditions of the Barents Sea for the period 1900–1983. They found that strong year classes more frequently occurred during warm periods, and that poor year classes were most frequent during periods of cold in the Barents Sea. Ellertsen et al. (1987) compared the mean temperature of the upper 30 m during March–April during spawning, and the number of cod at age 3, taken from the ICES working group on Arctic fisheries (Anonymous, 1985) (Figure 2.18), and found poor year classes to be correlated with low temperatures, supporting Saetersdal and Loeng. Ellertsen et al. (1987) also concluded that only poor year classes were formed when the temperature was low.

Discussion and Conclusion

The year-class strength of the Arcto-Norwegian cod shows large variations. Data on the year-class abundance of three-year-olds during 1946–1985 (Anonymous, 1985) show that the difference between the average of the smaller year classes and the largest year class (1970, 1,800 mill.) is a factor of 15. Year-class abundance at age two months (Bjørke and Sundby, 1987) does not show a considerable larger range of variation. During the period 1979–1985 in which data on age two months exist, the difference is a factor of 20. Bjørke and Sundby (1987) showed that postlarval indices were correlated to O-group indices, while Randa (1984) showed that O-group indices were correlated to the abundance at age three years. He concluded that the O-group index could be applied in forecasting general year-class strength. This implies that much of the variation in year-class abundance is determined within the first months of life by the

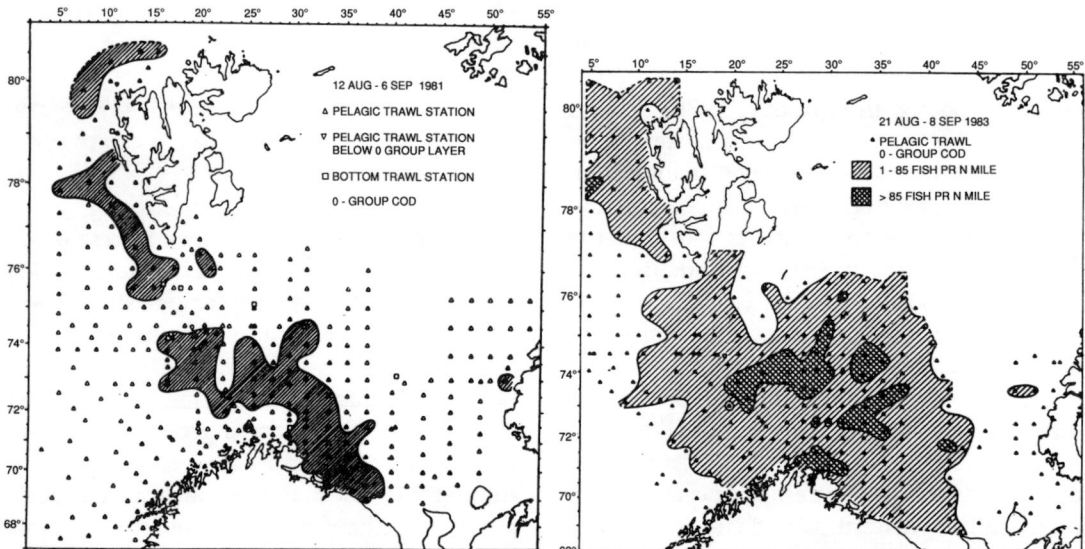

Figure 2.17. The distribution of O-group cod in August–September 1981 and 1983. (From Anonymous, 1985.)

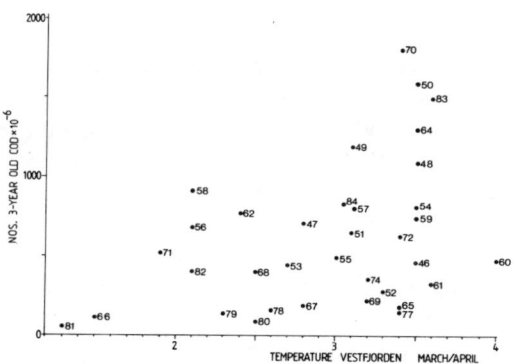

Figure 2.18. The relation between the size of the year class as a 3-year-old fish and the mean temperature in the upper 30 m of the sea in March–April in Lofoten.

varying effects of recruitment mechanisms.

A main purpose of the early larval field investigations was to monitor the feeding of larvae, their feeding conditions, and survival and compare these to later quantitative measurements of the year-class strength. It was expected that the causes of large or small year classes of the Arcto-Norwegian cod should also affect changes in the larval morphology and other biological parameters related to survival of the larvae, such as the level of nauplii concentration where the larvae were found, the feeding ratio and feeding incidence of the larvae, and the level of predator concentration.

The field investigations in 1976–1981 were inadequate for understanding the causes of the variation in year-class strength. There were several reasons for this: first, Mother Nature supplied us with only poor year classes during this period, so only small changes in the year-class strength were recorded. The number of individuals in each year class at age three only varied between 96 mill. and 303 mill. Second, due to field sampling capacity, the study of qualitative processes had to be emphasized at the cost of quantitative measurements. For instance, only one cohort of larvae was monitored in most years, and this appeared to be far from sufficient to reveal the most important processes that rule the formation of a year class.

Nevertheless, important processes related to survival of larvae were found. The investigations showed that the residence time of the water masses was longer at the spawning grounds in Vestfjorden and at the spawning banks farther north than in other areas near the spawning grounds. The nauplii investigations showed that the concentration level was higher at the spawning grounds, and that the spawning grounds coincided with the first feeding area. It was also documented that the feeding ratio of the cod larvae were higher in areas where the nauplii concentration was higher. All this showed that the first feeding areas of cod larvae had more favorable feeding conditions than other areas.

The presence of vertical microstructures of fish larvae and their food has been proposed as an important factor for recruitment during the early life stages. As previously stated, both eggs and larvae of the Arcto-Norwegian cod are exclusively distributed in the mixed layer and thus subjected to atmospheric forces. There are three main forces that govern the vertical plankton distribution in the mixed layer (Sundby, 1983): (i) buoyancy, (ii) wind-induced turbulence, and (iii) the ability to vertically migrate. When there is no wind mixing, the buoyancy forces may create high concentrations at the surface. Additionally, vertical migration or other behavior patterns can create concentration peaks or patches. Numerous vertical profile measurements with both the Mocness plankton trawl and pumps show that very sharp gradients are seldom found, though the use of in situ particle counters has revealed the occasional occurrence of dense patches of small planktonic organisms (Tilseth and Ellertsen, 1984a). The presence of frontal processes and Langmuir cells may generate vertical gradients, but these do not seem to frequently influence the vertical distribution in the first feeding areas of the Arcto-Norwegian cod. As stated earlier, the ability for the cod larvae to vertically migrate seems to be very limited except during extremely calm conditions. The nauplii, even though much smaller than the larvae, seem to have a larger ability to migrate vertically, but even their movement is counteracted by the turbulence at moderate wind speeds (5–8 m · s^{-1}). One may therefore question whether small-scale concentration peaks in the mixed layer can be an important factor for recruitment. The turbulence itself can, up to a certain level, bring about a favorable effect in plankton contact rates (Rothschild and Osborne, 1988), by increasing the probability of the larvae encountering a food item. A recent report indicates that the feeding ratio of cod larvae increased by a factor of 2.8 when the average wind speed increased from 2 m · s^{-1} to 6 m · s^{-1} (Sundby and Fossum, 1989). This increase corresponds rather well with the theory by Rothschild and Osborn (1988). However, it can be questioned if there exists an optimal level of turbulence with respect to the number of successful attacks on the prey.

For the Arcto-Norwegian cod, temperature seems to be the most important parameter related to recruitment. The very good 1983 year class was also exceptional with respect to larval condition, having the highest dry weight/standard length-ratio recorded. Also the postlarval investigations showed a high average larval length (Bjørke and Sundby, 1987). The most prominent environmental feature of 1983 was the high temperature of the coastal water masses during spawning — 3.6°C, the highest recorded since 1961. Only two years earlier, in 1981, eggs and early larvae experienced the lowest environmental temperature (1.3°C) ever recorded since temperature measurements started in 1936, and became a very poor year class. The relationship between year-class strength at age three years and the temperature during the egg/early larval stages indicates that temperature strongly influences the recruitment of the Arcto-Norwegian cod (Figure 2.18). The figure indicates that a high temperature is a necessary condition for the formation of a strong year class. It also implies that other processes are important, since far from all years with high temperatures produced strong year classes. At present, no other physical parameter or process seems to be more dominant than temperature. The significance of temperature is supported by the work of Saetersdal and Loeng (1987), who found that strong and medium year classes occurred most frequently in warm years. Since most of the temperature variations in the Norwegian Sea/Barents Sea are of a geographically large scale (Dickson and Blindheim, 1984; Blindheim et al., 1981), the same variations also appear in the Lofoten spawning area.

The temperature-dependent spawning of *C. finmarchicus* found in Lofoten, causing a delay in spawning of about 1.5 months in the coldest year (1981) compared to the warmest (1960), may be the most important process to cause variable larval survival. There are also other temperature-dependent processes related to recruitment, such as a rapid development through the egg and larval stage which would decrease mortality by predation and reduce the transport and dispersion out of favorable first feeding areas.

It is uncertain to what extent predation on eggs and larvae influences the recruitment success of cod. The mortality of cod eggs from spawning to hatching was 90% in 1983 and 1984, when strong year classes were formed. It has been documented that adult herring prey on cod eggs, but the quantitative importance of this is uncertain. The preying upon cod larvae by herring is probably negligible, as they overlap only to a small extent in time and space.

The analyses of data collected from 1976–1987 continue, with special emphasis on the temperature effects on match/mismatch between cod larvae and their prey, the effect of turbulence on cod larvae feeding, and other factors that are supposed to be of importance in setting year-class strength.

References

Anonymous. 1975. Report of a colloquium on a larval fish mortality studies and their relation to fishery research, January 1975. Circ. U.S. Dep. Commer., Nat. Oceanic and Atmos. Adm., Nat. Mar. Fish. Serv. (395):1–5.

Anonymous. 1983. "Torskelarvens oppvekstvilkår i kyststrommen."Årsrapport til NFFR for 1982. In Norwegian.

Anonymous. 1984. Report on the Working Group

on Larval Fish Ecology, Hirtshals, Denmark, 25–27 June 1984. ICES C.M./L:37.

Anonymous. 1985. Report of the Arctic Fisheries Working Group, Copenhagen, 25 Sept.-2 Oct. 1985. ICES C.M./G:4.

Anonymous. 1987. Report on the Working Group on Larval Fish Ecology, Hirtshals, Denmark, 17–19 June 1987. ICES C.M./L:28. 81 pp.

Baranenkova, A. S. 1965. Notes on the condition of formation of the Arcto-Norwegian tribe of cod of the 1959–1961 year classes during the first year of life. ICNAF Spec. Publ. 6:397–410.

Bjørke, H., and Sundby, S. 1984. Distribution and abundance of post larval Northeast arctic cod and haddock. Proceedings of the Soviet-Norwegian symposium on reproduction and recruitment of Arctic cod, Leningrad, 25–30 September 1983:72–98.

Bjørke, H., and Sundby, S. 1986. Abundance indices for the Arcto-Norwegian cod for the period 1979–1986 based on investigations in June/July. ICES C.M./G:75. 7 pp. 9 figs.

Bjørke, H., and Sundby, S. 1987. Abundance indices and distribution of postlarvae and O-group cod. Contribution to the third Soviet/Norwegian symposium in Murmansk, 25–30 May 1986: The effect of oceanographic conditions on distribution, and population dynamics of commercial fish stocks in the Barents Sea. 19 pp.

Bjørke, H., Hansen, K., Johannessen, M., and Sundby, S. 1988. Postlarve undersøkelser-juni/juli 1987. Havforskningsinstituttets egg- og larveprogram. Rapport nr. 14. 1988.

Blaxter, J. H. S. 1966. The effect of light intensity on the feeding ecology of herring. In Light as an ecological factor. pp. 393–409. Ed. by R. Bainbridge, G. C. Evans, and O. Rackham. Br. Soc. Symp. 6, Oxford, UK.

Blindheim, J., Loeng, H., and Saetre, R. 1981. Long term temperature trends in the Norwegian coastal waters. ICES C.M./C:19. 13 pp.

Dahlberg, M. D. 1979. A review of survival rates of fish eggs and larvae in relation to impact assessment. Mar. Fish. Rev. 41(3):1–12.

Davenport, J., Lønning, S., and Kjørsvik, E. 1981. Osmotic and structural changes during early development of eggs and larvae of the cod, Gadus morhua L. J. Fish. Biol. 19:317–331.

Dickson, R. R., and Blindheim, J. 1984. On the abnormal hydrographic conditions in the European Arctic during the 1970s. Rapp. P.-v. Reun. Cons. int. Explor. Mer 185:201–213.

Eggvin, J. 1932. Vannlagene pa fiskefeltene. Årsberetning vedkommende Norges fiskerier 1932 (2):90–95. In Norwegian.

Ellertsen, B., Fossum, P., Solemdal, P., and Sundby, S. 1987. The effect of biological and physical factors on the survival of Arcto-Norwegian cod and the influence on recruitment variability. Contribution to the third Soviet/Norwegian symposium in Murmansk, 26–30 May 1986: The effect of oceanographic conditions on distribution, and population dynamics of commercial fish stocks in the Barents Sea. 29 pp.

Ellertsen, B., Fossum, P., Solemdal, P., Sundby, S., and Tilseth, S. 1984. A case study on the distribution of cod larvae and availability of prey organisms in relation to physical processes in Lofoten. In The propagation of cod Gadus morhua L. pp. 453–478. Ed. by E. Dahl, D. S. Danielssen, E. Moksness, and P. Solemdal. Flødevigen rapportser. 1, 1984.

Ellertsen, B., Furnes, G. K., Solemdal, P., and Sundby, S. 1981a. Effects of upwelling on the distribution of cods eggs and zooplankton in Vestfjorden. In Proc. from Norwegian Coastal Current Symposium, Geilo, Norway, 9–12 September 1980. pp. 604–628. Ed. by R. Saetre and M. Mork. Univ. Bergen.

Ellertsen, B., Moksness, E., Solemdal, P., Tilseth, S., and Øiestad, V. 1976. The influence of light and food densities on the feeding success in larvae of cod (Gadus morhua L.), field and laboratory observations. ICES C.M./F:34. 31 pp.

Ellertsen, B., Solemdal, P., Strømme, T., Tilseth, S., and Westgård, T. 1980. Some biological aspects of cod larvae Gadus morhua L. Fisk. Dir. Skr. Ser. Havunders. 17:29–47.

Ellertsen, B., et al. 1979. Feeding and vertical distribution of cod larvae in relation to availability of prey organisms. ICES/ELH Symp./FM:Poster 1. 31 pp. 25 figs.

Ellertsen, B., et al. 1981b. Spawning period, transport and dispersal of eggs from the spawning area of Arcto-Norwegian cod (Gadus morhua L.). In The early life history of fish: Recent studies. Rapp. P.-v. Reun. Cons. int. Explor. Mer 178:260–267.

Ellertsen, B., et al. 1981c. Growth and survival of cod larvae in an enclosure. Experiments and mathematical model. Rapp. P.-v. Reun. Cons. int. Explor. Mer 178:45–57.

Fossum, P. 1986. A staging system for larval cod (Gadus morhua L.). Fisk. Dir. Skr. Ser. Havunders. 18:69–76.

Fossum, P. 1988. A tentative method to estimate mortality in the egg and early first larval stage with special reference to cod (Gadus morhua L.). Fisk. Dir. Skr. Ser. Havunders. 18:329–345.

Furnes, G. K., and Sundby, S. 1981. Upwelling and wind induced circulation in Vestfjorden. In

Proc. from Norwegian Coastal Current Symposium, Geilo, Norway, 9–12 September 1980. pp. 152–178. Ed. by R. Saetre and M. Mork. Univ. Bergen.

Gade, 1894. Temperaturmålinger i Lofoten 1891–1892. Carl C. Werner & Co., Christiania, Norway, 1899. 162 pp.

Godø, O. R., Nakken, O., and Raknes, A. 1984. Acoustic estimates cod off Lofoten and More in 1984. ICES C.M./G:47. 10 pp.

Godø, O. R., Raknes, A., and Sunnanå, K. 1985. Acoustic estimates cod off Lofoten and More in 1985. ICES C.M./G:66. 14 pp.

Harding, J. P., Marshall, S. M., and Orr, A. P. 1951. Time of egg-laying in the planktonic copepod *Calanus*. Nature 167:953.

Hjort, J. 1914. Fluctuations in the great fisheries of northern Europa viewed in the light of biological research. Rapp. P.-v. Reun. Cons. Perm. int. Explor. Mer 20:1–228.

Iles, T. D., and Sinclair, M. 1982. Atlantic herring: Stock discreteness and abundance. Science 215:627–633.

Kjesbu, O. S. 1989. The spawning activity of cod (*Gadus morhua* L.). J. Fish. Biol. 34:195–206.

Knutsen, G. M., and Tilseth, S. 1985. Growth, development and feeding success of atlantic cod larvae (*Gadus morhua* L.) related to egg size. Trans. Am. Fish. Soc. 114:507–511.

Kvenseth, P. G., and Øiestad, V. 1984. Large scale rearing of cod fry on the natural food production in an enclosed pond. In The propagation of cod *Gadus morhua* L. pp. 645–675. Ed. by E. Dahl, D. S. Danielssen, E. Moksness, and P. Solemdal. Flødevigen rapportser. 1, 1984.

Lasker, R. 1975. Field criteria for survival of anchovy larvae: The relation between inshore chlorophyll maximum layers and successful first feeding. Fish Bull. U.S. 73:453–462.

Lebour, M. V. 1919. The food of post-larval fish. J. Mar. Biol. Ass. U.K. 12:22–47.

Loeng, H., and Sundby, S. 1986. Hydrography and climatology of the Barents Sea and the coastal waters of northern Norway. Proceedings of workshop on comparative biology, assessment, and management of gadoids from the north Pacific and Atlantic oceans. Seattle, 24–28 June 1985.

Marr, J. C. 1956. The "critical period" of the early life history of marine fishes. J. Cons. int. Explor. Mer 21:160–170.

Marshall, S. M., and Orr, A. P. 1955. The biology of a marine copepod. Oliver & Boyd, Edinburgh, London. 188 pp.

May, R. C. 1974. Larval mortality in marine fishes and the critical period concept. In The early life history of fish. pp. 3–19. Ed. by J. H. S. Blaxter. Springer-Verlag, Berlin, Heidelberg, New York.

Melle, W. 1985. Predasjon på torskens egg og larver i Lofoten. Thesis in Fisheries Biology, Univ. Bergen. 142 pp. In Norwegian.

Melle, W., and Ellertsen, B. 1984. Predation on cod eggs and larvae: Potential predators in the spawning ground of the north east arctic cod. ICES Larval Fish Ecology Working Group, Hirtshals, Denmark, 25–27 June. 19 pp.

Mohus, I. 1981. "Micro-count" particle datalogger. Equipment manual. Sintef rep.STF 48 F 81018. 90 pp.

Moksness, E. 1978. Bassengstudier av torskelarvens naeringsvalg, vekst og overleving, fra klekking til metamorfose. Thesis, Univ. Bergen. In Norwegian.

Nordeng, H., and Bratland, P. 1971. Feeding of plaice (*Pleuronectes platessa* L.) and cod (*Gadus morhua* L.) larvae. J. Cons. int. Explor. Mer 34:51–57.

Pedersen, T. 1984. Variation of peak spawning of Arcto-Norwegian cod (*Gadus morhua* L.) during the period 1929–1982 based on indices estimated from fishery statistics. In The propagation of cod *Gadus morhua* L. pp. 301–316. Ed. by E. Dahl, D. S. Danielssen, E. Moksness, and P. Solemdal. Flødevigen rapportser. 1, 1984.

Pedersen, T. 1985. Variasjon i tidspunkt for maksimal gyteintensitet hos torsk (*Gadus morhua* L.) i Lofoten 1929–1983. Thesis in Fisheries Biology, Univ. Bergen. 121 pp. In Norwegian.

Randa, K. 1984. Abundance and distribution of 0-group Arcto-Norwegian cod and haddock 1965–1982. In The proceedings of the Soviet/Norwegian symposium on Reproduction and recruitment of arctic cod. Leningrad, 25–30 September 1983. pp. 189–210. Ed. by O. R. Godø and S. Tilseth.

Rothschild, B. J., and Osborn, T. R. 1988. Small-scale turbulence and plankton contact rates. J. Plankton Res. 10:465–474.

Saetersdal, G., and Loeng, H. 1987. Ecological adaptation of reproduction in Northeast Arctic cod. Fish. Res. 5:253–270.

Salzen, E. A. 1956. The density of eggs of *Calanus finmarchicus*. J. Mar. Biol. Ass. U.K. 35:549–554.

Solberg, T., and Tilseth, S. 1984. Growth, energy consumption and prey density requirements in first feeding larvae of cod (*Gadus morhua* L.). In The propagation of cod *Gadus morhua* L. pp. 145–166. Ed. by E. Dahl, D. S. Danielssen, E. Moksness, and P. Solemdal. Flødevigen rapportser. 1, 1984.

Solemdal, P. 1970. Intraspecific variation in size,

buoyancy and growth of eggs and early larvae of Arcto-Norwegian cod, *Gadus morhua* L., due to parental and environmental effects. ICES C.M./F:28. 11 pp.

Solemdal, P. 1982a. The spawning period of Arcto-Norwegian cod during the years 1976–1981. Report on the working group on larval fish ecology, Lowestoft, UK, 3–6 July 1981. ICES C.M./L:3. 21 pp.

Solemdal, P. 1982b. Sampling fish larvae with large pumps. Report on the working group on larval fish ecology, Lowestoft, UK, 3–6 July 1981. ICES C.M./L:3. 6 pp.

Solemdal, S. 1984. First feeding period of cod larvae from the Lofoten area during the years 1980–1983. Larval fish ecology ICES Working Group, Hirtshals, Denmark, 25–27 June 1984. L:37.

Solemdal, P., and Ellertsen, B. 1984. Sampling fish larvae with large pumps, quantitative and qualitative comparisons with traditional gear. *In* The propagation of cod *Gadus morhua* L. pp. 335–363. Ed. by E. Dahl, D. S. Danielssen, E. Moksness, and P. Solemdal. Flødevigen rapportser. 1, 1984.

Solemdal, P., and Sundby, S. 1981. Vertical distribution of pelagic fish eggs in relation to species, spawning behavior and wind conditions. ICES C.M./G:77. 26 pp.

Strømme, T. 1977. Torskelarvens lengde ved klekking og virkning av utsulting på larvens egenvekt og kondisjon. En eksperimentell undersøkelse pa norsk-arktisk torsk (*Gadus morhua* L.). Thesis, Univ. Bergen. In Norwegian.

Sund, O. 1925. Temperaturen i sjøen. Årsberetning vedkommende Norges fiskerier 1924 2:469–471. In Norwegian.

Sundby, S. 1980. Utviklingen innen oseanografisk forskning i Vestfjorden. Fisken Hav. 1980(1):11–25. In Norwegian.

Sundby, S. 1981. Vestfjordundersøkelsene 1978. A. Ferskvannsbudsjett og vindforhold. Fisken Hav. 1982(1). 16 pp. 18 figs. In Norwegian.

Sundby, S. 1983. A one-dimensional model for the vertical distribution of pelagic fish eggs in the mixed layer. Deep-Sea Res. 30(6A)645–661.

Sundby, S. 1984. Influence of bottom topography on the circulation at the continental shelf of northern Norway. Fisk. Dir. Skr. Ser. Havunders. 17:501–519.

Sundby, S. 1989. Factors affecting the vertical distribution of eggs. Contribution to ICES Symposium on "The ecology and management aspects of extensive mariculture." Nantes, France, 19–23 June 1989. ICES 1989/EMEM 59. 11pp.

Sundby, S., and Bratland, P. 1987. Kartlegging av gytefeltene for norskarktisk torsk i Nord-Norge og beregning av eggproduksjonen i arene 1983–1985 (Spatial distribution and production of eggs from Northeast-arctic cod at the coast of Northern Norway 1983–1985). Fisken Hav. 1987(1):1–58.

Sundby, S., and Fossum, P. 1989. Feeding conditions of Arcto-Norwegian cod larvae compared to the Rothschild-Osborn theory on small-scale turbulence and plankton contact rates. ICES C.M. 1989/G:19. 11pp.

Sundby, S., and Solemdal, P. 1984. Egg production of the Arcto-Norwegian Cod in the Lofoten area estimated by egg surveys. Proceedings of the Soviet-Norwegian symposium on Reproduction and recruitment of Arctic cod, Leningrad, 25–30 September 1983:113–135.

Sundnes, G., Leivestad, H., and Iversen, O. 1965. Buoyancy determination of eggs from the cod. J. Cons. int. Explor. Mer 14(3):249–252.

Tilseth, S., and Ellertsen, B. 1984a. The detection and distribution of larval Arcto-Norwegian cod, *Gadus morhua*, food organisms by an in situ particle counter. Fish. Bull. U.S. 82(1):141–156.

Tilseth, S., and Ellertsen, B. 1984b. Food consumption rate and gut evacuation processes of first feeding cod larvae (*Gadus morhua* L.). *In* The propagation of cod *Gadus morhua* L. pp. 167–182. Ed. by E. Dahl, D. S. Danielssen, E. Moksness, and P. Solemdal. Flødevigen rapportser. 1, 1984.

Tilseth, S., and Strømme, T. 1976. Changes in buoyancy and activity during starvation of cod larvae (*Gadus morhua* L.). ICES C.M./F:33. 12 pp.

Tilseth, S., Klungsøyr, J., Falk-Petersen, S., and Sargent, J. R. 1987. Fatty acid composition as indicator of food intake in cod larvae (*Gadus morhua* L.) from Lofoten, northern Norway. ICES C.M. 1987/L:31. 13 pp.

Westgård, T. 1988. A model for the vertical distribution of pelagic spheres in the wind-mixed upper layers of the ocean. A computer realization. Havforskningsinstituttets egg-og larveprogram. Rapport 1988.

Wiborg, K. F. 1948. Investigations on cod larvae in the coastal waters of northern Norway. Fisk. Dir. Skr. Ser. Havunders. 9(3):1–27.

Wiborg, K. F. 1950. Utbredelse og forekomst av fiskeegg og fiskeyngel på kystbankene i Nordnorge våren 1948 og våren 1949. Fisk. Dir. Småskr. 1950(1):1–26. In Norwegian.

Chapter 3

Fluctuation in the Cod Biomass of the West Greenland Sea Ecosystem in Relation to Climate

Holger Hovgård and Erik Buch

Abstract

Changes in the temperature conditions at West Greenland in this century coincide generally with the changes in the cod fishery, indicating the existence of a relatively strong climatic effect on the cod stock.

An analysis of the structure of the Sea cod stock since 1956 indicates that low recruitment in the last 20 years played an important role in the collapse of the cod fishery. Interpretations of various larval and young-fish data, as well as theoretical considerations, lead to a hypothesis that claims that the variations in recruitment to the West Greenland Sea cod stock may be caused by variable inflow of larvae from Iceland. According to this view, the climatic effect is not to be found in the temperature changes, per se, but rather in a change in hydrographical conditions in the Irminger Sea.

Since 1980 there has been a southern displacement in the cod fishery off West Greenland. During the 1980s, the size-at-age of cod has shown a considerable decrease. Both of these factors coincide with the latest cooling period, and a direct effect of temperature on distribution and growth might be postulated.

Introduction

During the present century, Greenland has gradually changed from a nation of hunters to a nation of fishermen. One reason for this radical change was the development of a rich cod fishery starting in the 1920s. It evolved quickly from a local, small-boat fishery to an international offshore fishery, primarily by trawlers, with peak catches between 400,000 and 500,000 tons (mt) annually. After the mid-1960s, the cod fishery decreased drastically, with negative consequences for the Greenland economy.

In the same period, Greenland, as did other parts of the Northern Hemisphere, experienced a climatic improvement which lasted until about 1967. Since then, climatic conditions have generally deteriorated.

This chapter will discuss possible causes for the decline in the cod fishery in relation to the change in climate.

Variations in Climate

The climate in the Greenland area has undergone great changes in the course of time. Paleoclimatic studies have revealed that one must go back 1,000 years in time (i.e., to when Eric the Red colonized Greenland) to find atmospheric temperatures as high as those observed in the present century.

Meteorological observations have been carried out at the Nuuk (Godthaab) meteorological station since 1876. The average temperatures from this station are shown in Figure 3.1. A general rise in temperatures started in the early 1920s and lasted until the middle of the 1960s, when a decline in temperatures took place. During the 1970s, the air temperatures improved for a short period, to be followed by a second, very strong cooling at the beginning of the 1980s,

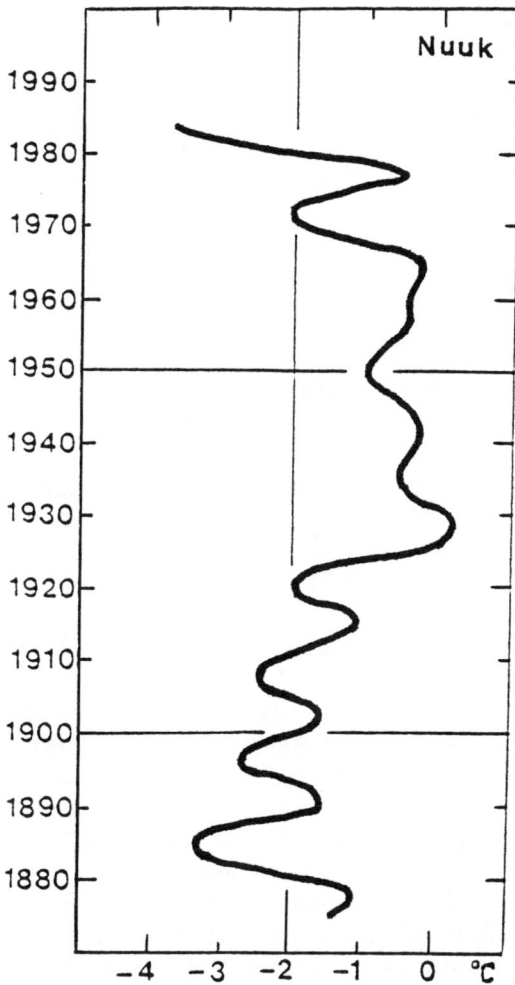

Figure 3.1. Average air temperatures at the Nuuk/Godthaab meteorological station, 1876–1984.

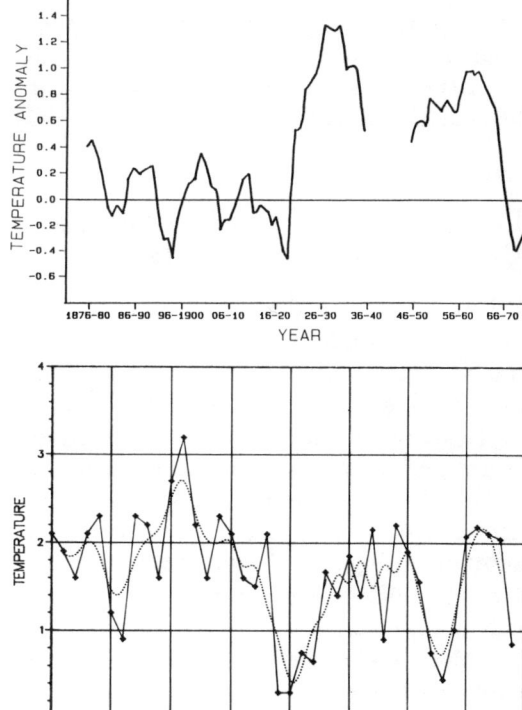

Figure 3.2. Ocean temperatures. (a) Sea surface temperature anomalies (5-year running mean) for West Greenland, 1876–1974. [Prepared by Jens Smed, ICES.] (b) Mean temperature of the upper 40 m. on Fylla Bank by the middle of June, 1950–1989. ♦—♦, actual temperature; ...., 3-year running mean.

during which Greenland experienced the coldest atmospheric conditions within the 100 years that meteorological observations have been carried out.

The principal changes in the air temperatures are reflected in a time series of sea surface temperatures from the West Greenland area (Figure 3.2a, b).

West Greenland Cod Fishery

The occurrence of cod at West Greenland has been of a periodic character. In the last century, two short cod periods are known viz., in the 1820s and in the late 1840s (Hansen, 1949). After having been nearly absent from the West Greenland fishing banks for a period of 50–70 years, cod returned to these waters during the 1920s.

The development of the West Greenland cod fishery in this century is illustrated in Figure 3.3. In the 1920s and 1930s, a good cod fishery developed, carried out mainly by non-Greenlandic long-liners and jiggers. During World War II, the fishery was reduced to what could be taken by the Greenland fishing fleet of small boats. After the war, when foreign nations gradually returned to the area with modern and effective vessels, the West Greenland Sea cod fishery experienced an explosive development, frequently yielding catches in excess of 300,000 tons per year.

Around 1970, the catches decreased drastically, although with a temporary slight improvement in the late 1970s, whereafter the fishery decreased further to the 1985 level of 13,000 mt.

Linking Changes in Fisheries to Changes in Temperature

When comparing the development in cod catches (Figure 3.3) with the changes in air and sea temperatures (Figures 3.1 and 3.2), an overall relationship between temperature and catches is indicated, as (i) when the warm period started, cod catches rose significantly; (ii) when the cooling started around 1970, catches decreased nearly instantaneously; and (iii) when a second very strong cooling occurred in 1982–1983, catches of cod declined to almost nil.

More elaborate correlations between catches and temperatures are not meaningful, as catches do not relate simply to stock size in the 1920–1980 period, due to major changes in fishing technology and effort.

The Development of the Cod Stock Since 1956

For the years after 1956, catch-at-age data for cod are available and an analytical as-

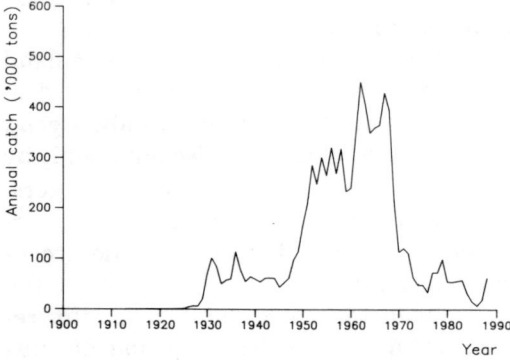

Figure 3.3. Catches of cod off West Greenland, 1900–1988.

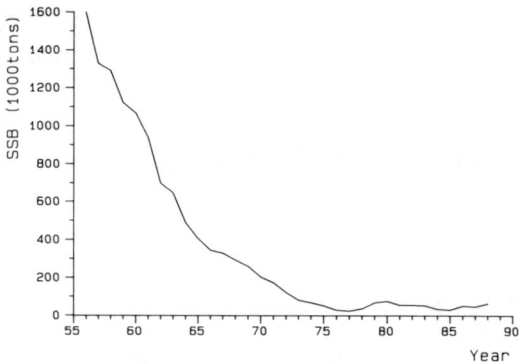

Figure 3.4. Spawning stock biomass ('000 tons) of the West Greenland cod stock, 1956–1984.

sessment of stock size can hence be made for that period. The spawning stock biomass (SSB) declined from 1.6 mill. tons in 1956 to less than 50,000 mt in 1984 (Figure 3.4). The decrease is exponential, with a yearly rate of decline between 13% (1956–1968) and 27% (1969–1977). The year-class strength, measured as the number of 3-year-old cod, revealed great variability during the same period (Figure 3.5). However, the data indicate that the recruitment was generally good during the decade 1953–1963, but since then, it has been poor.

The most reasonable explanation for the trends in the cod stock size therefore seems to be that an intensive fishery substantially reduced the stock size, but the actual collapse of the fishery was caused by a failure in recruitment. If, for instance, maintaining recruitment on the 1952–1963 level and using the mean fishing mortality of later years, the SSB would stabilize at approximately 180,000 mt. In reality, fishing mortality was high concurrently with a poor recruitment, leading to a doubling of the rate of decline of the SSB in the 1970s. Therefore, most considerations regarding the relationship between climate and the cod stock is given to the recruitment process.

However, climate might have other effects on the stock (i.e., affecting stock distribution and migrations or the growth of the individual species). These aspects will therefore also be discussed.

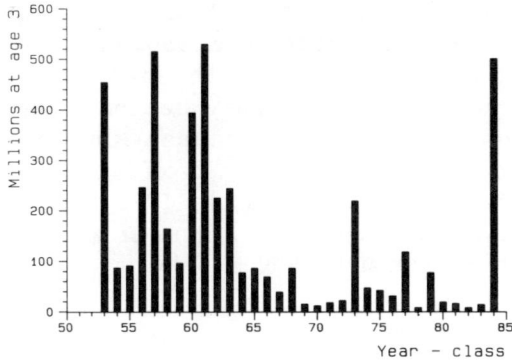

Figure 3.5. Year-class strength, measured as numbers at age 3, at West Greenland, 1953–1984.

Spawning Area and Larval Drift

Mature prespawning cod are found around Iceland and in waters southeast and southwest of Greenland. Spawning seems to occur in a more or less continuous belt over this region.

The currents in the area in question are shown in Figure 3.6. There are two currents of importance, the cold East Greenland Current of polar origin and the warm, more saline Irminger Current, which is a side branch of the North Atlantic Current. The two meet in the area southeast of Greenland and flow side by side under some mixing round Cape Farewell, turning northward along the west coast of Greenland.

This current pattern carries eggs and larvae in a clockwise direction around the southern part of Greenland. The actual distance carried will depend on the current velocity and duration of the pelagic phase of young cod. Our knowledge of both these factors is scarce. The mean velocity is of the order of 0.2 m/s, but great fluctuations, both seasonally and between years, do exist (Hansen and Buch, 1986). The length of the pelagic phase has not been examined, but from Icelandic young-fish surveys off Iceland and East Greenland, it is known that substantial numbers of cod are still pelagic at mid-August (i.e., four months after the peak spawning) (Wilhjalmsson and Magnusson,

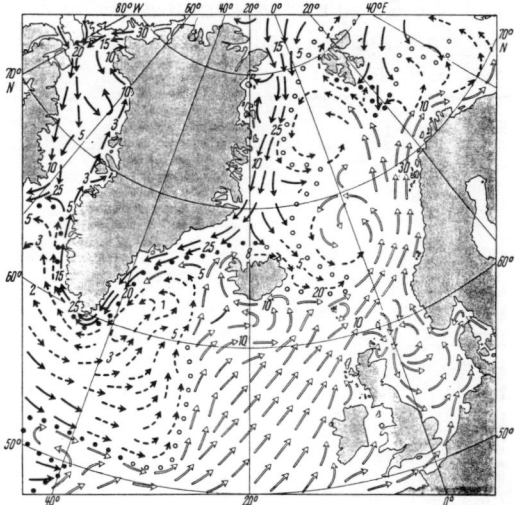

Figure 3.6. Surface currents in the northern part of the Atlantic Ocean. ⇒ Atlantic water; →, polar water; – →, mixed water; ● ● ● o o o, polar front. The numbers represent the current speed in cm/s.

1984; Jonsson, 1982). The pelagic phase of the Barents Sea cod, subjected to rather similar temperature conditions, is found to be about six months (Bergstad et al., 1987).

Assuming a current velocity of 0.2 m/s and a pelagic phase of four months leads to a transport distance of 2,000 km, approximately the distance from the Icelandic spawning area to the areas around the Fylla Bank at West Greenland. Correspondingly, the larvae from the West Greenland spawning grounds should be carried out somewhere in the Davis Strait.

These theoretical considerations on larval flow distances are supported by some empirical findings.

A recruitment of icelandic cod to Greenland has been assumed since Tåning (1937) showed young cod drifting in Irminger water between Iceland and Greenland. More systematic evidence of this drift has been archived by the Icelandic young fish survey conducted annually in August since 1971. These surveys indicate, however, that the magnitude of this larval drift varies significantly between years as large numbers of 0-group cod are only seen in 1973, 1984, and

Table 3.1. Abundance indices of O-group cod from the Icelandic O-group survey off east Greenland. (From Anonymous, 1987.)

Year	Index
1971	+
1972	No survey
1973	135
1974	2
1975	+
1976	5
1977	7
1978	2
1979	2
1980	1
1981	19
1982	+
1983	+
1984	372
1985	32
1986	+

1985 (Table 3.1). Since the cod year-classes of 1973 and 1984 are the only really large ones seen in the West Greenland fisheries since the mid 1960s it therefore seems as if good year-classes are correlated with the occurrence of an invasion of young cod from Iceland. A further, although indirect, support of the cod larval drift is the finding of relatively large numbers of haddock of the 1984 and 1985 year-classes by German bottom-trawl surveys off West Greenland (conducted on an annual basis since 1982). As haddock occurs very rarely at West Greenland and the closest known spawning site for haddock is off southwest Iceland, this shows that Irminger Water reached West Greenland in 1984 and 1985.

It is more difficult to find empirical evidence supporting the hypothesis that the larvae from the West Greenland spawning areas are carried out somewhere in the Davis Strait. Except for larvae surveys carried out in June/July since 1950, no regular surveys have been made prior to the 1980s. Scrutinizing charts of larval distribution from these surveys indicates that a large proportion often are found far offshore, and these larvae are probably lost from the Greenland cod stock. However, in the peak days of the West Greenland cod stock in the 1950s and 1960s the northern Store Hellefiske bank (67°N latitude) was a large nursery area for smaller cod. This might have been the settling area for the small cod spawned off Southwest Greenland. In the years after 1970, few cod have been taken in the offshore areas north of approximately 65°N latitude.

The Relationship Between the Cod Stock and Climate

The near collapse of the cod stock in the late 1960s can be traced back to recruitment failure. It further coincides with a marked change in climate, as revealed by the reduction in air and sea temperature.

It is possible that the low water temperature found since the mid-1960s directly has caused the reduction in recruitment, by influencing larval physiology, stability of water masses, and the dynamics of the plankton community. However, the clear relationship between the occurrence of large numbers of O-group cod off East Greenland with later strong year-classes off West Greenland indicates that recruitment is strongly coupled to variations in the number of young fish recruited from Iceland. These variations could in turn be caused by a change in the current pattern in the Irminger Sea. It is known that the circulation patterns in the whole of the North Atlantic Ocean are very closely related to the difference in air pressure between the subtropical high near the Azores and the subarctic low near Iceland. This pressure difference rules the strength of the westerlies, which in turn has decisive effects on the strength and temperature of ocean current patterns (Veronis 1973, 1981).

The complexity of mechanisms behind the transport of cod larvae from the Icelandic spawning grounds to the West Greenland area can be illustrated by the fact that in 1984, the transport of young cod from Iceland to West Greenland was due to a blocking of the Denmark Strait by cold Arctic

Table 3.2. Catch of cod by Greenland state-owned trawlers (500–1000 BRT), by year and division.

Year	Total catch (tons)	Percent of catch, by divisions				
		1B	1C	1D	1E	1F
1973	7.083	1.3	19.4	43.0	29.7	6.7
1974	9.876	0	32.4	26.5	34.4	6.7
1975	12.404	0.2	77.9	11.4	9.9	0.7
1976	11.020	0.1	25.5	27.5	46.9	0
1977	8.731	0	24.5	16.0	56.6	2.9
1978	16.217	0	67.5	12.8	18.7	1.0
1979	9.617	0	60.3	32.9	6.6	0.2
1980	6.670	24.0	26.6	28.5	20.5	0.4
1981	11.846	0	31.4	35.9	32.7	0
1982	18.932	0.6	19.2	35.9	44.3	0.1
1983	14.821	1.9	3.4	24.0	70.1	0.7
1984	4.060	0	0	41.9	53.8	4.3
1985	1.752	0	0	8.0	85.8	6.3

water, preventing the Irminger Water — and thereby also cod larvae — from following its normal path into the Iceland Sea. Instead, it was forced towards Greenland. The preceding year-class of Icelandic origin observed at West Greenland was the 1973 year-class, but in 1973, a blocking of the Denmark Strait by Arctic water was not observed; therefore, some other mechanism was responsible for the drift that year.

In general, it may be concluded that today there is no clear understanding of which processes are responsible for the drift of cod larvae from Iceland to Greenland, and research on this subject is one of the major objectives in the years to come.

The coupling between the atmospheric and the oceanic processes is very complex and far from being fully understood. This is the reason why great efforts are being devoted to establishing the World Ocean Circulation Experiment (WOCE), with the declared goals to develop models useful for predicting climate changes and to collect data necessary to test them and to find methods for determining long-term changes in the oceanic circulation. The Greenland Fisheries Research Institute is involved in WOCE, together with sister organizations from the other Nordic countries concentrating on a five-year project concerned with ocean circulation in the northern North Atlantic. The knowledge obtained will hopefully be helpful to the understanding of the West Greenland cod stock.

Changes in Distribution of Adult Cod

The spatial distribution of cod catches off West Greenland has shown marked differences since World War II. In the 1950s and 1960s high catches were taken in the northern areas (ICNAF/NAFO Division 1B) where only small catches are taken today.

A more formal treatment of changes in distribution is impeded by the lack of regular bottom-fish surveys and the often poor catch statistics of earlier years. The participation of the many different fleets in the fishery, each exerting variable effort both seasonally and between years, also makes any interpretation of total catch distribution very difficult. However, the catch distribution of the Greenland state-owned trawler fleet, which consists of very similar stern trawlers, might shed some light on the magnitude of changes in the distribution of the stock in recent years. The catch distribution is shown in Table 3.2 and in a somewhat simplified form in Figure 3.7. The mean latitude of catches is high in 1975 and in the years 1978–80. From 1981, the fisheries has gradually

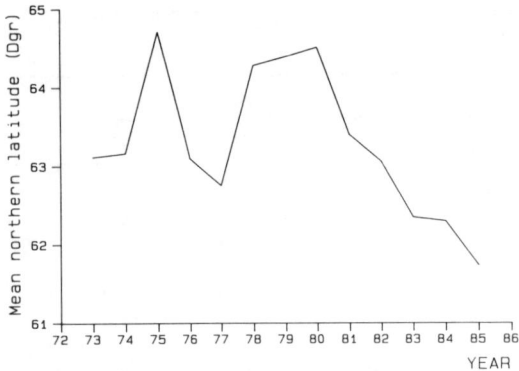

Figure 3.7. Mean northern latitude of catches taken in directed cod fishery by the Greenland Home Rule trawler fleet, 1973–1985.

Table 3.3. Mean length and weight of cod by age (age groups 4 to 8) off West Greenland in 1979 and 1984. (From Hansen, 1987.)

Year age	1979		1984	
	Length (cm)	Weight (kg)	Length (cm)	Weight (kg)
4	47.61	1.07	39.33	0.57
5	58.96	2.05	49.28	1.13
6	70.65	3.51	59.80	1.98
7	78.04	4.77	65.59	2.62
8	87.23	6.53	72.18	3.46

moved southward to the present very southern distribution.

Temperature conditions during the years 1973 to 1980 were moderately warm, showing only relatively small year-to-year variations (Figure 3.2b), and the annual changes of the importance of various fishing area is therefore not easily explained by changes in temperature. On the other hand, the southern movement in the fisheries in the beginning of the 1980s correlates well with the pronounced cooling during that period.

Tagging experiments have shown that cod tagged in southwestern Greenland more often migrate to Iceland than do more northerly tagged cod (Hansen, 1949; Riget and Hovgard, 1989). A southern displacement of the stock might therefore give rise to a larger emigration of cod, leading to a reduction in the population off West Greenland.

Changes in Size at Age

A dramatic change in size at age of cod in the West Greenland area between 1979 and 1984 has been found by Hansen (1987). During this period, length and weight at age for the most important age-groups declined by roughly 15% and 45%, respectively (Table 3.3).

This decrease in size at age correlates well with the overall change in temperature regime, and it is possible that the reduction in size reflects a lower growth caused either by a direct temperature effect or by temperature-induced changes in the production of food species.

However, strictly speaking, the observations refer to change in size at age and along with other observations, point to the possibility that especially the larger individuals have left the area. This, of course, will show up as a reduction in mean size at age for the remaining population.

References

Anonymous. 1971. Report of the north western working group 1970. ICES C.M.1971/F:2.

Anonymous. 1983. Report of the working group on cod stocks off East Greenland. ICES C.M.1983/Assess:8.

Anonymous. 1987. Report of the working group on cod stocks off East Greenland. ICES C.M.1987/Assess:10.

Bergstad, O. A., Jorgensen, T., and Dragesund, O., 1987. Life and ecology of the Gadaid resources of the Barents Sea. Fish. Res. 5:119–163.

Hansen, P. M. 1949. Studies on the biology of the cod in Greenland waters. Rapp. P.-v. Reun. Cons. int. Explor. Mer 123:1–83.

Hansen, P. M. 1968. Report on cod eggs and larvae. Spec. Publ. Int. Comm. Northw. Atla. Fish. 7:127–137.

Hansen, H. H., and Buch, E. 1986. Prediction of year-class strength of Atlantic cod (*Gadus morhua*) off West Greenland. NAFO Sci. Coun. Studies 10:7–11.

Hansen, H. H. 1987. Changes in size at age of cod off West Greenland, 1979–1984. NAFO Sci. Coun. Studies 11: 37–42.

Jonsson, E., 1982. A survey of spanning and reproduction of the Icelandic cod. Rit Fiskideildar 6(2):1–41.

Riget, F., and Hovgard, H., 1989. Recaptures by year-class of cod in East Greenland/Icelandic waters from tagging experiments off West Greenland. NAFO SCR. Doc. 89/24.

Taning, Å. V. 1937. Some features in the migration of cod. J. Cons. Perm. int. Explor. Mer. 12:1–35.

Veronis, G. 1973. Model of world ocean circulation: I. Wind-driven, two layer. J. Mar. Res. 31:228–288.

Veronis, G. 1981. Dynamics of large-scale ocean circulation. Evolution of Physical Oceanography, Ed. by B. A. Warren and C. Wunsch. MIT Press, Cambridge, MA.

Wilhjalmsson, H., and Magnusson, J. V. 1984. Report on the O-group fish survey in Icelandic and East Greenland waters, August 1974. ICES C. M. 1984/H:66.

Chapter 4

The Caribbean Sea
A Large Marine Ecosystem in Crisis

William J. Richards and James A. Bohnsack

Abstract

The Caribbean Sea is one of the most beautiful on earth, with blue water surrounded by green islands with white sand beaches on the north and east, and by continental mountain ranges on the south and west. But beneath this beauty lies a troubled ecosystem showing signs of stress, especially on its important shallow-water coral reefs. Many of the local artisanal fisheries show signs of chronic stress or have collapsed. The bleaching and dying of corals on massive scales is unexplained, as is an epizootic disease that has almost eliminated *Diadema*, a dominant sea urchin and an important grazer. Fishery resource demands of growing coastal populations, land development, and tourism are increasing the stress on a limited ecosystem that is characterized by low net productivity. Although 40% larger than the neighboring Gulf of Mexico to the North, the Caribbean produces a catch of only 30% of the U.S. Gulf fishery. The most productive waters are off the northern coast of South America where nutrient input from rivers, estuaries, and wind-induced upwelling is greatest. The remaining area is characterized by relatively clear, nutrient-poor waters. Despite the problems, considerable biological oceanographic information exists. Most fishery populations appear linked at a Caribbean-wide level, making coordinated management desirable but difficult because of the high number of political entities (38). International cooperation through planned cooperative management could provide solutions to these problems.

Introduction

The Caribbean Sea (Figure 4.1) — the second largest in the world — is located in the Western Hemisphere between North and South America (roughly bounded by 61°W longitude on the east, 89°W on the west, 22°N latitude on the north, and 8°N on the south). It has an area of 2,515,900 km^2 and is noted for its many islands, especially the Leeward and Windward Islands on the eastern boundary, Cuba and Hispaniola to the north, Jamaica and Cayman Islands in the northwest, and additional continental islands in the south. Besides the islands there are numerous banks and breaking shoals surrounding the basin. The Caribbean is comprised of four deep basins — the Venezuelan Basin in the east, which is partially separated by a ridge from the Colombian Basin in the west; in the northwest, the Cayman Trough lies north of Jamaica and a series of shallow banks between Jamaica and Honduras; and north of the Trough is the Yucatan Basin. The entrances to the Caribbean are quite narrow and most are relatively shallow, with only the Yucatan Channel greater than 1,800 m. Half of the water area is over 3,600 m and 75% is over 1,800 m. The deepest part of the Atlantic lies just outside the Caribbean (the Puerto Rican Trench at 9,200 m). Circulation is influenced by the trade winds and is generally a westward-flowing surface current with some upwelling along the South American continent. Surface water temperatures are warm, with little variation between winter lows of 25.5°C and summer highs of 28°C.

Other areas are often included in the Caribbean because of close proximity and similarities of marine fauna. These areas in-

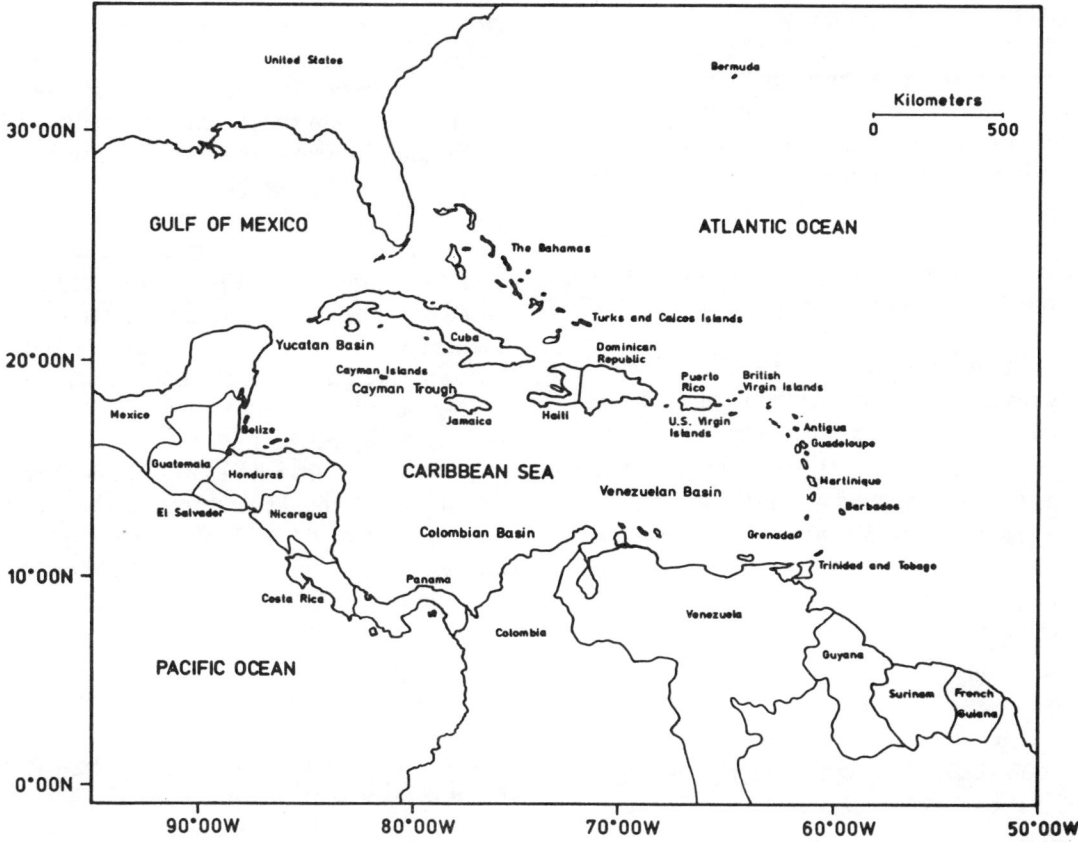

Figure 4.1. The Caribbean Sea and adjacent areas.

clude the Bahamas, Gulf of Mexico, Straits of Florida, and even Bermuda, which all lie to the north, plus the northern coast of South America to the mouth of the Amazon River to the east. It is this area that was included in a major research effort in the 1970s entitled Cooperative Investigations in the Caribbean Sea and Adjacent Areas (CICAR). Extensive background information is available in the two FAO Fisheries Reports that emanated from this multidisciplinary international study (FAO, 1969, 1977). Of special interest is the excellent summary of the extensive research activities (Bayer, 1969) which have been conducted in the Caribbean.

The purpose of this chapter is to describe some recent events in this area that may indicate serious problems for the natural resources. The area is known for its exquisite natural beauty, fascinating coral reefs with myriad life forms, and warm climate. As a result, people are drawn to the area to enjoy these resources. Recent events that indicate that this ecosystem may be stressed include signs of overfishing of reef resources, an epizootic of the spiny sea urchin *Diadema antillarum* Philippi, widespread bleaching of corals, plus white band and black band diseases, fish kills, oil pollution, mining, land erosion associated with deforestation, and population declines of several species of sea turtle. These events are reviewed and prognoses made about the future.

Fishery Resources

The fishery resources of the Caribbean and adjacent areas have been reviewed in the past by FAO (1969), Idyll (1971), Klima (1977), Wolf and Rathjen (1974), Munro

(1983), Goodwin *et al.* (1985), and Goodwin (1987), to list just a few. The fisheries consist of near-shore trap and hook-and-line fisheries that are mostly artisanal over the large reef areas, trawl fisheries in selected areas for shrimp and demersal fishes where bottom topography is suitable, and pelagic fisheries for tunas, billfishes, and flying fish. Extensive fisheries for invertebrates — spiny lobster, conch, and oysters — also exist, with limited fisheries for sea turtles and whales.

The total estimated catch of marine fishery resources of the Caribbean and adjacent regions for 1985 (the most recent data available from FAO, 1987) was 499,473 metric tons (mt). We arrived at this value by including the catch from Caribbean and adjacent areas with Mexico's catch limited to Quintano Roo (data from Ojeda Paullada, 1986), and the U.S. catch limited to South Florida. These data are given in Table 4.1. Of this catch, Venezuela accounts for 53.5%, Cuba for 16%, Guyana for 8%, and the Dominican Republic for 3%, with the remaining 34 countries for 2% or less. In 1976, Venezuela's catch was 40%; thus, upward trends of the total catch have been due to Venezuela, as most of the other countries' catches have been rather steady over this 10-year period. Since much of the fishing is artisanal, there is probably a substantial amount of unreported catch. However, overall, the fishery resources are not large when compared to the smaller Gulf of Mexico to the north, where the U.S. commercial landings were 1.089 million metric tons (mmt) in 1985 and the recreational catch was estimated at 109,000 mt in 1985 (Richards and McGowan, 1989). Adding to this Mexico's catch in the Gulf of 260,754 mt results in a total catch of 1.459 mmt from a sea that is 60% the size of the Caribbean. The smaller Caribbean catch is the result of lower productivity and small shelf size. The only Caribbean area with high productivity is the large shelf off Venezuela which, along with that country's expanded fishery for large pelagics, especially for yellowfin tuna and swordfish, accounts for its high catch figure.

Spiny lobster

The spiny lobster (*Panulirus argus*) is one of the most valuable species in the Caribbean. On Jamaica, lobster accounted for only 3.7% of sample catch by numbers, but represented 25% of the total value (Munro, 1983). The 1985 catch was 30,645 mt from the western central Atlantic. This is the highest catch reported over the past seven years, from a low of 21,883 mt in 1981 (FAO, 1987). This catch may well be underestimated but not to the degree indicated by Suman's (1987) report that Cuban fishermen take 64,000 tons of spiny lobster from the Gulf of Batabano annually in artificial structures, as this must be a misprint. In 1985, Cuba reported a total catch of 13,578 mt. The United States imports almost 20,000 mt, while its domestic landings were 3,000 mt in 1986 (Thompson, 1987), which also indicates that the catch is underreported since the United States does not import from Cuba. Concern also exists about long-term stock persistence with increasing effort. Although catch has been roughly the same, the catch per unit effort and total effort in Florida has increased dramatically (Jones *et al.*, 1985). In many areas the minimum legal size is well below the size of reproductive maturity. The spiny lobster has a long and complicated larval life (Richards and Potthoff, 1981). This larval life is estimated to be 6 to 9 months in duration and the pelagic larval stages travel great distances (Richards and Goulet, 1977). Consequently, distant stocks may be responsible for local recruitment. Presumably if these distant reservoir stocks are harvested at similar high levels as in Florida and before reproductive maturity, a population crash would result.

Coralline reef fishes

Fishing in coralline reef areas is the major fishery in the Caribbean area. Coral reefs are a major component of the shallow-water Caribbean ecosystem. The coral reefs are centered in the Caribbean and range to adjacent areas of southern Florida and the Bahamas. Fringing reefs occur around most Caribbean islands and local fisheries depend mostly on shallow-water reef resources. Milliman (1976) and Glynn (1976) described the geology, biology, and history of Caribbean coral reefs. The living coral reefs are made predominantly by hermatypic (calcium-secreting) corals and require clear, oceanic, shallow, low-nutrient waters, with sunlight and warm temperatures. Coral growth can be limited by high turbidity, exposure to fresh water or air, extreme temperatures, depth, pollution, and excess nutrients. Of the three main types of reefs, fringing reefs dominate in the Caribbean (Glynn, 1976). A few atolls exist and a true barrier reef extends 240 km along the coast of Belize and north of Providencia Island in the southwestern Caribbean (Milliman, 1976; Glynn, 1976).

The fishery is fundamentally artisanal. Traps and handlines are the principal modes of fishing, with some netting and spearfishing. Munro (1983) estimated that production from coralline shelves should be 4–6 mt $\cdot$ km^{-2} $\cdot$ yr^{-1}. Munro (1977) estimated that there are 129,620 km^2 of coralline shelf in the Caribbean which, based on his production estimate, should then yield 518,400–777,720 mt $\cdot$ yr^{-1}. Since current total production is less than this, there must be a large amount of unreported catch (which is typical for artisanal fisheries), catch lost because of ciguatera poisoning (a persistent problem with tropical fisheries), or it may be that production estimates are too high. However, Munro (1983) reviewed many recent estimates and indicates that production may be much higher — 8 to 24 t $\cdot$ km^{-2} $\cdot$ yr^{-1} from some Pacific areas. Further research is needed, but since several areas show evidence of growth overfishing (Appeldoorn and Lindeman, 1985), management of the fisheries is needed. In Puerto Rico, the greatest diversity of fishery production occurred in depths less than 30 fathoms (Anonymous, 1979). This zone accounts for 70% of the fish landings. More than 170 species of finfish are captured for commerce, although the bulk of the catch is made up of less than 50 species. The U.S. Caribbean Fishery Management Council recognized 65 shallow-water species with economic importance. On Jamaica, fish traps captured 97 species, of which 21 accounted for more than 1% of the catch and none more than 12% (Munro, 1983). Traps are the primary means for capturing reef fishes and crustaceans, and management may be accomplished by regulating mesh sizes of these traps (Munro, 1983). Recruitment of fish and crustaceans is interdependent throughout the area, and management practices (or lack of) may impinge on other areas. Richards (1984) and Richards and Potthoff (1981) showed the wide distribution of early life history stages throughout the region.

Turtles

Sea turtles, once abundant and ubiquitous around the Caribbean, have been extensively exploited and are now considered threatened or endangered in many areas (Bacon et al., 1984). Local management actions range from total protection to uncontrolled harvesting.

Conch

Berg and Olsen (1989) reviewed the Caribbean queen conch *(Strombus gigas)* fishery. Conch fisheries have collapsed in many areas (e.g., Cuba, Belize, Florida, U.S. Virgin Islands) and are under various levels of protection or regulation in many areas. U.S. imports, believed to approximate one-half the total catch, averaged 500,000 kg from 1972 through 1985 (Berg and Olsen, 1989). The ability to maintain this catch is in doubt based on a decline in CPUE.

Recent Critical Events

Several widespread epizootic events have occurred in the Caribbean in the past eight

Table 4.1. Estimates of catch of fish, crustaceans, and molluscs from the Caribbean Sea and adjacent waters. Data mostly from FAO (1987), with pelagic catches of non-Caribbean countries deleted and catches of Mexico and the U.S. restricted to Quintana Roo and South Florida, respectively.

Country/area	1976	1977	1978	1979	1980	1981	1982	1983	1984	1985
Anguilla	0	0	0	0	0	0	0	0	0	0
Antigua Brd	1727	1673	2204	1632	1601	1777	2004	2246	2246	2246
Aruba	650	700	770	770	770	770	770	770	770	770
Bahamas	2917	3741	3762	4029	5026	4372	4686	5211	5341	8188
Barbados	5162	3166	3683	4342	3735	3411	3480	6522	5787	3915
Belize	1588	1529	1497	1415	1324	1325	1345	1515	1299	1388
Bermuda	468	496	451	451	547	442	422	481	486	506
British Virgin Islands	318	318	318	318	318	318	318	318	318	318
Cayman Islands	0	0	0	0	0	0	0	0	0	0
Colombia	5880	7621	5837	5320	5155	9363	6367	3058	7491	9924
Costa Rica	456	254	342	110	132	141	228	254	549	399
Cuba	80482	72361	71248	66576	68898	60040	68718	74437	76869	79394
Dominica	1024	1047	1070	642	1445	1514	1545	1000	500	446
Dominican Republic	6435	4235	4573	6845	8199	9167	11448	10951	12845	15814
French Guiana	1207	1266	1240	1186	1150	1436	1992	2500	2679	3035
Grenada	1700	2958	3111	2175	2021	943	1221	1856	1281	1243
Guadeloupe	5060	9525	9000	8500	8000	8300	8240	8826	8940	9000
Guatemala	100	100	100	100	50	50	150	150	150	150
Guyana	20956	24299	19813	26051	22929	22691	25124	26830	31581	41295
Haiti	3700	3850	3700	3750	3800	3800	3900	4000	4100	4100
Honduras	4234	4953	5478	6333	6031	4550	3906	7621	6113	7426
Jamaica	10100	10100	9600	9600	9000	7757	7741	8428	9191	9500
Martinique	4233	2432	4375	4564	4891	4700	5500	5107	5174	5200
Mexico (Quintana Roo)	4400	4400	4400	4400	4400	4400	4400	4343	4420	4400
Montserrat	120	120	120	120	120	120	120	120	120	120
Netherlands Antilles	950	1000	1010	1010	1090	990	1030	1030	1030	1030
Nicaragua	6507	7753	7676	5637	5025	3385	2701	3018	2767	2179
Panama	7366	5690	6267	1217	3558	1445	4972	547	2995	2995
Puerto Rico	2563	2881	3197	3597	2557	1784	2202	2655	2354	1496
St Chris Nev	1600	1600	1700	1700	1850	1880	1880	1100	1300	1500

Table 4.1 *(continued)*

Country/area	1976	1977	1978	1979	1980	1981	1982	1983	1984	1985
Saint Lucia	2200	2500	2600	2600	2400	2404	2404	2635	1182	1044
St Vincent	379	581	698	547	547	547	547	547	547	547
Suriname	4340	4833	3444	3401	2689	3284	2808	3375	3957	3184
Trinidad-Tobago	4417	4303	4823	3840	4461	3804	4574	4240	3593	2862
Turks Caicos	1269	946	1128	1385	1190	989	1061	1236	1212	1349
US (South Florida)	4548	4460	4450	4445	4445	4400	4400	4400	4400	4400
US Virgin Islands	500	502	557	547	669	634	883	611	680	600
Venezuela	139184	143033	154858	151152	160352	155193	188257	197913	220823	265510
TOTAL	347765	341326	349120	340493	349575	332126	401344	399854	435030	499473

years starting with a massive fish kill in 1980 which closely followed a powerful hurricane (Atwood, 1981). Since then, there has been mass mortality of the long-spined sea urchin, a white banding disease of coral, and a widespread bleaching of coral. These events are briefly summarized below.

Mass mortalities of fishes

Unexplained mass mortalities of reef fishes occurred around the Caribbean in August and September 1980 following Hurricane Allen (Atwood, 1981). Unfortunately, specimens were not saved to allow determination of possible causes for this event (Williams and Williams, 1987).

Mass mortalities of sea urchins

In 1983, a mass mortality of the sea urchin (*Diadema* spp.) began off Panama and within one year had spread throughout the Caribbean (Lessios et al., 1984b). Mortality reached approximately 98% in many areas (Lessios et al., 1984a; Hughes et al., 1985). This species was the most important algal grazer on Caribbean reefs, particularly on reefs subjected to intense fishing (Jackson, 1987). Many reefs are now carpeted by macroalgae that are smothering corals as deep as 10 to 15 m (Jackson, 1987). The long-term effect on the ecosystem is unknown. Also unknown is the cause of this catastrophic disturbance, but Bauer and Ageter (1987) isolated *Clostridium*, an anaerobic bacteria from sea urchins, and demonstrated its lethal effect. Kills of two other sea urchins in the Caribbean have been reported, but are unrelated to the *Diadema* mortalities (Williams et al., 1986).

Coral mortalities

Considerable concern exists about possible major disruptions of reef ecology by unexplained recent, widespread bleaching and possible mass mortalities of corals in the Caribbean (Roberts, 1987; Williams and Williams, 1987). Bleaching occurs when corals expel resident symbiotic algae (Zooxanthellae) in their tissues, resulting in a white appearance of the coral. High temperatures and other environmental stresses often result in bleaching events. Corals often recover after such events, but not always. The algae symbionts function in nutrient recycling, calcification, and energy fixation. Death of corals can stop reef growth and increase reef erosion and degradation. An earlier episode of this bleaching was associated with the 1983 El Niño event.

In addition to the coral bleaching phenomena, a disease called white band disease afflicts coral with serious consequences. This disease has killed up to 90% of *Acropora palmata* off Buck Island, St. Croix, U.S. Virgin Islands (Ogden, 1987). It is not known for wide-ranging impacts, but rather local events. Another disease termed black-ring disease also affects corals.

General reef degradation

Survey results suggest that Caribbean and western Atlantic coral reefs and associated fisheries have deteriorated significantly in the last 10 years (Rogers, 1985; Dustan and Hales, 1987). Associated with reef degradation is coastal alteration, deforestation, and drainage of mangrove forests and wetlands (Woodley, 1987; Jackson, 1987).

Endean (1976) reviewed disturbance and recovery of coral reefs. Causes of reef degradation include sedimentation from coastal alterations due to development and deforestation (Risk et al., 1980; Hudson, 1981; Jackson, 1987), anchor damage (Davis, 1977), excess nutrients (Kinsey and Davies, 1979), predation (Endean, 1976), ship groundings, storms, hurricanes, and diver contact. Atwood et al. (1987) and Jackson (1987) noted problems of petroleum pollution in the Caribbean. Hurricanes are frequent events that can be devastating locally to coral reefs and other communities (Woodley et al., 1981; Porter et al., 1981).

Conclusions

In this brief review we have touched on some of the problems affecting the natural resources of the Caribbean. Many of the countries are poor and suffer overpopulation problems, but because of the natural beauty and mild climate they are actively pursuing growth through tourism expansion. Already tourism is a large industry, as evidenced by the number of cruise ships operating. The February 7, 1988, issue of the *Miami Herald* lists 58 cruise ships with a capacity of about 53,000 passengers operating in the Caribbean from the ports of Miami, Ft. Lauderdale, Tampa, and San Juan. The *Miami Herald* reports that additional cruise ships are planned because this industry can expand significantly. Besides this single aspect of tourism, it can be expected that additional expansion will occur in other tourist activities together with associated development, resource demands, and habitat alteration. We believe that the Caribbean as a large marine ecosystem may be showing signs of environmental stress, but the real crisis lies in the lack of coordinated support among the 38 nations to monitor the system. Coordinated regional management is desirable because efforts of individual countries may not be effective. Coordination is difficult because the Caribbean is divided geographically and politically into at least 38 countries or territories with different cultures and varying levels of public education and economic development. A number of marine laboratories located in the Caribbean are doing excellent research on local problems, some of which have system application (Ogden 1987). But a fully funded coordinated effort such as the CICAR work in the 1960s is lacking and does not appear to be forthcoming.

Bohnsack has proposed that these fishery resources can best be maintained by creating marine fishery reserves, areas with no consumptive usage, which offer a potential way to manage reef fish fisheries by protecting species composition, population age structure, spawning potential, and intraspecific genetic variability. Reserves will insure against recruitment failure due to environmental variability. A minimum of 20% of shelf habitat is recommended for reserve protection. With this approach, each country contributes in proportion to its shelf area and potential biotic wealth.

The Caribbean is not the only large marine ecosystem with these problems. Solving these problems can be achieved only through coordination among the countries involved, with someone exercising leadership. Nationalistic concerns are justified, but at the same time, it must be recognized that strong system interactions exist.

Acknowledgments

We thank Dr. Bradford E. Brown, Southeast Fisheries Center, for critically reviewing the manuscript. Dr. Carl J. Berg, Florida Department of Natural Resources, kindly provided information on conch. We also appreciate the help rendered by Mr. Lewis Fraser, Sr. and his staff at Poseidon Maritime Services, Inc. on cruise ship information. Finally we thank Dr. Kenneth Sherman for inviting us to participate in this outstanding symposium.

References

Anonymous. 1979. Fisheries resources of Puerto Rico and the United States Virgin Islands. SEFC Contribution No. 79–54F. National Marine Fisheries Service, Southeast Fisheries Center, Miami.

Appeldoorn, R. S., and Lindeman, K. C. 1985. Multispecies assessment in coral reef fisheries using higher taxonomic categories as unit stocks, with an analysis of the artisanal haemulid fishery. Proc. 5th Int. Coral Reef Congr., Tahiti, Vol. 5:507–514.

Atwood, D. K. (Editor). 1981. Unusual mass fish

mortalities in the Caribbean and Gulf of Mexico. Published and distributed by the Atlantic Oceanographic and Meteorological Labs., NOAA, Miami. 46 pp.

Atwood, D. K., et al. 1987. Petroleum pollution in the Caribbean. Oceanus 30(4):25–32.

Bacon, P., et al. 1984. Proceedings of the western Atlantic turtle symposium. Center for Environmental Education, Washington, DC. 3 vols.

Bauer, J. C., and Agerter, C. J. 1987. Isolation of bacteria pathogenic for the sea urchin Diadema antillarum (Echinodermata: Echinoidea). Bull. Mar. Sci. 40(1):161–165.

Bayer, F. M. 1969. A review of research and exploration in the Caribbean Sea and adjacent areas. FAO Fish. Rep. No. 71, Vol. 1:41–91.

Berg, C. J., Jr., and Olsen, D. A. 1989. Conservation and management of queen conch (Strombus gigas) fisheries in the Caribbean. In Marine invertebrate fisheries: Their assessment and management. pp. 421–422. Ed. by J.F. Caddy. John Wiley & Sons, New York. 752 pp.

Davis, G. E. 1977. Anchor damage to a coral reef on the coast of Florida. Biol. Conserv. 11:29–33.

Dustan, P., and Hales, J. C. 1987. Changes in the reef- coral community of Carysfort Reef, Key Largo, Florida: 1974 to 1982. Coral Reefs 6:91–106.

Endean, R. 1976. Destruction and recovery of coral reef communities. In Biology and geology of coral reefs. Vol. 3, Biology 2. pp. 215–253. Ed. by O. A. Jones and R. Endean. 435 pp.

FAO. 1969. Symposium on investigations and resources of the Caribbean Sea and adjacent regions. FAO Fish. Rep. No. 71 (vols. 1 and 2): 165 pp. and 347 pp.

FAO. 1977. Symposium on progress in marine research in the Caribbean and adjacent regions. FAO Fish. Rep. No. 200:547 pp.

FAO. 1987. 1985 yearbook of fishery statistics. Food and Agriculture Organization of the United Nations. Vol. 60:463 pp.

Goodwin, M. 1987. Changing times for Caribbean fisheries. Oceanus 30(4):57–64.

Goodwin, M., Orbach, M., Dandifer, P., and Towle, E. 1985. Fishery sector assessment for the eastern Caribbean: Antigua/Barbuda, Dominica, Grenada, Montserrat, St. Christopher/Nevis, St. Lucia, St. Vincent and Grenadines. Contract No. 38-0000-C-5011. U.S. Agency for International Development. Island Resources Foundation, Red Hook Center Box 33, St. Thomas, VI.

Glynn, P. W. 1976. Aspects of the ecology of coral reefs in the western Atlantic region. In Biology and geology of coral reefs. pp. 271–324.

Hudson, J. H. 1981. Growth rates in Montastraea annularis: A record of environmental change in Key Largo Coral Reef Marine Sanctuary, Florida. Bull. Mar. Sci. 31:444–459.

Hughes, T. P., Keller, B. D., Jackson, J. B. C., and Boyle, M. J. 1985. Mass mortality of the echinoid Diadema antillarum Philippi in Jamaica. Bull. Mar. Sci. 36:377–384.

Idyll, C. P. 1971. The potential for fishery development in the Caribbean and adjacent seas. Univ. Rhode Island Int. Center Mar. Res. Devel. 16 pp.

Jackson, J. B. C. 1987. Protection of the tropics. Oceanus 30(4):87–88.

Jones, A. C., et al. 1985. Ocean habitat and fishery resources of Florida. In Florida aquatic habitat and fishery resources. pp. 437–543. Ed. by W. Seaman, Jr. Florida Chapter of the American Fisheries Society, Kissimmee, FL. 543 pp.

Kinsey, D. W., and Davies, P. J. 1979. Effects of elevated nitrogen and phosphorus on coral reef growth. Limnol. Oceanogr. 24(5):935–940.

Klima, E. F. 1977. An overview of the fishery resources of the west central Atlantic region. FAO Fish. Rep. (200):231–252.

Lessios, H. A., et al. 1984a. Mass mortality of Diadema antillarum on the Caribbean coast of Panama. Coral Reefs 3:173–182.

Lessios, H. A., Robertson, D. R., Cubit, J. D. 1984b. Spread of Diadema mass mortality through the Caribbean. Science 226:335–337.

Milliman, J. D. 1976. Caribbean coral reefs. In Biology and geology of coral reefs. Vol. 1, Geol. 1. pp. 1–50. Ed. by O. A. Jones and R. Endean. Academic Press, Orlando, FL.

Munro, J. L. 1977. Actual and potential fish production from the coralline shelves of the Caribbean Sea. FAO Fish. Rep. (200):301–321.

Munro, J. L. (Editor). 1983. Caribbean coral reef fishery resources. ICLARM Studies and Reviews. No. 7:276 pp.

Ogden, J. C. 1987. Cooperative coastal ecology at Caribbean marine laboratories. Oceanus 30(4):9–13.

Ojeda Paullada, P. 1986. Agenda estadistica pesquera. Secretaria de Pesca. Mexico. 99 pp.

Porter, J. W., et al. 1981. Population trends among Jamaican reef corals. Science 294:249–250.

Richards, W. J. 1984. Kinds and abundance of fish larvae in the Caribbean Sea and adjacent areas. NOAA Tech. Rep. NMFS SSRF-776. 54 pp.

Richards, W. J., and Goulet, J. R., Jr. 1977. An operational surface drift model used for studying larval lobster recruitment and dispersal. *In* Cooperative investigations of the Caribbean and adjacent regions-II. pp. 363–374. Ed. by H. B. Stewart, Jr. FAO Fish. Rep. (200):547 pp.

Richards, W. J., and McGowan, M. F. 1989. Biological productivity in the Gulf of Mexico: Identifying the causes of variability. *In* Biomass yields and geography of large marine ecosystems. pp. 287–325. Ed. by K. Sherman and L. M. Alexander. AAAS Selected Symposium 111. Westview Press, Boulder, CO. 493 pp.

Richards, W. J., and Potthoff, T. 1981. Distribution and seasonal occurrence of larval pelagic stages of spiny lobsters (Palinuridae, *Panulirus*) in the western tropical Atlantic. Proc. Gulf Caribb. Fish. Inst. 33:244–252.

Risk, M. J., Murillo, M. M., and Cortes, J. 1980. Observaciones biologicas preliminares sobre el arrecife coralino en el Parque Nacional de Cahuita, Costa Rica. Rev. Biol. Trop. 28(2):361–382.

Roberts, L. 1987. Coral bleaching threatens Atlantic reefs. Science 238:1228–1229.

Rogers, C. S. 1985. Degradation of Caribbean and western Atlantic coral reefs. Proc. 5th Int. Coral Reef Cong. Tahiti 6:491–496.

Suman, D. O. 1987. Intermediate technologies for small-scale fishermen in the Caribbean. Oceanus 30(4):65–68.

Thompson, B. G. 1987. Fisheries of the United States, 1986. Cur. Fish. Stat. (8385): 119 pp.

Williams, E. H., and Williams, L. B. 1987. Caribbean mass mortalities. Oceanus 30(4):69–75.

Williams, L. B., Williams, E. H., Jr., and Bunkley, A. G., Jr. 1986. Isolated mortalities of the sea urchins *Astropyga magnifica* and *Eucidaris tribuloides* in Puerto Rico. Bull. Mar. Sci. 38(2):391–393.

Wolf, R. S., and Rathjen, W. F. 1974. Exploratory fishing activities of the UNDP/FAO Caribbean Fishery Development Project, 1965–1971: A summary. Mar. Fish. Rev. 36(9):1–8.

Woodley, J. D. 1987. Managing marine resources. Oceanus 30(4):85–86.

Woodley, J. D., *et al.* 1981. Hurricane Allen's impact on Jamaican reefs. Science 214:749–755.

Chapter 5

Productivity and Fisheries Potential of the Banda Sea Ecosystem

Jenne J. Zijlstra and Martien A. Baars

Abstract

Wyrtki (1958, 1961) postulated that in eastern Indonesian waters, periods of upwelling and downwelling alternate in response to the monsoonal seasons. New data for this area, obtained during the Indonesian-Dutch Snellius II expedition, supported this hypothesis and extended the knowledge about the biological consequences of this change in hydrography in the area studied, the eastern Banda Sea and northern Arafura Sea. Various forms of upwelling were observed in August 1984 (southeast monsoon), whereas conditions were generally oligotrophic during February 1985 (northwest monsoon). During upwelling, stocks and productivity in the ecosystem were generally enhanced by a factor of 2 to 3, as evidenced by data on chlorophyll, primary production, and meso- and macrozooplankton. The changing conditions also influenced the phytoplankton and zooplankton species composition. Pelagic fish resources, estimated by acoustic methods, appear to be significantly higher during the upwelling period and are probably greatly underexploited.

Introduction

Studies on marine ecosystems in tropical waters are relatively few compared to those of temperate and even polar regions. The study reported on here concerns the ecosystem of the Banda and Arafura Sea in southeastern Indonesian waters. The area has a complex geomorphology, with the Aru Basin forming a transition between the shallow (< 100 m) semi-enclosed Arafura Sea in the east and the deep Banda Sea (> 4000 m), fringed by numerous islands to the west (Figure 5.1). The area, about 400,000 km^2 in size, is under monsoonal influence, with southeast winds prevailing between April and November and northwest winds between December and March. Surface currents, mainly westerly during the southeast monsoon and easterly during the northwest monsoon, conform to this change in wind direction. Wyrtki (1958, 1961) postulated the presence of an alternating monsoon-related system of up- and downwelling, both for the eastern Banda Sea and along the nearby northwest coast of Australia (Figure 5.2). He expected upwelling during the southeast monsoon among others because of a decline in sea surface temperatures and sea level in that period, and more important, because of seasonal changes in temperature/salinity profiles in the Banda Sea. Temperature profiles, collected at six stations at the end of the southeast monsoon in October 1929, suggested an upward displacement compared to profiles collected at about the same positions at the end of the northwest monsoon period in March 1957. Rochford (1977), studying the situation off northwest Australia, failed to detect any signs of upwelling in that region. Wyrtki's hypothesis for the eastern Banda Sea evoked several Indonesian studies in the area, demonstrating the coincidence of lower surface temperatures and higher nutrient levels in the photic zone at the end of the southeast monsoon as compared to the situation at the end of the northwest monsoon (Nontji, 1975; Birowo and Ilahude, 1977). Relatively

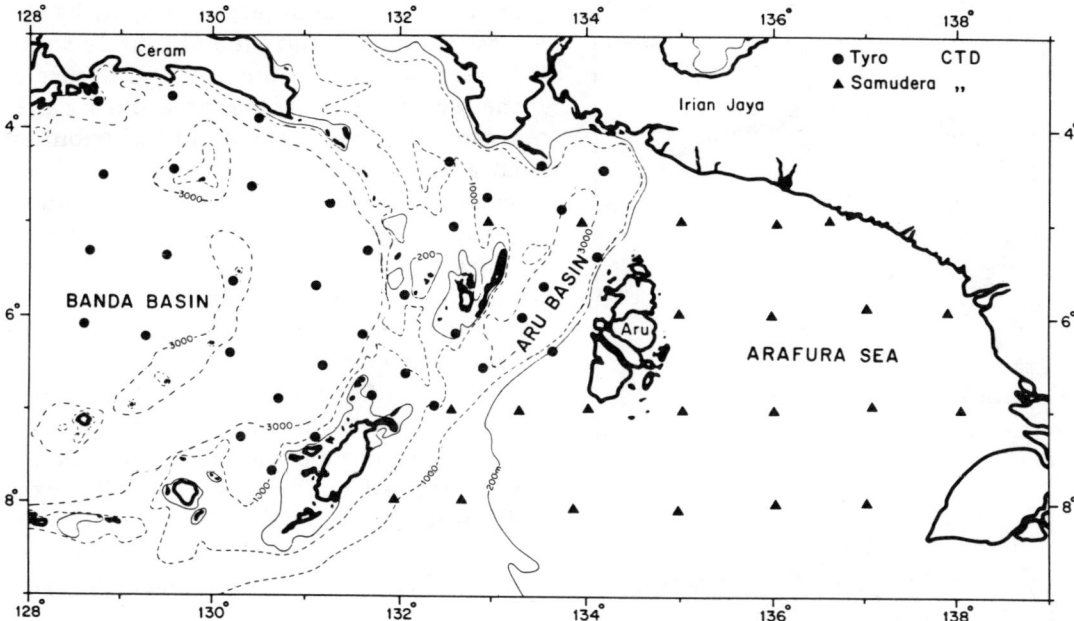

Figure 5.1. Map of the eastern Banda Sea and the northern Arafura Sea, indicating geomorphology and isobaths, with stations sampled by RV *Tyro* (circles) and RV *Samudera* (triangles). RV *Tenggiri*, for which no cruise track is shown, worked in the same area as RV *Tyro*.

high nutrient concentrations were also observed in the southern part of the Arafura Sea during the southeast monsoon, indicating a pronounced slope upwelling from the Banda Sea into this shallow area (Rochford, 1962, 1966). However, surface temperature observations in the Banda Sea (Boely et al., 1990), together with temperature profiles for a limited number of years, suggest the presence of large, interannual variations in surface temperature and temperature profiles in the area, probably partly related to the ENSO phenomenon (Nicholls, 1984; Zijlstra et al., 1990). Such variations would affect conclusions on vertical water movements derived from temperature profiles collected in different years. Therefore, to test Wyrtki's hypothesis, new studies on the hydrographic regime in the area that measured vertical water movements within a season were required. At the same time, the biological consequences of seasonal changes in the hydrography of the area, which had so far not been considered, could be investigated. In this context, an Indonesian-Dutch study, forming part of a larger

cooperation in the ocean sciences between the two countries under the name of the Snellius II Expedition (the first Snellius Expedition took place in 1929–1930), ventured to investigate the monsoon-related physical, chemical, and biological changes in the area, including its pelagic fish resources. Three ships, two Indonesian and one Dutch, participated in the project during two periods of one month each (August 1984, southeast monsoon; February–March 1985; northwest monsoon), covering both the eastern Banda Sea and the northern part of the Arafura Sea (Baars and Zijlstra, in press) (Figure 5.1).

Two vessels, RV *Tyro* (Dutch) and RV *Samudera* (Indonesian) concentrated mainly on hydrographical features, nutrient distribution, and lower food-chain characteristics, dividing the area between them. The third vessel, RV *Tenggiri* (Indonesian), covered the eastern Banda Sea and Aru Basin with an acoustic survey using an echo-integrator system in an attempt to estimate fish resources. The Dutch vessel used a simpler acoustic system in the same area.

This chapter gives a summary of the re-

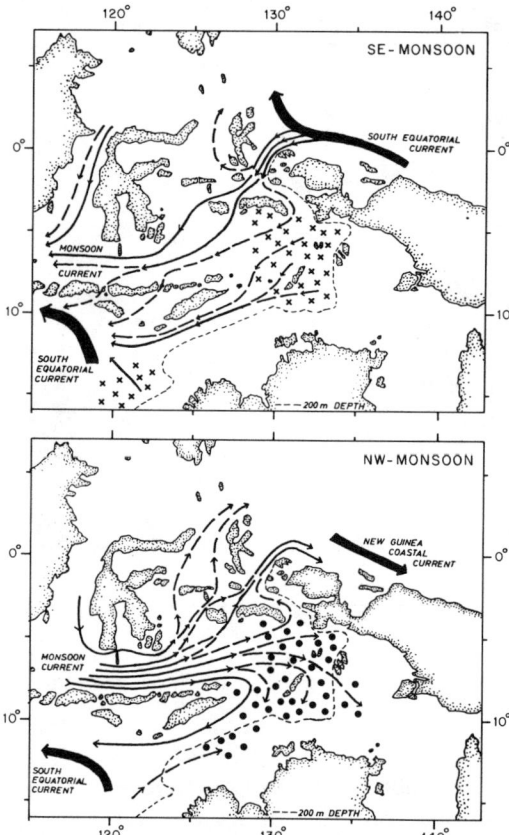

Figure 5.2. Surface current patterns during the southeast and northwest monsoon and the location of areas of upwelling (x) and downwelling (●) during both periods, according to Wyrtki (1958).

sults of the studies in which some 30 Indonesian and Dutch scientists participated. The majority of the basic papers, presented during a symposium in Jakarta in November 1987, will appear in a special volume of the *Netherlands Journal of Sea Research*: 1990.

Hydrography and Nutrients

Surface temperatures in August were some 3°–4°C lower than in February (26°–30°C), conforming to Wyrtki's observations (Figure 5.3). During both periods, low-salinity surface water was present in the Arafura Sea, originating from river outflow from Irian (New Guinea), carried westward in particular during the southeast monsoon by the westerly surface current and causing stratification, shallow in the east and deeper in the western part of the area during that season. Temperature/salinity distribution on a west-east section in the area did not show a clear upturn of the isotherms towards the east during the southeast monsoon, so characteristic for upwelling conditions in the Eastern Boundary Current systems, so that the hydrographic structure in the area provides no clear evidence for upwelling in that season (Figure 5.3). The salinity distribution in particular indicates that during the southeast monsoon, the shallow Arafura Sea was stratified, but was generally fully mixed down to the bottom during the northwest monsoon. As will be shown later, this may have affected the fertility in that area during the northwest monsoon.

A more dynamic picture of the vertical movements in the water masses was obtained in the Banda Sea and Aru Basin by comparing the depth of isopycnals and isotherms during successive visits to stations during each cruise. The time interval between visits was 6–21 days. The results shown in Figure 5.4 for three groups of stations, two in the eastern Banda Sea and one in the Aru Basin, indicated upward water movements in the upper 100–150 m in all three areas during the southeast monsoon and downward movements during the northwest monsoon. Upwelling rates during the southeast monsoon were strongest in the Aru Basin, with rates up to 300 cm $\cdot$ d^{-1}; more to the west, rates declined to 30–40 cm $\cdot$ d^{-1}. However, downwelling with rates up to 100 cm $\cdot$ d^{-1} was only present in the eastern Banda Sea during the northwest monsoon (Zijlstra et al., 1990). At depths below 100–150 m, no significant changes in depths of isopycnals could be observed. The distribution of nutrients (P, N, Si) and oxygen concentrations in the water column agreed with this picture, with higher free nutrient concentrations in the euphotic zone in August (southeast monsoon) (Table 5.1) and an upward displacement of the nitra-

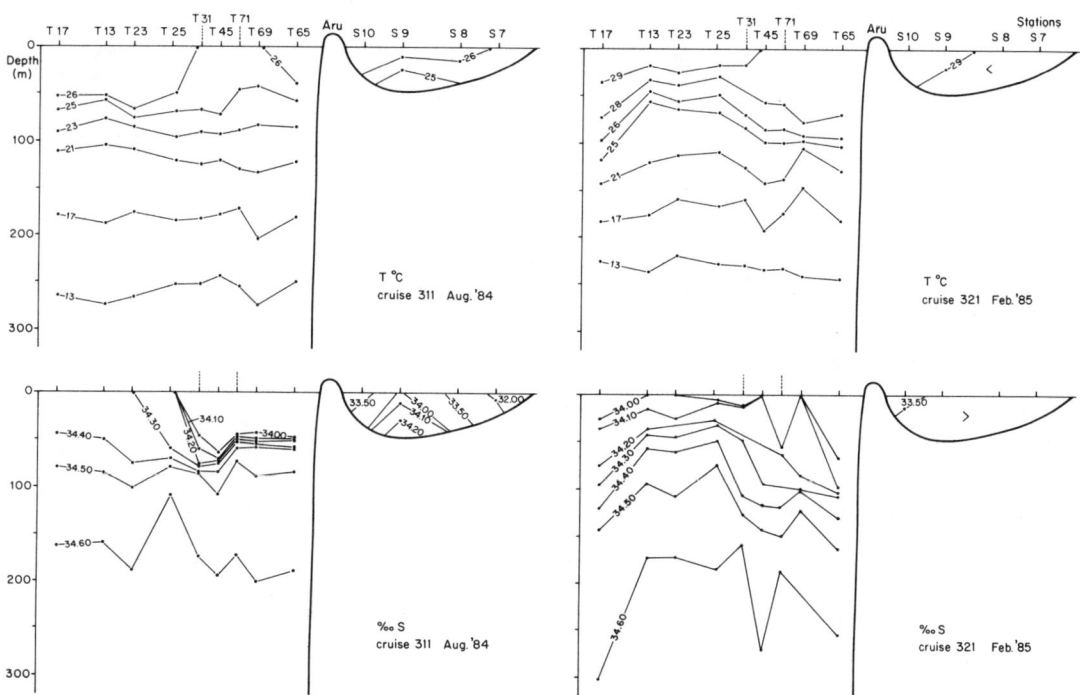

Figure 5.3. West-east section of the area at about 7°S, showing temperature (**upper figures**) and salinity distribution (**lower figures**) during both seasons. (From Zijlstra et al., 1990.)

cline (Figure 5.5) and the oxygen depth curve in that season (Wetsteyn et al., 1990; Tijssen et al., 1990). In the surface water, the free nitrate concentration, most likely the limiting nutrient in the area, was below detection during the northwest monsoon, but not during the southeast monsoon (Figure 5.6). In accordance with the observations of Rochford (1962, 1966) in the Arafura Sea, evidence was found for an upslope, eastward transport of colder, more saline, nutrient-rich bottom water during the southeast monsoon (slope upwelling) (Figure 5.7), which probably fertilized the upper waters of that area by a mixing process with the westward moving surface layers (Zijlstra et al., 1990; Wetsteyn et al., 1990). Whether the Arafura Sea was also nutrient enriched by fresh-water outflow from Irian (New Guinea) is an open question (Wetsteyn et al., 1990; Gieskes et al., 1988), but the moderate contribution of river water to the upper water layers (euphotic zone) (1–5%) seems to make this unlikely, unless nutrient concentrations in the river water were very high.

Plankton System

The biological response to the change in the physical regime was unambiguous, with higher biomasses and productivity at almost all levels of the food chain in August (Baars and Zijlstra, in press). For instance, chlorophyll a concentrations were on average two times higher in August than in February (Figure 5.8). Highest chlorophyll concentrations in August were found in the northern part of the Aru Basin and in the northern Arafura Sea, where the highest upwelling rates had been recorded (Gieskes et al., 1988). Chlorophyll a concentrations in February 1985 were most strongly reduced in the western part of the area (Banda Sea), but not so much in the Arafura Sea. As mentioned earlier, the latter area was probably

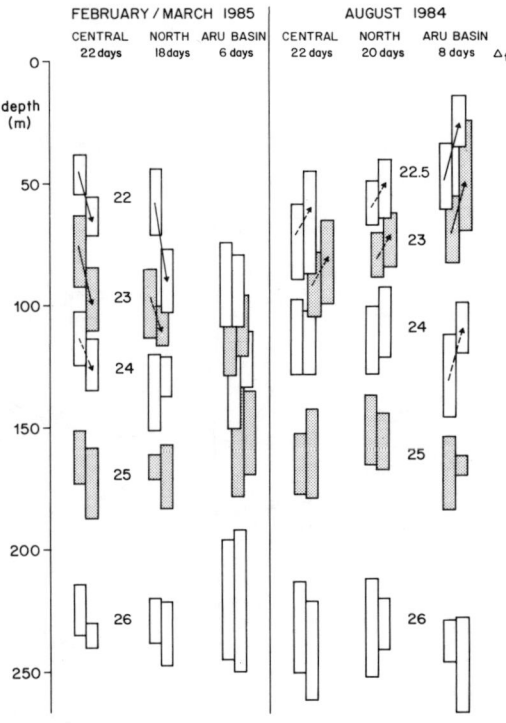

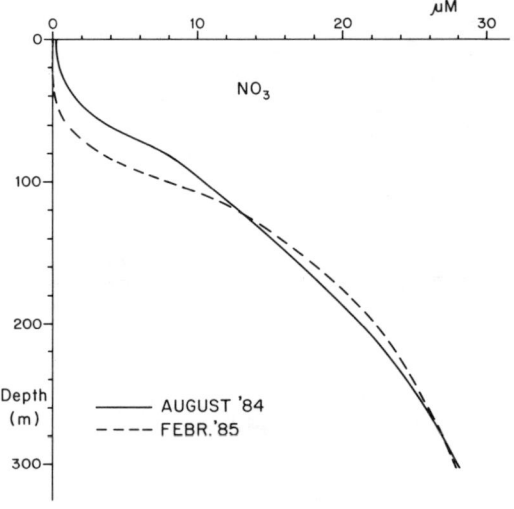

Figure 5.5. Nitrate depth distribution during August 1984 and February–March 1985. The distribution gives an average of all Banda Sea and Aru Basin stations sampled. (From Wetsteyn et al., 1990.)

Figure 5.4. Changes in depth of sigma-t-planes in two areas of the eastern Banda Sea and one in the Aru Basin, which were visited twice during both cruises. Time lapse between visits was 6–21 days. Data of 3–9 stations were pooled, with depth ranges indicated by blocks. Depth ranges during first visit are to the right, the second visit to the left. Arrows indicate significant changes. Data by RV Tyro. (From Zijlstra et al., 1990.)

well mixed in most parts during February–March, so that nutrients were available in the euphotic zone by remineralization and upward transport.

The vertical distribution of chlorophyll *a* showed a marked difference between the two seasons. During August the highest concentrations were usually found close to the surface, in the upper water layers, whereas in February a deep chlorophyll maximum generally was present at the bottom of the mixed layer, just above the nutricline (Gieskes et al., 1990). Such a distribution is often indicative of nutrient limitation in the euphotic layer (Figure 5.9).

Primary production, estimated as daytime production with the ^{14}C method (deck-incubation and in situ measurements), measured at 15–16 stations in the Banda Sea and Aru Basin was on average 1.9 g C · m^{-2} · d^{-1} (range 0.51–6.83) in August and 0.9 g C · m^{-2} · d^{-1} (range 0.31–1.45) in February, with the highest estimates in August 1984 in the northern part of the Aru Basin (Gieskes et al., 1990). About half the difference in primary production between the two seasons can be explained in terms of "new production," when the rate of upwelling in

Table 5.1. Concentrations of inorganic nutrients at three depth levels in August 1984 and February 1985. (From Baars and Zijlstra, in press.)

Nutrient (μm)	0 m		50 m		100 m	
	August	February	August	February	August	February
Phosphate	0.1	0.04	0.3	0.1	0.7	0.6
Nitrate	0.2	0.0	2.4	0.5	10.4	8.2
Silicate	1.5	1.0	4.6	1.8	12.3	9.7

Productivity of the Banda Sea Ecosystem / 59

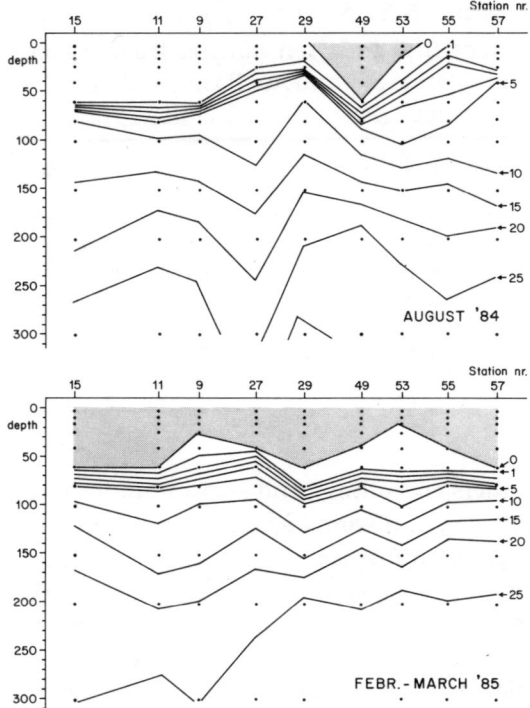

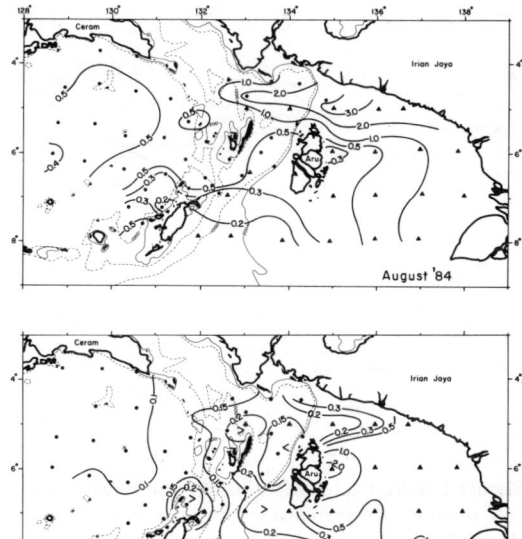

Figure 5.8. Distribution of chlorophyll a (mg · m^{-3}) during August 1984 and February 1985. (From Gieskes et al., 1988.)

Figure 5.6. West-east section showing the nitrate concentration at various depth levels during both seasons. The area above the zero-line indicates that nitrate concentrations were below detection. (From Baars and Zijlstra, in press.)

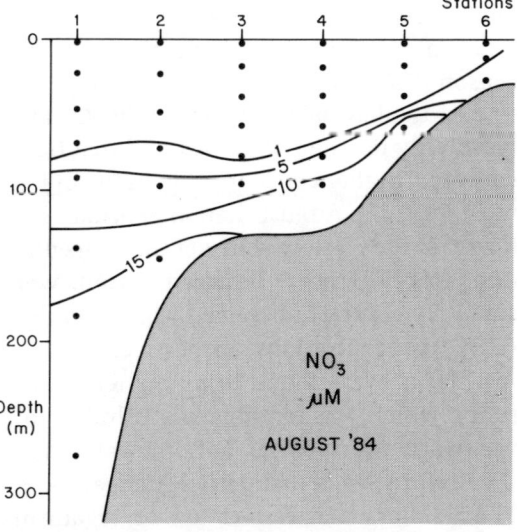

Figure 5.7. West-east section over the slope of the Arafura Sea at about 8°S, showing the isolines for nitrate concentrations. (From Wetsteyn et al., 1990.)

August and the nutrient concentration of the source water is taken into account and the Redfield ratio is applied (Wetsteyn et al., 1990). The production estimates obtained by the ^{14}C method were in good agreement with those obtained by other methods (e.g., increase in oxygen concentration in the euphotic layer during daytime, increase in ATP-concentrations and POC in euphotic layer during daytime period) (Tijssen et al., 1990; Vosjan et al., 1990; Cadée, 1988). Accepting the August and February estimates of primary production as maximum and minimum estimates of a full annual cycle, the estimates would suggest an annual primary production of 400–500 g C · m^{-2} · yr^{-1}, which is about twice that for the North Sea (Fransz and Gieskes, 1984) and also higher than the estimates for most areas on the northwestern Atlantic Shelf (O'Reilly and Busch, 1984). The change in the production rate between the two seasons could be related experimentally to changes from a mainly light-limited situation (August) to a mainly nutrient-limited one (February) (Zevenboom and Wetsteyn, 1990).

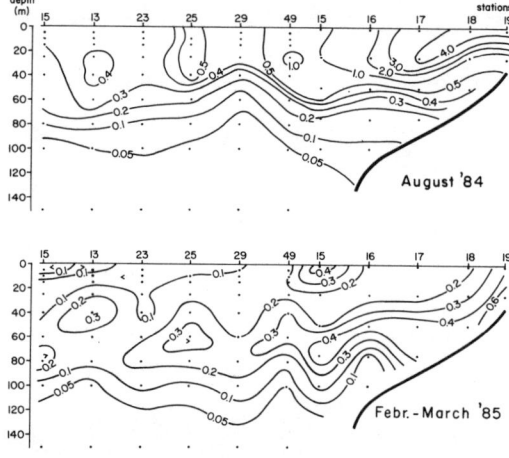

Figure 5.9. Vertical distribution of chlorophyll *a* during both seasons, as shown in a vertical (east-west) section. (From Gieskes et al., 1988.)

Table 5.3. Means of pigment concentrations ($\mu g \cdot dm^{-3}$) in samples from the eastern Banda Sea and Aru Basin, as determined by HPLC. (From Gieskes et al., 1988.)

Pigment	Aug. 1984	Feb.–March 1985
Chlorophyll *a*	0.279	0.077
Fucoxanthin	0.065	0.003
19'hexanoyloxyfucoxanthin	0.064	0.009
Zeaxanthin	0.025	0.051
Chlorophyll *b*	0.031	0.012

Differences between the two seasons were not restricted to biomass and production, but also concerned the size and species composition of the autotrophs. As shown in Table 5.2, a large part of chlorophyll *a* in August was contained in particles $> 8\,\mu m$, whereas in February, much of the chlorophyll was found in the size fraction $< 1\,\mu m$ (Gieskes et al., 1988). Moreover, changes in the relative abundance of several algal pigments, as determined by HPLC-analysis, indicated major differences in the algal assemblages in the two seasons (Gieskes et al., 1988). As shown in Table 5.3 an analysis of samples from the upper 25 m showed higher concentrations of chlorophyll *a*, fucoxanthin, 19'-hexanoylfucoxanthin, and chlorophyll *b*, whereas zeaxanthin was more abundant in February. As fucoxanthin tends to be characteristic for diatoms, 19'-hexanoyloxyfucoxanthin for Prymnesiophycea (among others coccoliths), and zeaxanthin for cyanobacteria, it would appear that the area under the influence of the monsoon changes from a mainly diatom-prymnesiophycea system in August to a Prymnesiophycea-cyanobacteria system in February. Diatoms in particular (fucoxantin) were dominant in the northern part of the Aru Basin in August, where the highest upwelling rates were found, while prymnesiophycea dominated in that period in the western part of the area. In February, cyanobacteria (zeaxanthin) dominated in the western part of the area and prymnesiophycea in the eastern part of the area investigated (Aru Basin). A larger contribution of cyanobacteria in February is confirmed by the small size ($< 1\,\mu m$) of the chlorophyll *a* containing particles in that season (Table 5.2).

Particulate organic carbon (POC) analyzed only for the Banda Sea and Aru Basin showed a higher concentration in the uppermost 100 m in August than in February by a factor of 1.5; below that depth, concentrations and differences between seasons were small. By contrast, dissolved organic carbon (DOC) concentrations, in particular in the Aru Basin, were found to be higher in February, which was tentatively attributed to a westward transport of bottom water from the Arafura Sea in that period (Cadée, 1988). The biomass and respiration of organisms with a size between 0.5–50 μm (filter size) were estimated by ATP and ETS-analysis, respectively (Vosjan and Nieuwland, 1987;

Table 5.2. Percentage of chlorophyll *a* in different size fractions of suspended matter in the eastern Banda Sea and Aru Basin. (Modified from Gieskes et al., 1988.)

Particle size (μm)	August (%)	February (%)
0.4–1	14.6	46.5
1–8	36.4	28.5
> 8	48.1	14.9

Vosjan et al., 1990). In the upper 40 m, the biomass (ATP) was highest in August, but integrated over the whole upper 300-m water column, the total biomass was similar in the two seasons, due to the fact that the distribution of ATP biomass reached deeper in February than in August. The (potential) respiration of these small organisms was again higher in August.

The same was found for mesozooplankton, where displacement volume in August, during the upwelling season, exceeded that in February by a factor of 2–3 (Schalk, 1987; Baars et al., 1990), decreasing from 23 ml $\cdot$ m^{-2} in August 1984 to 9 ml $\cdot$ m^{-2} in February 1985 (Baars and Zijlstra, in press). The distribution of the zooplankton resembled that of chlorophyll a (Figure 5.10), showing highest concentrations in August in the eastern part of the Aru Basin and Arafura Sea. Compared to chlorophyll a, the distribution of the zooplankton in August appears to shift somewhat westward. As in phytoplankton, the monsoon-related changes in the zooplankton were not restricted to abundance, but also concerned composition. During August larger calanoid species typical for upwelling conditions, such as Calanoides philippinensis and Rhincalanus nasutus, were found abundantly, particularly in the northeastern part of the area, but were virtually lacking or far less abundant in February. However, fine-filter feeders like Undinula darwini and U. vulgaris were equally numerous in both seasons. Zooplankton biomass (0.5–1 g C $\cdot$ m^{-2}) was on average comparable to estimates of about 0.75 g C $\cdot$ m^{-2} from the nearby Gulf of Carpentaria (Rothlisberg and Jackson, 1982). These estimates are somewhat lower than those for temperate seas (North Sea: $\pm$ 2 g C $\cdot$ m^{-2}), but productivity may have been similar because of short generation times due to high temperatures in the mixed layer (26°–30°C), where most of the zooplankton was found (Schalk, 1987; Baars et al., 1990).

Micronekton and Fish

Whereas differences in mesozooplankton between seasons could probably be attributed mainly to changes in abundance, this does not seem to be the case for macrozooplankton and micronekton collected with an RMT8 net up to depths of 500 m (Schalk et al., 1990a). In this group, retained by a 5 mm mesh size and consisting mainly of pelagic crustacea and mesopelagic fishes, a near doubling of the biomass between the poor and the rich season was largely effectuated by higher individual weights during August (Table 5.4). The weight differences were partly due to the size composition, but also to changes in the weight per size, probably indicating poorer feeding conditions in February 1985. Although some changes in species composition were observed, for instance in pteropods, they tended to be less pronounced than in phyto- and zooplankton. The distribution of micronekton in August, shown in Figure 5.11 for both seasons, does not conform to the general pattern encountered in other groups in showing no particularly high biomass in the Aru Basin.

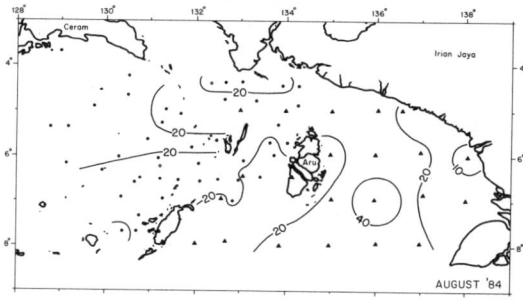
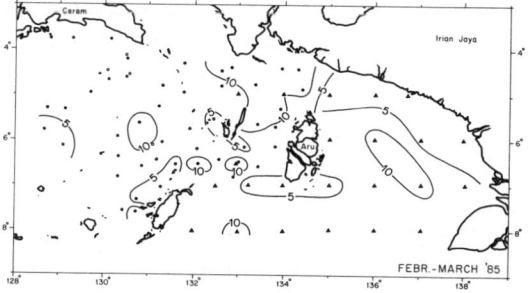

Figure 5.10. Distribution of zooplankton biomass (displacement volume, ml $\cdot$ m^{-2}) during August 1984 and February 1985 in the upper 150 m. Data by RV Tyro (●) and RV Samudera (▲). (From Baars and Zijlstra, in press.)

Table 5.4. Biomass (gram wet weight per 10,000 m^3) and abundance (n · 10,000 m^{-3} in parentheses) of micronektic Crustacea and fishes, caught by RMT1+8-sampling in August 1984 and February 1985 in the eastern Banda Sea and the Aru Basin. (From Schalk et al., 1990a.)

	Aug. 1984	Feb. 1985
Crustacea	82.6 (1833)	33.2 (1168)
Fishes	37.4 (67)	25.8 (80)

As indicated in the introduction, attempts to assess the pelagic fish resources in the area had to be restricted to acoustic surveys carried out by two ships. Acoustic recordings of RV *Tyro* concerned two types of traces — vague, more-or-less continuous ones and distinct, isolated traces. The continuous, vague traces were similar to deep-scattering-layer recordings (DSL), encountered in open oceans and probably reflecting (parts of) the mesopelagic fauna (Schalk et al., 1990b). The distinct isolated traces were thought to represent larger schooling organisms such as fish or squid. They showed clear diurnal migration, with distinct schools during daytime at depths up to 300–400 m, distintegrating at night in surface waters. Both ships participating in the acoustic surveys noted considerable differences in echo-abundance between the two seasons (Figure 5.12). RV *Tyro* observed twice as many traces in August as in February, whereas RV *Tenggiri* recorded 3–4 times more backscattering in August. Fish abundance was estimated by the echo-integrator system of *Tenggiri* at 600–900 thousand tons in August and at about 150–250 thousand tons in February, which should be compared to an actual annual catch of up to 30,000 tons of pelagic fish (Amin and Nugroho, 1990). Both ships observed the highest trace abundance in the eastern part of the area studied during August (Aru Basin), where upwelling and biological activity was most pronounced in that period. Identification of the organisms responsible for the distinct traces was not possible, but mesopelagic fishes such as myctophids can probably be excluded as the responsible acoustic backscatterers because of their distribution pattern in August, which was very different from that of the traces (Schalk et al., 1990a,b). Judging from the composition of the pelagic fish catches in the area, the fish traces might consist mainly of small pelagic fish like clupeids, engraulids, and carangids. The large changes in abundance between seasons might also indicate a dominance of small pelagics in the fish resources observed, because these are known for their short life cycles (sometimes two generations within a year) (Dwiponggo et al., 1986; Dalzell and Pauly, 1990). Such short life cycles would allow populations to adapt their size to the seasonal variations in biological productivity occurring in this area, where periods with poor feeding conditions cannot be bridged by inactivity and energy conservation at low temperatures, as in temperate waters during winter. Another consequence of the presumably short life cycles would be that P/B ratios would be

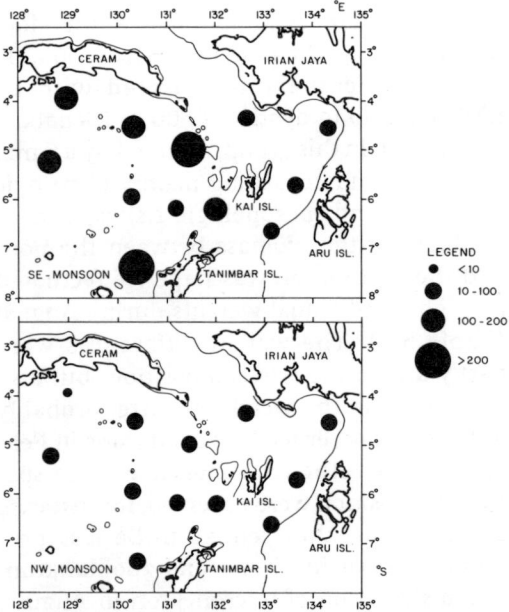

Figure 5.11. Distribution of micronekton (e.g., Crustacea, fish) biomass (gram wet weight per 10,000 m^3) in August 1984 and February 1985. (From Schalk et al., 1990a.)

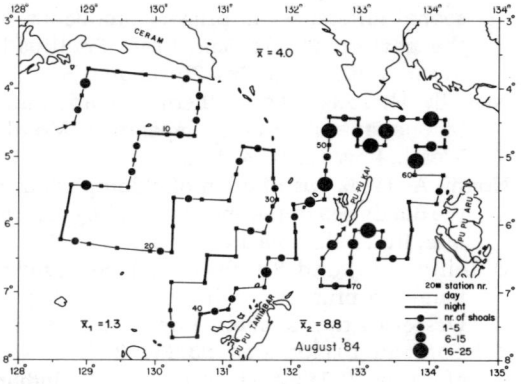

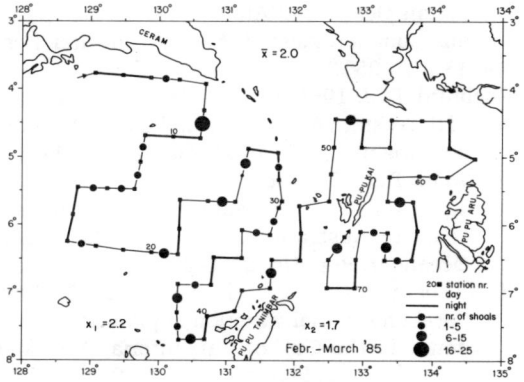

Figure 5.12. Distribution of distinct echo traces during August 1984 and February 1985, as recorded by RV *Tyro*. (From Schalk *et al.*, 1990b.)

high as compared to those of fish stocks in temperate waters, so that the potential fishing yield of the area might be considerable, much more than the present catch.

Another possible explanation of large seasonal differences in "fish" abundance would be that part of the traces observed were caused by tuna, known to be present in the area (Anonymous, 1985). In that case the difference in echo abundance between the two seasons could be explained by migrations.

Conclusions

The overall picture evolving from this study in a tropical area is one of marked changes in the abundance, activity and composition of almost all components of the pelagic system, related to a monsoon-induced alternation between periods of up and downwelling. The seasonal response was observed in phytoplankton, mesozooplankton, micronekton, and depending on the interpretation of the distinct echo traces, in fish. Noteworthy is the relatively high primary production estimate for the area of 400–500 g C $\cdot$ m^{-2} $\cdot$ yr^{-1}, a higher estimate than those for the temperate waters of the North Sea and northwest Atlantic shelf, but probably not as high as that for the Gulf of Thailand. Moreover, acoustic recordings covering the eastern Banda Sea suggest that fisheries resources in the area might be considerable, particularly if the recorded backscattering was caused by small pelagics with a short life cycle and a consequently high P/B-ratio. It seems likely that the events recorded for the Banda and Arafura Sea are also present in other areas under monsoonal influence, as evidenced for instance by similar seasonal changes in zooplankton abundance in the Gulf of Thailand and the South China Sea (Brinton, 1963, 1975) and in parts of the Indian Ocean (Tranter and Kerr, 1969).

Acknowledgments

This study was carried out as part of the Snellius II Expedition, under the auspices of the Netherlands Council of Sea Research (NRZ) and the Indonesian Institute of Sciences (LIPI). We gratefully acknowledge the excellent cooperation of our Indonesian colleagues during all stages of the expedition and the efforts of captains and crews of the three research vessels involved.

References

Amin, E. M., and Nugroho, D. 1990. Acoustic surveys of pelagic fish resources in the Banda Sea during August 1984 and February-March 1985. Proc. Snellius-II Symp., Neth. J. Sea Res. 26:(in press).

Anonymous. 1985. Fisheries statistics of Indonesia 1982, no. 12. Directorat Jenderal perikanan, Departemen Partanian, Jakarta, Indonesia.

Baars, M. A., and Zijlstra, J. J. In press. Monsoon-induced changes in the pelagic ecosystem of the Banda Sea: Preliminary result of the Indonesian-Dutch Snellius II expedition 1984/1985. Proc. Westpac Symposium, December 1986, Townsville, Australia.

Baars, M. A., Sutomo, A. B., Oosterhuis, S. S., and Arinardi, O. H. 1990. Zooplankton abundance in the eastern Banda Sea and northern Arafura Sea during and after the upwelling season, August 1984 and February 1985. Proc. Snellius-II Symp., Neth. J. Sea Res. 26:(in press).

Birowo, S., and Ilahude, A. G. 1977. On the upwelling of the eastern Indonesian waters. Papers presented at the 13th Pacific Science Congress, Vancouver, Canada, August 1975. Published by LIPI, Jakarta:69–89.

Boely, T., Gastellu Etchegorry, J. P., Potier, M., and Nurhakim, S. 1990. Seasonal and interannual variations of the sea surface temperatures in the Banda and Arafura Sea area. Proc. Snellius-II Symp., Neth. J. Sea Res. 26:(in press).

Brinton, E. 1963. Zooplankton abundance in the Gulf of Thailand and South China Sea. Naga Rep 1(4):53–58.

Brinton, E. 1975. Euphausiids of the southeast Asian waters. Naga Rep 4(5). 287 pp.

Cadée, G. C. 1988. Organic carbon in the upper 100 m and downward flux in the Banda Sea: Monsoonal differences. Neth. J. Sea Res. 22(2):109–121.

Dalzell, P., and Pauly, D. 1990. Assessment of the fish resources of southeast Asia, with emphasis on the Banda and Arafura Seas, Indonesia. Proc. Snellius-II Symp., Neth J. Sea Res. 26:(in press).

Dwiponggo, A., Hariati, T., Banon, S. Palomares, M. L., and Pauly, D. 1986. Growth, mortality and recruitment of commercially important fishes and penaeid shrimps in Indonesian waters. ICLARM, Tech. Rep. 17:91 pp.

Fransz, H.G., and Gieskes, W.W.C. 1984. The unbalance of phytoplankton production and copepod production in the North Sea. Rapp. P.-v. Reun. Cons. int. Explor. Mer 183:218–225.

Gieskes, W. W. C., Kraaij, G. W., Nontji, A., Setiapermana, D., and Sutomo, 1988. Monsoonal alternation of a mixed and layered structure in the phytoplankton of the euphotic zone of the Banda Sea (Indonesia): A mathematical analysis of algal pigment fingerprints. Neth. J. Sea Res. 22(2):123–137.

Gieskes, W. W. C., Kraaij, G. W., Nontji, A., Setiapermana, D. and Sutomo, 1990. Monsoonal differences in primary production in the eastern Banda Sea. Proc. Snellius-II Symp., Neth. J. Sea Res. 26:(in press).

Nicholls, N. 1984. The southern oscillation and Indonesian sea surface temperature. Month. Weath. Rev. 112:424–432.

Nontji, A. 1975. Distribution of chlorophyll a in the Banda Sea by the end of upwelling season. Mar. Res. Indonesia 14:49–59.

O'Reilly, J.E., and Busch, D.A. 1984. Phytoplankton primary production in the Northwestern Atlantic shelf. Rapp. P.- v. Reun. Cons. int. Explor. Mer 183:255–268.

Rochford, D. J. 1962. Hydrology of the Indian Ocean. II. The surface waters of the south east Indian Ocean and Arafura Sea in the spring and summer. Aust. J. Mar. Freshwater Res. 13:227–251.

Rochford, D. J. 1966. Some hydrological features of the eastern Arafura Sea and the Gulf of Carpentaria in August 1964. Aust. J. Mar. Freshwater Res. 17:31–60.

Rochford, D.J. 1977. Upwelling off the northwest coast of Australia. CSIRO Aust. Dir. Fish. Oceanogr. Rep. No. 85:1–25.

Rothlisberg, P. C., and Jackson, C. J. 1982. Temporal and spatial variation of plankton abundance in the Gulf of Carpentaria, Australia 1975–1977. J. Plankton Res. 4:19–40.

Schalk, P. H. 1987. Monsoon-related changes in zooplankton biomass in the eastern Banda Sea and Aru Basin. Biol. Oceanogr. 5:1–12.

Schalk, P. H., Witte, J. IJ., Budihardjo, S., and Hatta, A. M. 1990a. Monsoon influences on micronekton of the Banda Sea (Indonesia). Proc. Snellius-II Symp., Neth. J. Sea Res. 26:(in press).

Schalk, P. H., Zijlstra, J. J., and Witte, J. IJ. 1990b. Spatial and seasonal differences in acoustic recordings of the Banda Sea (Indonesia) obtained with a 30 KHz echosounder. Proc. Snellius-II Symp., Neth. J. Sea Res. 26:(in press).

Tijssen, S. B., Wetsteyn, F. J., and Mulder, M. 1990. Production and consumption rates in the upper 300 m in the eastern Banda Sea during and after the upwelling season, August 1984 and February 1985. Proc. Snellius-II Symp., Neth. J. Sea Res. 26:(in press).

Tranter, D.J., and Kerr, J. D. 1969. Seasonal variation in the Indian Ocean along 110°E. V. Zooplankton biomass. Aust. J. Mar. Freshwater Res. 20:77–84.

Vosjan, J. H., and Nieuwland, G. 1987. Microbial biomass and respiratory activity in the surface waters of the East Banda Sea and northwest Arafara Sea (Indonesia) at the time of the southeast monsoon. Limnol. Oceanogr. 32:767–775.

Vosjan, J. H., Nieuwland, G., and Ruyitno. 1990. Monsoonal differences in microbial biomass respiratory activity and bacterial numbers in the Banda Sea. Proc. Snellius-II Symp., Neth. J. Sea Res. 26:(in press).

Wetsteyn, F. J., Ilahude, A. G., and Tijssen, S. B. 1990. Nutrient distribution in the upper 300 m of the eastern Banda Sea and northern Arafura Sea during and after the upwelling season, August 1984 and February 1985. Proc. Snellius-II Symp., Neth. J. Sea Res. 26:(in press).

Wyrtki, K. 1958. The water exchange between the Pacific and the Indian Oceans in relation to upwelling processes. Proc. 9th Pacific Sci. Congr. 16:61–66.

Wyrtki, K. 1961. Physical oceanography of the southeast Asian waters. Naga Report 2:1–195.

Zevenboom, W., and Wetsteyn, F. J. 1990. Growth limitation and growth rates of (pico)phytoplankton in the Banda Sea during two different monsoons. Proc. Snellius-II Symp., Neth. J. Sea Res. 26:(in press).

Zijlstra, J. J., et al. 1990. Monsoonal effects on the hydrography of the upper waters (< 300 m) of the eastern Banda Sea and northern Arafura Sea, with special reference to vertical transport processes. Proc. Snellius-II Symp., Neth. J. Sea Res. 26:(in press).

Part Two:
Biodynamics of Large Marine Ecosystems

Introduction

Forced to describe the great diversity of marine biota in only a few words, we draw upon the observation that the size, density-per-unit volume, and lifespans of the viruses, bacteria, larger plankton, fish, and mammals that live in the sea vary over many orders of magnitude. This is the problem of scale. How can we think about the interrelationships between bacteria that are only a few microns in diameter and adult fish that might be a meter in length? Or how can we compare the vital rates of phytoplankton and zooplankton when the average life span of a zooplankton might be 100 times greater than that of the phytoplankton? Finally, how do we consider the interaction of the physical environment with these biological properties of populations, particularly when physical variability can exist on scales as large as the major ocean basins themselves or as small as the microscale remnants of turbulence, of the order of only a few millimeters?

The conventional approach is descriptive. A 15-micron phytoplankton is assigned to organisms that are 15 microns in diameter, an 800-micron zooplankton is classified as an 800-micron zooplankton, and so on. The major gyres of the ocean are thought of in terms of the large scale, eddies associated with these gyres are thought of in terms of the mesoscale, and phenomena such as langmuir circulation are considered to be small scale. The physics and biology can now be put on the same descriptive diagram.

The mere descriptions of large and small, long time and short time are, however, not particularly interesting. What we are really concerned with is dynamic interaction among the large and small and long and short events. How does the cascade from large-scale physical events to microscale physical events drive the dynamics among populations of the sea?

These questions are what biodynamics is about. Biodynamics is exciting because it focuses on a very difficult problem that involves generalities that can be drawn from the interaction of the wide array of organisms with their physical environment, which is likewise also variable.

Aspects of biodynamics are considered in different ways by different authors. In Chapter 6, Thomas Osborn and myself set the stage for developing criteria for reducing the very high dimensionality of the marine ecosystem by setting forth notions on how kinetic variability affects trophodynamics and hence population dynamics. In Chapter 7, Thomas Dickey continues this theme by showing how particular scales are important and how events at the various scales have been measured. Taking another tack in Chapter 8, Osborn, Hidekatsu Yamazaki, and Kyle Squires provide the groundwork for experiments that show the biodynamic effects of turbulent flow or its remnants.

If we are to be able to measure long-term changes in the response of populations to the environment, we will need to understand the genetics of the populations. New techniques for studying population genetics and other biochemical properties of organisms are described by Dennis Powers, Fred Allendorf, and Thomas Chen in Chapter 9. Another set of techniques critical to analyzing samples of organisms is image identification. Techniques for image identification, such as those described by Mark Berman in Chapter 10, will do much to end the decades of undersampling of marine biota.

Putting some of these notions and techniques together enables a fresh view of the marine ecosystem, which Geoffrey Laurence attempts to articulate in Chapter 11. Finally, Christopher Taggart and Kenneth Frank in Chapter 12 discuss the biodynamic setting involving the interaction of scales with fish larvae.

This section of the volume represents the beginnings of a new approach to the study of the sea, an approach that integrates physics and biology at biologically meaningful scales. As the study of marine biodynamics moves ahead, a more fundamental understanding of the interaction between production, population, dynamics, and trophodynamics will be achieved.

—*Brian J. Rothschild*

Chapter 6

Biodynamics of the Sea
Preliminary Observations on High Dimensionality and the Effect of Physics on Predator-Prey Interrelationships[1]

Brian J. Rothschild and Thomas R. Osborn

Abstract

The difficulty in predicting fluctuations in marine organism abundance is related to a need to simultaneously account for variability induced by density-dependent population-dynamic processes and variability induced by the physical structure of the sea measured on appropriate time-and-space scales. Combining two major sources of variability generates a situation where the number of parameters or dimensions exceeds the number of variables available for their estimation. Hence procedures for either reducing the number of parameters or increasing the number of observations need to be considered. Taking account of the interaction of population dynamics and physical oceanography at the most fundamental level is a prerequisite to efficient evaluation of the trade-off between the number of parameters or dimensions and the number of observations. In this regard, this chapter takes into particular account the effect of small-scale water motion on predator-prey relations in the plankton. Preliminary calculations show that the velocities of the water in which plankton are ambient approximate the velocities of the plankton, testifying to the potential importance of the effect of water motion on the "food signal" and suggesting that much of the energy used in search for food or avoiding predation is derived from the motions of the water in which the plankton are entrained.

Introduction

Fluctuations and trends in marine plant and animal population abundance are evident from even a casual perusal of the literature. These fluctuations range from changes in intra-annual periodicities, such as those that depend on production cycles (e.g., Cushing, 1967); to population collapses (e.g., Murphy, 1966) or explosions (e.g., Cushing, 1984); or to changes in the character of the entire ecosystem (Hempel, 1978).

The exciting questions regarding these fluctuations involve their cause or causes. If the causes were known, then it would be possible to predict the fluctuations and to more fully understand the effects of long-term, climatically induced temperature changes (see, e.g., Kelly, 1984), pollution, and fishing, as well as the still-vexing problems of multiple-species interactions.

But at present there seems to be little hope for making such predictions, as environmental correlations are not generally reliable; the population dynamics effects of pollution are not understood (except, of course, at high pollutant concentrations); and while certain yield-per-recruit considerations in fisheries are fairly well understood, those that relate to the recruitment-stock re-

---

[1] This chapter is based upon Rothschild and Osborn (1986). For subsequent work the reader is referred to Rothschild and Osborn (1988) and Rothschild *et al.* (in press).

lation are not at all well known.

A major difficulty in answering these questions is that a theory does not exist that is capable of linking the population dynamics of marine organisms with environmental variability, whether naturally or anthropogenically induced. Such a theory, a *biodynamic theory of the sea*, needs to be developed.

This chapter provides preliminary observations on specifications for a biodynamic theory in terms of its high-dimensionality context and micro- and fine-scale kinetics, a likely major source of biodynamic variability. A problem set in a high-dimensionality context implies that the number of *important* variables or dimensions for any nontrivially long time series will be much greater than the number of observations, mostly because of shifts in the relative importance of subsets of independent variables among the total set of independent variables. Identifying which set of variables is important, in which years, is thus a key problem requiring criteria to identify particularly critical dimensions or variables.

Since the setting for most population dynamic problems is highly dimensional and most analyses are based on only a few dimensions or variables, often for practical reasons, the researcher must have selected from many dimensions the few dimensions that were analyzed. How was the selection made? Was the procedure arbitrary? What are the consequences of such a selection in terms of inferences that are either extrapolative into the future, or beyond the data set at hand?

A frequent approach to high-dimensionality settings is to examine the data and select the data sets for analysis and eventual publication that seem most correlated. But in a highly dimensional system, the magnitude of the correlation among a few variables must, by definition, be a function of the unconsidered variables. When the unconsidered variables change, the correlation breaks down and some other correlation operates. Is it valid to say that the original correlation is invalid? Probably not; the problem was poorly framed in the sense that the controlling variables were not taken into account.

High Dimensionality

The high-dimensionality problem arises from the fact that the population dynamics of organisms is affected by many different variables, in contrast to simpler problems where there are only a few dominant sources of variability. As an example, consider the time series of abundance for any fish stock over the last century. It would not be difficult to think of more than 50 or 60 variables that might affect the abundance of the stock, even though only four or five may be important in any single year. Such a model suggests that the study of causes for population fluctuation needs to concentrate on determining how the major sources of variability associated with each data point change rather than either identifying short-term correlations as being important or to the contrary suggesting that a jumble of data points represents no relationship.

Mathematical Features of High-Dimensionality Problems

High-dimensionality problems represent a special class of inverse problems. An "inverse problem" is simply a problem in which a squared data-kernel matrix is inverted to solve for coefficients. Such solutions are routinely obtained using least-squares procedures under circumstances when the number of observations exceeds the number of parameters. Inverse methodology generalizes the problem to the case where the number of observations is less than the number of parameters.

Following Menke (1984), perhaps the simplest way of describing these interrelationships is by the matrix equation;

$$\mathbf{Gm} = \mathbf{d} \qquad (1)$$

where $\mathbf{G}$ is an m × n matrix, the data kernel; $\mathbf{m}$ is an m-element column vector of parameters, and **d** is an n-element data vector. In other words, there are m parameters and n observations.

We immediately recognize the least-squares regression problem when we invert the squared data-kernel matrix equation solving for **m**. The ordinary least-squares inversion process is possible only when n > m. If these conditions do not apply, then a least-squares solution is not available. However, restricted solutions are available for the situation where m < n. This requires computing a generalized inverse. Whereas the least-squares inverse when m < n is $[\mathbf{G}^T\mathbf{G}]^{-1}$, the minimum length inverse m > n is $\mathbf{G}^T[\mathbf{G}\mathbf{G}^T]^{-1}$. But the problem is only partially solved, because while we have a solution, it is not unique. This means that having a solution, it is still necessary to agree on the procedure used to derive the solution and to further reduce the alternatives among the non- unique solutions. This requires information external to the problem. But how should one choose the external information?

One of the ways of selecting appropriate external information relates to identifying those dimensions or variables that are particularly important, but seem to be ignored. The broadest possible classification would need to include population dynamic variability, physical-oceanographic variability, and genetic variability. The idea is captured in Figure 6.1, despite the fact that only a few dimensions can be displayed on a two-dimensional surface. The figure shows a dependent variable "population response" as a function of indices of relevant physical oceanographic and population density variables. The figure is specifically drawn to indicate interaction between the two indices, but the results are the same (aside from special cases) even if interaction does not exist. The figure shows that while there is a deterministic relation among the three variables, samples taken more or less at random and projected on either the Z-X plane or the Z-Y plane, ignoring the variability due to either Y or X, would result in a jumble of points.

Interactions among major well-known sources of variation are displayed in Figure

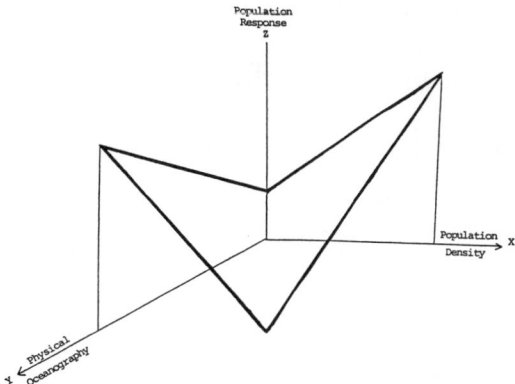

Figure 6.1. Response of population abundance to an index of physical-oceanographic variability and an index of density dependence. The particular hypothetical relation shows the response is "twisted" and that there is no variation about the surface. If we did not account for physical factors and projected all variability on the X-Z plane, then we would say there is no relationship, despite the fact that a strong relationship exists. Such a situation is probably quite common.

6.2. The notion of *transformation* is the essence of Figure 6.2. In Figure 6.2 we can see that, in effect, a predatory act transforms the prey of a particular genotype into predator metabolic products, some of which eventually become somatic and reproductive tissue, the latter creating a new set of genotypes.

In interpreting Figure 6.2, it is important to recognize the difference between predator-prey contact and the ingestion of a prey by a predator. Generally speaking, not all "contacted" prey are ingested. Our primary concern is the ratio between prey and predator that actually exists and the ratio between prey and predator apparent to the predator: that is, the ratio of actually ingested prey to predator or the ratio of transformed prey to predator. Hence we make the distinction between a "contact," which is when a prey is within a distance where it could be perceived and ingested under ideal sensory conditions, and "effective contact," where a prey is actually ingested.

Physical Environment

Figures 6.1 and 6.2 make the point that in considering a theory for population fluctua-

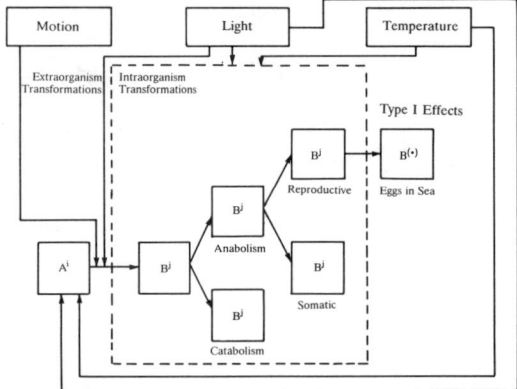

Figure 6.2. Example of linkages between a schematic of conceptual physical model and a biological transformation model. The biological transformation refers to a single organism A^i. The superscript refers to its genetic subpopulation. A^i is destroyed by predator B^j. A^i is thus transformed into anabolic, catabolic, reproductive, and somatic products, and eventually into eggs in the sea. The eggs in the sea have a variety of genetic attributes. The physical affect both the extra- and intra-organism transformations.

tion, it is necessary to at least think about major sources of variation, particularly since the sources of variation are often interactive. Figure 6.2 makes the point that it is not sufficient to consider only the interaction and population dynamics; rather it is necessary to appreciate the various modalities within which the variables operate. There are at least three ways that contact rates can be modified as a function of physical variables or population density. Type I variables involve the physical variables that affect the behavior and physiology and contact rates of the predator and prey directly; Type II variables involve the physical variables that affect the effective contact rates between predator and prey, even given constant predator-prey ratios; and Type III variables are the density-dependent variables that change effective contact rate as a function of predator or prey density.

In order to study the manifestation of Type I, II, III effects, we use the notion of a "food signal." Because a signal can be considered as any time-dependent function conveying information, a food signal can be thought of as the time-dependent availability of food as perceived by a predator. In a sense, this formulation is quite similar to placing the prey in a queue where the predator is the server and the prey are the customers (see, e.g., Beyer and Laurence, 1980).

The construction of the food signal or food queue depends upon the geometry of the predator-prey system. Geometries can be thought of in term of D_0, D_1, D_2, and D_3 occupancy volumes (see Rothschild, 1985).

D_0 volume. The D_0 volume is the track taken by fish larvae through the water. If the track is considered as being in one dimension, then it has no volume. But if the track is in fractal dimension (Mandelbrot, 1977) or in the dimension swept clear by the larval corpus, a volume exists. In actuality, we are more concerned by the path and whether it is straight or convoluted in the presence or absence of food, for example (see Hunter and Thomas, 1974).

D_1 volume. This is the volume specified by Rosenthal and Hempel (1970) and Blaxter and Staines (1971) and consists of the perceived volume swept clear by the fish larvae. However, this volume is only useful for considering prey that do not move because if prey have a particular velocity, more will be contacted as their velocity increases. In other words, as the fish larvae move through the D_1 volume, they may contact more or less prey depending on the velocity of the prey.

D_2 volume. If we consider that prey have velocity, we need to construct a D_2 occupancy volume, the diameter of which is a function of not only the perceptive capabilities of the predator but of the relative velocity of both, a well-known theory of search strategy (e.g., Koopman, 1956; Gerritsen and Strickler, 1977). The difficulty with the D_2 volume is that it still does not account for the motions in which the predator exists.

D_3 volume. The D_3 volume is intended to take account of the dynamics of the sea because it considers th perceptive fields and the velocity of the prey as affected by the physics of the sea. For example, small-scale turbulence and shear affect the relative velocity of prey and predator; differences in irradiance as a function of the intensity of light

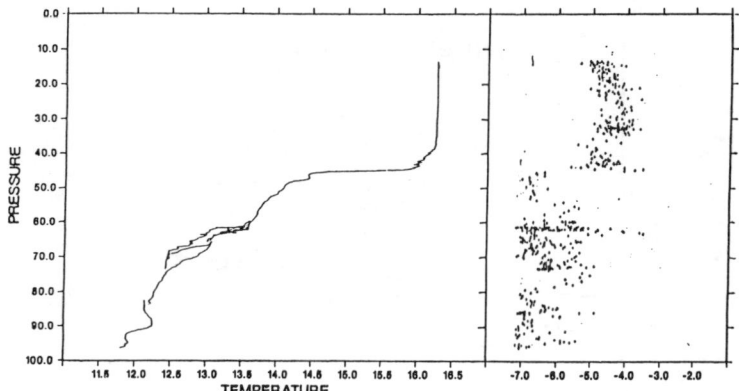

Figure 6.3. The calculated dissipation rate for the 4 December 1980 profile of the *Dolphin*. The dissipation rate is plotted on a logarithmic scale in units of W · m^{-3}. The value represents the average of the estimates from the vertical and the horizontal velocity shear data over 8-second intervals. [Reprinted from Osborn and Lueck, J. Phys. Oceanogr. 15:1502–1520, with permission. Copyright 1985, American Meteorological Society.]

striking the sea surface, turbidity, and depth all effect the perceived area swept clear as well; and finally, temperature affects the D_3 space in terms of, for example, its effect upon the length of the D_0 dimension.

The classification of occupancy volumes suggests that various sources of variability might contribute to the trophic interrelationships. While not necessarily independent, the D_0 volume pertains to the volume swept by the predator, the D_1 volume to the perceived volume; the D_2 volume, to volume encapsulating prey; and the D_3 volume, to the modification of the D_2 volume by physical variables.

It is remarkable that D_2 and D_3 volumes are not generally considered, even though they probably represent large sources of variation. If this is the case, then it is difficult to interpret many experiments on planktonic food requirements that do not take into account D_2 and D_3 considerations. The need to consider the D_2 and D_3 volume is obviated to some extent by important advances in understanding the role of small-scale turbulent fluctuations in modifying planktonic contact rates. Results from the last 10 years allow us to set the range of turbulent intensities for the time scales of variation that are relevant to planktonic communities.

Direct measurements of the rate of turbulent energy dissipation are available from the equatorial oceans, frontal regions, eastern and western boundary currents, warm core rings, and mid-ocean gyres. While the oceans are still sparsely sampled for turbulence, a reasonable picture is emerging for the energy dissipation that scales turbulence intensity. As an example, consider Figure 6.3, which shows a vertical profile of dissipation rate derived from a submarine-mounted instrumentation suite by Osborn and Lueck (1985). The upper layer is strongly turbulent, as is the top of the thermocline. There is a thin entrainment region between the two where the turbulence drops off substantially. The dissipation stops in the thermocline and only occurs in intermittent patches below that. The strong turbulence in the thermocline is often seen and thought to be due to the local shear.

With substantial averaging, the energy dissipation rates show a monotonic increase with stratification (Gargett and Holloway, 1984; Lueck et al., 1983). There is no consensus (Gregg et al., 1986) as to the nature of the relationship (Figure 6.4), but the trend has been seen in several data sets and between different investigators. This suggests that the degree of stratification is an important variable in the three-dimensional turbulence structure.

Measurements in the mixed layer (Dillon et al., 1981; Oakey and Elliott, 1982; Gargett et al., 1984; Gregg et al., 1986) all gauge the response of the upper ocean to forcing by wind or buoyancy flux. Dissipation increases with the cube of the wind speed (Oakey and Elliott, 1982) and falls off inversely with depth in the first few meters

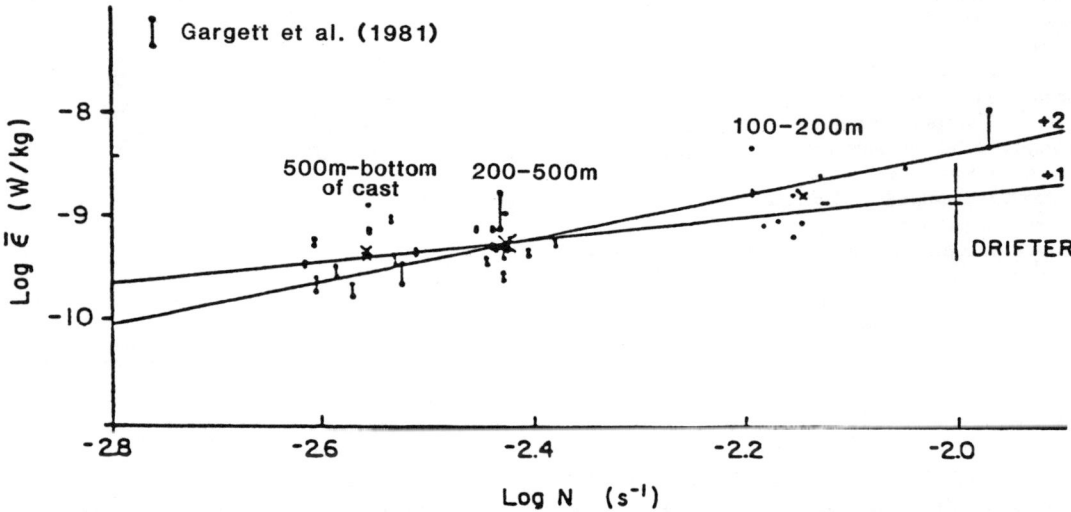

Figure 6.4. Average dissipation rates during the Drifter cruise, as a function of N, compared to the summary in Lueck et al. (1983). It is evident that few averages exceed 1×10^{-9}. The vertical line for the Drifter data shows the range of burst averages, and the cross bar is the ensemble average. The Drifter data are somewhat lower than the projected lines for slopes of +1 and +2. In view of the vastly different numbers of samples in the estimates, the variety of data systems, differences in processing algorithms, and possible seasonal effects, a well-defined dependence of dissipation rate on stratification should not be expected. [Reprinted from Gregg et al., J. Phys. Oceanogr. 16:856–885. Copyright 1986, American Meteorological Society.]

(Dillon et al., 1981). The diurnal effect can be dramatic. Some progress on modelling has been reported by Price et al. (1986).

Stratification limits the vertical scale of turbulent fluctuations, although not necessarily the vertical extent of the patches. Detailed measurements of the three turbulent velocity components (Gargett et al., 1984) from the Pisces submersible show the suppression of velocities at scales larger then $(E/N^{**}2)^{**}1/2$ where E is the rate of turbulent energy dissipation and N is the Vaisala frequency. This scale is usually on the order of 0.5 m. The appropriate scaling collapses the low wave number portion of the spectra (Figure 6.5) and can provide estimates of the turbulent velocity fluctuations as a function of wave length.

These introductory remarks enable us to consider food signals in more detail, particularly their interaction with Type I, Type II, and Type III variables. For the purposes of

Figure 6.5. Typical streamwise (ϕ_{11}) and vertical (ϕ_{33}) velocity component spectra from each of the four classes discussed, scaled by Kolmogoroff variables k_s and ϕ_s. [Reprinted from Gargett et al., J. Fluid Mech. 144:231–280, with permission. Copyright 1984, Cambridge Univ. Press.]

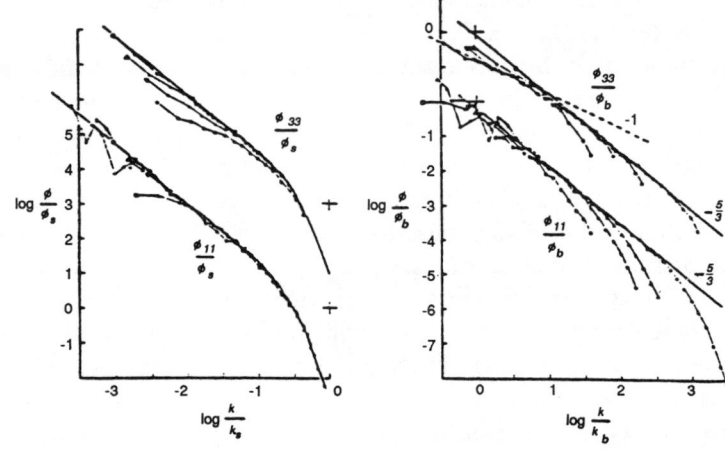

this chapter, we are particularly interested in the small-scale physical effects in the context of Type II variation.

Food Signals and Type I, II, and III Effects

The food signal, then, is the temporal distribution of a prey encountered by a predator. This is a function of at least the (i) relative density of predator and prey, (ii) relative velocities of predator and prey, and (iii) effects on these velocities of the kinetic structure of the ambient medium.

To consider the problem in more detail and the ways in which the Type I, Type II, and Type III variables operate, let us define a food signal over a 12-hour period. Divide the period into n small, equal intervals of length t so that nt = T. The length of t is such that only one food particle can occur in the interval (Figure 6.6). Let us further suppose that all food particles are identical. The food signal is then each predator's Lagrangean view of the occurrence of food as a function of time. The distribution of perceived food particles as a function of time can be characterized by both the density of food and its variance.

Type I effects

As suggested above, Type I effects are quite well known. However, one aspect of a Type I effect that is often not considered is how changes in temperature, for example, will affect a potential average predator-prey transaction. The complexity of the problem can be shown with a simple example. Suppose predator and prey performance are difference functions of temperature (Figure 6.7). It can be seen then that it is important to

Figure 6.6. The food signal or food queue as perceived by a predator. Each dot is a prey. The food signal is distributed in one dimension, time.

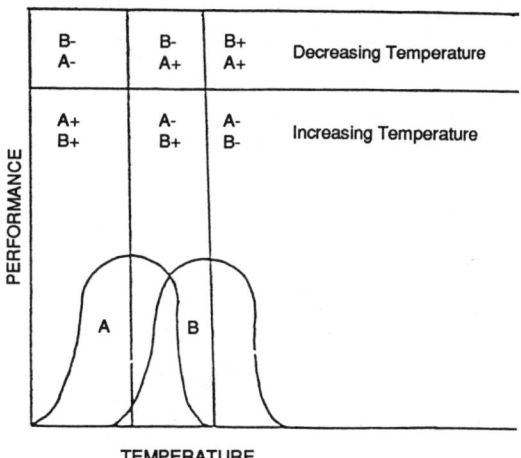

Figure 6.7. Performance as function of temperature for two populations, A and B. As temperature increases and decreases, A and B are variously advantaged and disadvantaged. For example, at *increasing* temperatures, in the low range A and B are advantaged; in the medium range, B is advantaged over A; and in the higher range, both A and B are disadvantaged. Hence the advantage in performance of one species over another does not depend entirely on the absolute value of temperature, but on its derivative.

know whether temperature is increasing or decreasing. For identical ranges of temperature, species A can be advantaged over B, and B can be advantaged over A, and hence it is important to know not only the absolute value of an independent variable, but the derivative of temperature change. In other words two equivalently warm years might have opposite biodynamic effects.

In the context of food signal, Type I effects modify velocities of feeding behavior of predator and prey. By increasing or decreasing the activity of either the predator or the prey or both, the frequency of the signal is modified (Figure 6.8). The greatest increases or decreases involve situations where the performance of the predator is enhanced at the expense of the prey or vice versa.

Type II effects

Type II effects concern the effects of changes in the relative velocity of predator and prey induced by changes in the velocity of the ambient medium. While Type I effects modify

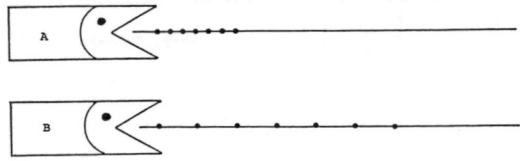

Figure 6.8. Temperature as a Type I effect increases contact rate even though the densities of food in A and B are identical. The same result would occur if turbulence increases contact rate, but this would be a Type II effect.

the performance (e.g., swimming speed), behavior, or physiology of the predator or the prey, Type II effects operate to affect the contact rate between predator and prey.

For example, increasing turbulent motion should increase the contact rates between predator and prey. It is well known from the kinetic theory of gas that increasing the velocity of particles increases their contact rates. Further modifying the relative direction of predator and prey particles also affects the contact rates. Meyer (1899) has pointed out the functional relation between the number of contacts and the relative velocities of predator and prey particles as

$$C = \frac{\rho^2 \pi}{\lambda^3} 2v \sin \tfrac{1}{2}\theta \qquad (2)$$

where λ is the mean distance between prey, ρ is the radius of action between predator and prey, v is the relative velocity of predator and prey, and θ is the angle between the predator and prey. Equation 2 shows that contact rates increase with relative velocity, all other things being equal. Now, if there is a symmetry in motion, equation 2 can be integrated so that

$$C' = \frac{4}{3} \frac{v\rho^2 \pi}{\lambda^3} \qquad (3)$$

Therefore, the addition of turbulence increases the probability of contact among parcels of water and if these velocities pertain to the organisms entrained in the water, then the contact of the organisms is increased. Inasmuch as water motion may not be symmetrical over all spherical coordinates — that is, vertical turbulence may be greater than horizontal turbulence or vice versa —

the influence of C on C' becomes important.

In order to obtain a rough idea of the effects of kinetics, we can compare the turbulence velocities of the water with the velocities of the predator and prey. Consider, as an example, the plot of temperature and turbulent dissipation rate in watts $\cdot$ m^{-3} in Figure 6.3. The figure shows that turbulence is relatively high in the mixed layer, with a dissipation rate of about 3×10^{-5} watts $\cdot$ m^{-3}. This rate is comparable to other measurements in the upper ocean. It is not unusual to obtain much higher rates.

To make these observations relevant to the prey-predator problem, it is necessary to estimate turbulent velocity fluctuations as a function of scale. To do that, we compute a velocity shear spectrum (from the same data used to compute the turbulent dissipation rate between 31.5–36.3 d bars). The spectrum (Figure 6.9) is, as conventionally computed, where K is wave number in (cycles per meter), v is velocity, and s is a horizontal component. Now to obtain the velocity squared for any range of wave number, integrate between k_1 and k_2 cycles per meter

$$\varphi(k) = \left[\text{F.T.} \left(\frac{\partial v}{\partial \chi} \right) \right]^2 \qquad (4)$$

where K is wave number in (cycles per meter), v is velocity, and χ is a horizontal component.

Now to obtain the velocity squared for any range of wave number, integrate between k_1 and k_2 cycles per meter

$$v^2 = \int_{k_1}^{k_2} k^{-2} \varphi(k)\, dk. \qquad (5)$$

For the plankton problem, two scales seem appropriate: the 1 m → 10 cm scale and the 10 cm → 1 cm scale. Integrating for both the horizontal and vertical scales yield:

Scale	1 m → 10 cm	10 cm → 1 cm
r.m.s. horizontal velocity fluctuation	1.2 cm $\cdot$ s^{-1}	0.5 cm $\cdot$ s^{-1}
r.m.s. vertical velocity fluctuation	1 cm $\cdot$ s^{-1}	0.5 cm $\cdot$ s^{-1}

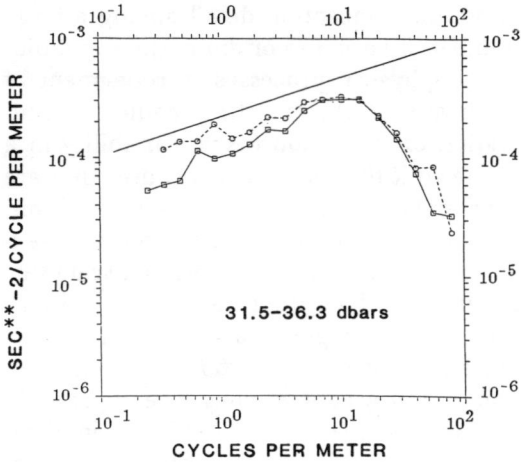

Figure 6.9. Spectra of the velocity shear data from a well-mixed portion of the upper layer and from the entraining region at the top of the thermocline. This figure was prepared using the procedures described by Osborn and Lueck (1984). The boxes are $\partial w/\partial x$ and the circles are $\partial v/\partial x$. [Reprinted from Osborn and Lueck, J. Phys. Oceanogr. 15:1502–1520, with permission. Copyright 1985, American Meteorological Society.]

The potential importance of these particular turbulent velocity fluctuations on predator-prey interactions can now be assessed. The r.m.s. 0.5 cm · s^{-1} velocities on the 10 cm to 1 cm scale are comparable in amplitude and scale to the feeding motions of plankton, but how important is the effect? Gerritsen and Strickler (1977) show the rate of particle interactions is proportional to

$$A = \left(\frac{v^2 + 3u^2}{u} \right) \quad (6)$$

where v is the mean speed of the prey and u the mean velocity of the predator. (This equation holds when $u > v$.) Adding an uncorrelated 0.5 cm · s^{-1} for each velocity increases the velocity-induced component of the contact rate by a large amount when predator *and* prey velocities are small relative to the added velocities to a factor of about 1.2 when the predator *and* prey are both large. Such an increase would not be at all insignificant if survival of predators was proportional to their food contact rate.

Our treatment of Gerritsen and Strickler's approach can only be used to estimate the magnitude of the effect. The point is clear that there is a substantial contribution to the interaction by small-scale turbulent motions, since they are comparable in size — often slightly larger than the planktonic motions on the same scale. While turbulent velocities are correlated over distance scales, making our calculations a rough approximation, the order of magnitude seems correct.

The magnitude of the estimate encourages further investigation because it would appear that not only are the D_3 considerations important, but that many feeding energetics calculations may be in error because the energy used in predation and avoiding predation may derive to a large extent from the environment.

The preceding argument considered only the apparent density of particles, assuming rather implicitly that they are uniformly distributed on the time line. The variance or spacing among the particles is important. Spacing owing to variance is a function of the relative velocities of the predator and the prey, as predators at relatively slow velocities, for example, reduce ability to detect patchiness and vice versa. At any rate, given a finite handling time for each prey, predators cannot ingest all prey that are closely adjacent. All other things being equal, increased turbulence should break up patches of prey and thereby increase the apparent density of prey (Figure 6.10).

Type III effects

Density dependence is a necessary prerequisite for population regulation. The above discussion has nothing to do with density de-

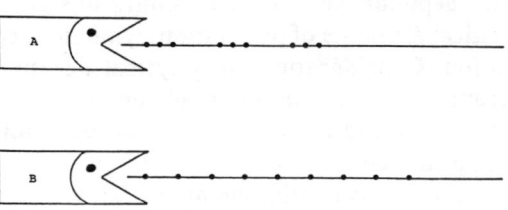

Figure 6.10. The effects of variance. The densities of food particles in A and B are identical. Under A, the particles are so close that the fish can only eat one particle in three, but in B, the particles are distributed by turbulence and the fish can eat every particle. Turbulence increases the apparent density three-fold.

pendence. In order to invoke density dependence, it is necessary to invoke changes in predation, mortality, and growth *rates* as a function of predator or prey density, not merely an increase in predation as predators increase (for example).

The critical issue is whether one or more predators consuming prey sample the prey with or without replacement. If predation is sufficiently intense, a single predator can sample the prey population without replacement, although this is unlikely. The more likely case is that a certain number of predators relative to the number of prey are required before the sampling process can be considered to operate without replacement.

The basic point is that when sampling shifts from a sampling-with-replacement mode to a sampling-without-replacement mode, then density dependence is operating and regulation is possible.

The importance of Type I and Type II effects is that they can induce Type III effects even though the relative density of predator and prey is constant.

Discussion

There has been little success in predicting fluctuations of organisms in the sea. This is not surprising since theoretical treatments do not take account of both sources of variability. Usually, either population dynamic variables are considered or oceanographic variables are considered, not both in concert. Furthermore, from an empirical point of view, considerations of population dynamics often do not take into account density dependence and in so doing miss the critical elements of population dynamic variation. Considerations of physical oceanographic variables are not made on appropriate scales and likewise cannot account for important components of variability.

From this qualitative appraisal, we can proceed to a more formal analysis of which variables and dimensions need to be taken into account. The inverse-problem formulation provides a setting for allocating sampling and conceptual detail among a large number of variables or dimensions and biological-physical processes. A refinement of the inverse problem setting requires consideration of major sources of variability in a framework that emphasizes causes that are proximal and not distal to the effects of concern. As an example, we have begun to examine the relation between small-scale turbulence and contact rates. Preliminary calculations suggest that small-scale turbulence can have an important effect on predator-and-prey contact rates. The importance of such a suggestion relates to its implications that (i) if predator survival is proportional to contact rate, small changes in contact rate at very young stages could have a dramatic effect at the time of recruitment, and (ii) significant components of energy required for contacting prey or avoiding predators may be derived from the physical environment rather than from metabolic sources.

References

Beyer, J. E., and Laurence, G. C. 1980. A stochastic model of larval fish growth. Ecol. Modelling 8:109–132.

Blaxter, J. H. S., and Staines, M. 1971. Food searching potential in marine fish larvae. *In* Fourth European marine geological symposium. pp. 467–485. Ed. by D. J. Crisp. Cambridge Univ. Press, Cambridge, UK.

Cushing, D. H. 1967. The grouping of herring populations. J. Mar. Biol. Assoc. U. K., N. S. 47:193–208.

Cushing, D. H. 1984. The gadoid outburst in the North Sea. J. Cons. int. Explor. Mer 1:159–166.

Dillon, T. M., Richman, J. G., Hansen, C. G., and Pearson, M. D. 1981. Near-surface turbulence measurements in a lake. Nature 290:390–393.

Gargett, A. E., *et al.* 1981. A composite spectrum of vertical shear in the upper ocean. J. Phys. Oceanogr. 11:1258–1271.

Gargett, A. E., and Holloway, G. 1984. Dissipation and diffusion by internal wave breaking. J. Mar. Res. 42:15–27.

Gargett, A. E., Osborn, T. R., and Nasmyth, P. W. 1984. Local isotropy and the decay of turbulence in a stratified fluid. J. Fluid Mech. 144:231–280.

Gerritsen, J., and Strickler, J. R. 1977. Encounter probabilities and community structure in zooplankton: A mathematical model. J. Fish. Res. Board Can. 34:73–82.

Gregg, M. C., D'Asaro, E. A., Shay, T. J., and Larson, N. 1986. Observations of persistent mixing and near-inertial internal waves. J. Phys. Oceanogr. 16:856–885.

Hempel, G. 1978. North Sea fisheries and fish stock — A review of recent changes. Rapp. P.-v. Reun. Cons. int. Explor. Mer 173:145–167.

Hunter, J. R., and Thomas, G. L. 1974. Effect of prey distribution and density on the searching and feeding behavior of larval anchovy *Engraulis mordax* Girard. *In* The early life history of fish. pp. 559–574. Ed. by J. H. S. Blaxter. Springer Verlag, Berlin.

Kelly, P. M. 1984. Recent climatic variations in the North Atlantic sector. Rapp. P.-v. Reun. Cons. int. Explor. Mer 185:226–233.

Koopman, P. M. 1956. The theory of search. I. Kinematic bases. Operations Research 4(3):324–531.

Lueck, R. G., Crawford, W. R., and Osborn, T. R. 1983. Turbulent dissipation over the continental slope off Vancouver Island. J. Phys. Oceanogr. 13:1809–1818.

Mandelbrot, B. B. 1977. Fractals: Form, chance, and dimension. Freeman, San Francisco.

Menke, W. 1984. Geophysical data analysis: Discrete inverse theory. Academic Press, Orlando, FL. 260 pp.

Meyer, O. E. 1899. The kinetic theory of gases. London, 1899.

Murphy, G. I. 1966. Population biology of the Pacific sardine *(Sardinops caerulea)*. Proceedings, Calif. Acad. Sci. 34:1–84.

Oakey, N. S., and Elliott, J. A. 1982. Dissipation within the surface mixed layer. J. Phys. Oceanogr. 12:171–185.

Osborn, T. R., and Lueck, R. G. 1984. Oceanic shear spectra from a submarine. In Internal gravity waves and small-scale turbulence, Proc., Hawaiian winter workshop, pp. 25–50, Ed. by P. Muller and R. Pujalet, Hawaii Inst. Geophysics, Honolulu.

Osborn, T. R., and Lueck, R. G. 1985. Turbulence measurements with a submarine. J. Phys. Oceanogr. 15:1502–1520.

Price, J. F., Weller, R. A., and Pinkel, R. 1986. Diurnal cycling: Observation and models of the upper ocean response to diurnal heating, cooling, and wind mixing. J. Geophys. Res. 91(C7):8411–8427.

Rosenthal, H., and Hempel, G. 1970. Experimental studies in feeding and food requirements of herring larvae *(Clupea harengus* L.). *In* Marine food chains. pp. 344–364. Ed. by J. H. Steele. Univ. Calif. Press, Berkeley.

Rothschild, B. J. 1985. Feasibility of relating recruitment to environmental variation. ICES, C. M. 1985/Biological Oceanography Committee L:38.

Rothschild, B. J. (Editor). 1988. Toward a theory on biological-physical interactions in the world ocean. NATO ASI Series. Series C: Mathematical and Physical Sciences, vol. 239. Kluwer, Dordrecht, The Netherlands. 650 pp.

Rothschild, B. J., and Osborn, T. R. 1986. Biodynamics of the sea: Preliminary observations on high dimensionality and the effect of physics on predator-prey interrelationships. ICES C.M.1986/L:25.

Rothschild, B. J., and Osborn, T. R. 1988. Small-scale turbulence and plankton contact rates. J. Plankton Res. 10(3):465–474.

Rothschild, B. J., Osborn, T. R., Dickey, T. D., and Farmer, D. M. In press. The physical basis for recruitment variability in fish populations. J. Cons. Cons. int. Explor. Mer.

Chapter 7

Physical-Optical-Biological Scales Relevant to Recruitment in Large Marine Ecosystems

Thomas D. Dickey

Abstract

The abundance and diversity of marine populations and their variations are affected by a variety of physical, optical, and biological processes along with their interactions. The particular problem of recruitment in large marine ecosystems entails the consideration of many processes which, as a whole, can span scales up to nine orders of magnitude. This chapter will consider pertinent time and space scales, as well as observational and modelling studies.

Introduction

Many interesting and important oceanographic problems require multidisciplinary data sets and modelling approaches. Included among these problems are biomass determination, primary and secondary productivity, biogeochemical fluxes and cycles, optical variability, bioluminescence, and larval fish recruitment. The problem of fluctuations of fish and invertebrate populations and recruitment on various scales has been the subject of considerable study and has been stimulated by both scientific and societal interests. Here the term recruitment refers to the process through which the young are "recruited" into the adult class. Recruitment depends on many physical, optical, and biological factors that control larval feeding and predation. Many of the important aspects (e.g., interactions with the benthos, species succession, genetic variations) of this complex problem are beyond the scope of the present review; however, more encompassing works include those by Sherman and Alexander (1986), Rothschild and Rooth (1982), and Rothschild (1986, 1988). In addition, descriptions of several regional programs dedicated to the recruitment problem have been presented by Sherman (1988). A conceptual model (Figure 7.1, after Dickey, 1988) may be used to illustrate how external physical forcing drives the ecosystem by redistributing energy and oceanic properties. There are many factors that affect phytoplankton, zooplankton, and fish populations. Ultimately, motion and light are the sources of energy for the ecosystem.

This chapter will (i) summarize some of the more important time and space scales relevant to the problem of recruitment in large ecosystems, (ii) review observational methods pertinent to the recruitment problem that may be useful for sampling relevant parameters on these scales, and (iii) describe modelling approaches that may be useful for the recruitment problem.

Time and Space Scales

Oceanographers have commonly studied phenomena by utilizing time series and spatial data sets and have used spectral analysis

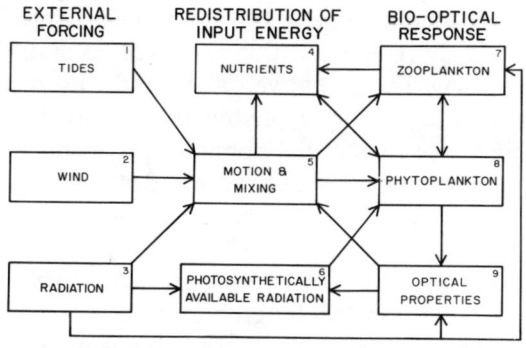

Figure 7.1. A conceptual model depicting some of the more important elements of the oceanic ecosystem, along with interactions (indicated by arrows) among the individual elements. (After Dickey, 1988.)

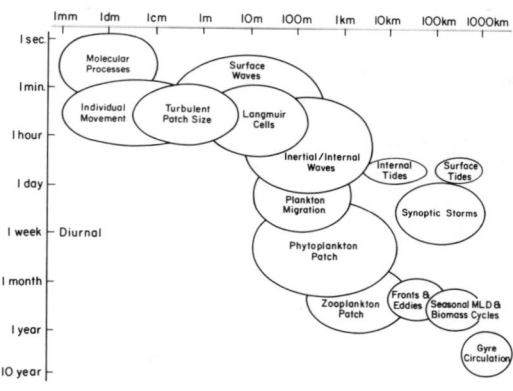

Figure 7.2. A schematic illustrating the relevant time and space scales of several physical, optical, and biological processes.

techniques to ascertain variance distributions of properties in time and space. Information on time and space scale variability can also be used to design experimental sampling plans and to identify and quantify relationships between the forcing and response of the ecosystem. Many in-depth reviews of ecologically relevant time and space scales have been written (e.g., Angel, 1976; Haury et al., 1978; Steele, 1978; Haury, 1982; Denman and Gargett, 1983; Denman and Powell, 1984; Legendre and Demers, 1984; Tett and Edwards, 1984; Klein and Steele, 1985; Mackas et al., 1985; Bakun, 1986; Boucher, 1988; Haury and Pieper, 1987; Steele, 1988).

Clearly, the recruitment of larval fish is affected by a broad range (up to nine orders of magnitude) of time and space scales of oceanic parameters. Representations of some of the more important physical, optical, and biological time and space scales are shown in Figure 7.2. Physical processes of relevance include molecular and turbulent diffusion, tides, storm mixing events, Langmuir cells, inertial motions, internal waves, diurnal and seasonal incident irradiance and heating cycles, and interannual phenomena (e.g., El Niño). Phytoplankton distributions are related in time to the physical forcing through light and nutrient availability and motion, particularly on storm event, tidal, internal gravity wave, diurnal, and seasonal scales (Figure 7.2).

Time scales of zooplankton are in part dependent upon phytoplankton time scales. Biological processes such as growth, phytoplankton photoadaptation, behavioral effects including food perception and feeding selectivity, and diel vertical migration of zooplankton populations are also relevant. In addition, the doubling times of organisms and their ambits are important aspects. The relationships between doubling time and particle diameters of phytoplankton, zooplankton, invertebrate carnivores and omnivores, and fish are indicated in Figure 7.3 (after Sheldon et al., 1972). The relative importance of particular physical versus biological processes is to first approximation determined by these two characteristic measures. There are at least some general relationships between physical time and space scales and trophic scales as demonstrated by Steele (1978), who noted that an empirical equation presented by Okubo (1971) between the standard deviation of dye concentration or a characteristic mixing scale and time, $\sigma \sim t^{1.17}$, generally conforms with biological trophic scales (see Figure 7.7). Denman and Powell (1984) assert that specific physical processes with time and space scales comparable to those of biological processes should have dominant ecological importance (e.g., since plankton are roughly passive).

One of the complicating aspects of the

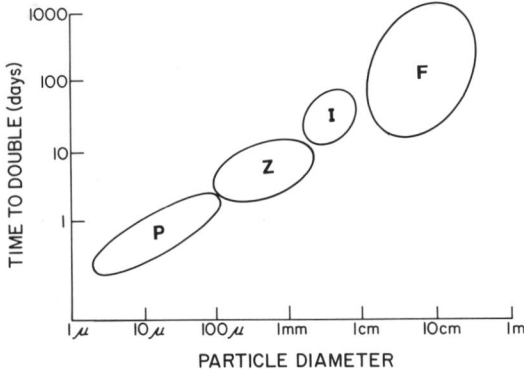

Figure 7.3. The general relationships between doubling time scales for phytoplankton (P), zooplankton (Z), invertebrate carnivores and omnivores (I), and fish (F) and particle diameter. (After Sheldon et al., 1972.)

recruitment problem is that the ecosystem is highly complex and generally nonlinear in nature. For example, it is likely that many of the more important phases of a given fish larva's life are critically affected by episodic physical forcing events. This important aspect must be considered for both observational and modelling studies. Another related vexing issue is the relationship between integrated regional effects and the global state. Ultimately, the relationships and couplings of biological as well as physical scales must be addressed.

It is well known that small time-scale processes correlate well with small spatial-scale processes (Figure 7.2) and that vertical scales are smaller than horizontal scales. Horizontal plankton patches have been studied extensively. Further, coastal regions with jets, fronts, and eddies tend to have richer spatial structure than open ocean regions. The subject of horizontal variability in abundances of phytoplankton and zooplankton (e.g., patchiness) has received considerable attention (Herman and Dauphinee, 1980) and is inherently important for modelling zooplankton swarms (Okubo, 1980) and fish recruitment (Sherman and Alexander, 1986).

The general problem of scales of the oceanic ecosystem is considered here. However, a few special considerations are warranted for the recruitment problem (Houde, 1982). First, egg and larval phases are relatively short stages within the lifetimes of organisms. Thus, relatively short and episodic events (e.g., storms, phytoplankton blooms) may have substantial impacts upon population dynamics. Fish eggs (on order of 1 mm, $O(1 \text{ mm})$, in diameter) are typically found in the upper 200 m, are positively buoyant, and generally drift with the currents. Larvae ($O(1-10 \text{ mm})$ in length) are often visual feeders, are usually negatively buoyant, and generally drift but are capable of short swimming bursts $O(1$ body length) with search rates of 0.01 to 10.0 liters $\cdot$ d^{-1} or daily ambits of $O(100$ to $1,000 \text{ m})$ according to Haury et al. (1978). It is evident that for individual eggs and larvae, scales on the order of 100 m and one day are quite important. However, the mesoscale ($O(10$'s to 100's km)) is important for the community as a whole, as larvae can be either retained in or advected from favorable nursery grounds. Another important aspect is the tendency toward dispersal or aggregation of larvae and their potential predators and prey. Thus, physical scales of the organisms and the fluid environment must be carefully analyzed.

In the following, some of the more important scales relevant to the fish larval recruitment problem are summarized. The standing stock time scales of primary importance are seasonal and interannual. For an individual, the diurnal light cycle, internal gravity wave (i.e., minutes to hours), advection (i.e., minutes to months), and mixing event time scales (minutes to a few days for effects with events occurring at intervals of days to a few weeks) are relevant in addition to the time scale associated with its motility. The spatial scales relevant for the larval community are often associated with specific oceanic regions and thus particular topographic features and circulation patterns (e.g., off northeastern United States; see Sherman et al., 1984). The horizontal larval patch scale is the lateral extent of a given larval aggregation and may be associated with the phytoplankton patch scales (Haury, 1982) and the larval prey scales. It is likely that larval patch sizes range from roughly

O(10 m) to O(10 km) and that interpatch scales range from O(100 m) to O(100 km). Vertical larval patch scales probably correspond to some degree with the scale of phytoplankton variability, O(< 10 m). The group encounter probability depends upon the respective group patch sizes and the interpatch distances along with the relative dispersion (or aggregation) rates. For the individual, the larva/food encounter probability scale may range from the length of the larva, O(millimeters), to the distance between the larva and its prey, O(millimeters to meters).

One of the areas of critical concern is the interaction of predator and prey. The terms predator and prey are often defined in a generic sense, but in order to quantify scales, it is necessary to be more specific. For the sake of the present development, the predator is considered to be a copepod zooplankton of scale O(1 mm) and the prey a diatom of scale O(0.01 mm). The encounter probability at the individual level is affected by the relative motions of the predator and the prey, both of which are controlled by the turbulence spectrum of the water and the properties of the organisms (e.g., size, shape, density). It is probable that motions with scales up to approximately a meter and a minute are most important. These scales are based upon the physical properties of the organisms, their motility and perception capability, and the turbulent flow field. The small-scale fluid environment of direct relevance to the predator and prey may be represented in general form as a turbulence energy spectrum, $E(k,t)$, for a homogeneous fluid (Figure 7.4, after Hinze, 1959). This spectrum is presented in wavenumber space (where wavenumber k is defined as $2\pi/\lambda$ where λ is the wavelength) and indicates a

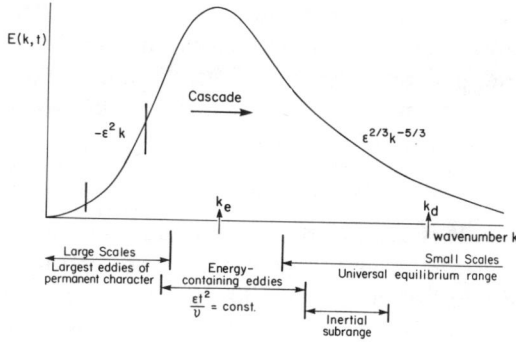

Figure 7.4. A representation of a spectrum for homogeneous turbulence. (After Hinze, 1959.)

spectral peak at the wavenumber of the most energetic turbulent eddies, k_e, and a small-scale Kolmogorov wavenumber, k_d, which marks the transition to the smallest scale motions governed by viscous rather than inertial effects.

There is considerable information now available concerning certain aspects of turbulence in the upper ocean. One of the important parameters obtained from ocean measurements is the dissipation rate of turbulent kinetic energy, ε. Summaries of values obtained under various oceanic conditions have been presented by Lueck and Reid (1984) and Gregg (1987). A few representative values of ε are given in Table 7.1. Using these values, the Kolmogorov length scale, η, and the Kolmogorov velocity scale, v_k, ranges have been computed (Table 7.1) according to the relations

$$\eta = \left(v^3/\varepsilon\right)^{1/4} \quad (1)$$

$$v_k = (v\varepsilon)^{1/4} \quad (2)$$

where v is the kinematic viscosity of seawater (~ 1 mm$^2 \cdot$ s^{-1}).

One of the most useful nondimensional

Table 7.1. Turbulence scales as related to predator-prey problem.

Region	ε (mm$^2 \cdot$ s^{-1})	η (mm)	v_k (mm $\cdot$ s^{-1})	Re$_z$	Re$_d$
North Pacific[1]	10^{-4}-1	10-1	0.1-1	0.1-1	0.001-.001
Scotia Shelf[2]	0.1-1	2-1	0.6-1	0.6-1	0.006-0.01
Seymour Narrows[3]	0.1-100	2-0.3	0.6-3	0.6-3	0.006-0.03

[1]Lueck and Osborn, 1982. [2]Oakey and Elliot, 1982. [3]Grant et al., 1962.

parameters in fluid mechanics is the Reynolds number. The Reynolds number is defined as the ratio of inertial to viscous force terms in the scaled Navier-Stokes equation for a fluid or

$$\text{Re} = \text{inertial force} / \text{viscous force}. \qquad (3)$$

It is used to delineate conditions under which laminar (low Re) versus turbulent (high Re) conditions prevail and thus facilitates the application of flow analysis. Turbulence is characterized by high mixing and diffusion rates. Unfortunately, the transition from laminar to turbulent conditions occurs over a range of Reynolds numbers. For example, transition occurs in the regime from Re~100 to 400 for a smooth sphere in a uniform flow (e.g., Clift et al., 1978). Reynolds numbers can also be defined with respect to organisms in small-scale flows. Although a spectrum of velocities acts upon the organisms, for convenience, a Reynolds number for the small Kolmogorov velocity scale for a copepod with a characteristic diameter of D_z may be defined as

$$\text{Re}_z = v_k D_z / \nu \qquad (4)$$

and the Reynolds number for a diatom with a characteristic diameter of D_d may be defined as

$$\text{Re}_d = v_k D_d / \nu. \qquad (5)$$

The characteristic Reynolds numbers are quite small, O(1) or less, with the Reynolds numbers of the zooplankton being two orders of magnitude greater than those of the diatoms (Table 7.1). An interesting depiction of Reynolds number dependency upon body length has been presented by Okubo (1987), who defined a swimming Reynolds number as

$$\text{Re}_s = V_s L / \nu \qquad (6)$$

where V_s is the animal swimming speed and L is the body length (Figure 7.5).

The empirical fit to their plot results in the following relations

$$\text{Re}_s = 269 \, L^{1.86} \qquad (7)$$

$$\text{and } V_s = 2.69 \, L^{0.86} \qquad (8)$$

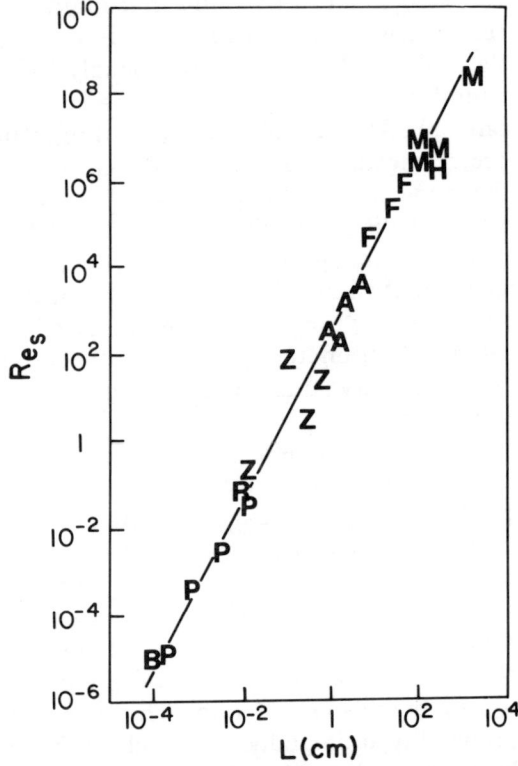

Figure 7.5. A relationship between Reynolds number, Re_L, for organisms of characteristic scale L and the scales of the organisms that include bacteria (B), phytoplankton (P), protozoa (R), zooplankton (Z), amphipod (A), fish (F), mammal (M), and man (H). (Figure after Okubo, 1987.)

where L is in cm and V_s is in cm $\cdot$ s^{-1}. The Reynolds number and the swimming velocity thus increase systematically with length over several orders of magnitude. If one considers the swimming speed of the zooplankton copepod (0.1 to 10 cm $\cdot$ s^{-1}), then Reynolds numbers for swimming copepods range between O(1) and O(100). One of the important points to be noted here is that Reynolds numbers relevant to the predator-prey problem vary by as much as five orders of magnitude (Table 7.1). Furthermore, the drag coefficients for these organisms also vary by up to five orders of magnitude. On this basis alone, the motion of the predator and the prey are expected to differ, resulting in substantial relative motion.

Other complicating factors include behavioral effects, the biomechanics of feed-

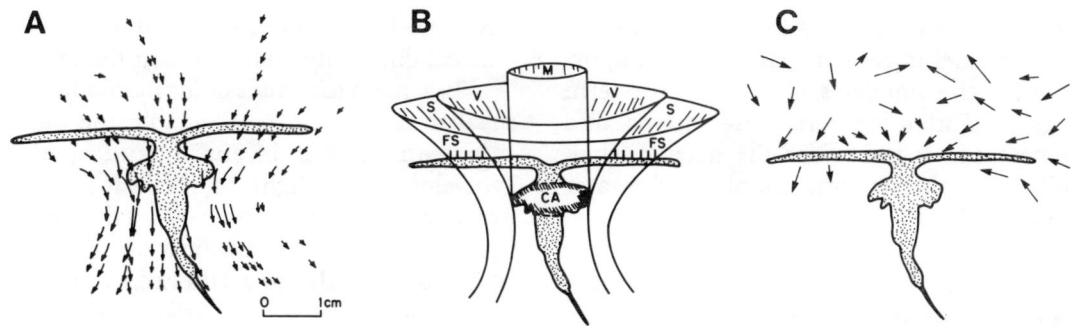

Figure 7.6. A depiction of a copepod (*Eucalanus crassus*). (a) Feeding currents are set up by the copepod and pathlines of algae are indicated by the arrows. (b) The various regions important to the copepod's feeding activity are indicated as the motion core (M), the viscous core (V), and the sensory core (S). Also shown are the copepod's capture area (CA) and field of sensing (FS) near antennae. (c) Finally, a hypothetic depiction of the copepod within a turbulent flow resulting in a disruption of the self-induced feeding currents. (Figure after Strickler, 1985.)

ing, and feeding currents. The biomechanics of aquatic animal locomotion is treated in detail by Lighthill (1975). The study of feeding currents has been done primarily in the laboratory (see review by Strickler, 1985). These studies suggest that organisms such as copepods are capable of sensing nearby food either visually or chemically at distances of O(1 mm) or less, of orienting themselves with respect to the flow field, and of setting up optimal feeding currents (Figure 7.6). The perturbation of feeding currents by in situ turbulent motion is being studied at present by Strickler and co-workers (pers. commun.). Particular feeding activities for a herbivorous copepod, along with their relevant scales, have been summarized according to Strickler (1985) as follows: (i) search for food — 10's cm and 10's sec; (ii) encounter and food recognition cm and sec; (iii) food capture, handling, and ingestion — mm and msec. It is clear that questions concerning encounter probabilities will require consideration of these small-scale processes as well as the larger scale effects associated with mesoscale and basin-scale circulation patterns.

Observations

The two primary methods that will be available to ocean scientists for observing the oceanic ecosystem within the foreseeable future are in situ sampling and remote sensing. The relative merits of various platforms (e.g., ships, buoys, satellites, airplanes) for studying particular temporal and spatial scales have been reviewed by Esaias (1981) and are illustrated in Figure 7.7. It should be emphasized that the very small as well as the great time and space scales are especially important for the global recruitment problem. Unfortunately, these extreme scales have been most inaccessible to ocean scientists. Nonetheless, technological advances should soon

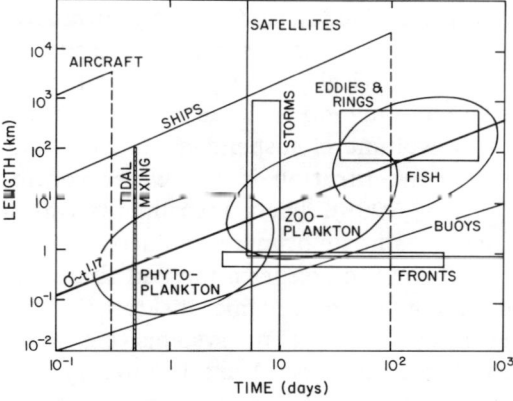

Figure 7.7. The space and time domains of (i) excursions of phytoplankton, zooplankton, and fish, (ii) some of the important physical processes, and (iii) several observational platforms (areas above and to the left of delineate domains undersampled by the particular platform). The line $\sigma R \sim t^{1.17}$ denotes the standard deviation time dependence for a diffusive process (Okubo, 1971). (Figure based upon diagrams by Steele, 1978; Esaias, 1981; and pers. commun. with Donald Collins of the Jet Propulsion Laboratory.)

enable measurements at these scales. In situ measurements remain particularly important for the smaller scale and longer term high-resolution temporal observations, and remote sensing provides the needed capabilities for the synoptic sampling of greater horizontal (primarily surface) scales.

In situ instrumentation and systems

A detailed review of recently developed and promising future multidisciplinary measurement systems has been presented by Dickey (1988). Therefore, only those instruments and systems most directly applicable to the recruitment problem are emphasized here. The development of multidisciplinary instrumentation systems has been highly dependent upon the availability of specific sensors. Some of the devices developed for use in upper ocean current studies are vector-measuring current meters, acoustical current-measuring systems, drifters, and drogues. Some of these instruments may be used in either moored or vertical profiling modes and can resolve time scales from minutes to several months or a few meters to a few hundred meters in the vertical depending on deployment strategy. Velocity fields have also been measured using several acoustical methods with vertical resolutions of O(m's). Acoustic tomography involves the measurement of the field of sound speed fluctuations within a control volume by transmitting acoustic signals along several diverse paths (Munk and Wunsch, 1979). One of the attractive features of this technique is that a relatively large (O(100's km) in the horizontal) volume of the ocean can be sampled synoptically. Vertical velocities have been most difficult to measure; however, shipboard measurements of vertical velocities have been done by using an acoustic Doppler system. Vertical velocity measurements have been taken with a modified vector-measuring current meter (Weller et al., 1985). Special velocity and temperature microstructure devices capable of resolving vertical scales on the order of 2 and 1 cm, respectively, have been developed and can be used to estimate vertical diffusivity and turbulent fluxes.

Two main functions of in situ bio-optical measurements are to (i) enable the determination of the intensity and quality (wavelength) of light available for photosynthesis at depth, and (ii) facilitate the identification and quantification of phytoplankton populations (including growth rates) and their products. Several of the sensors described in the following may be used for one or both of these purposes and provide virtually continuous sampling with vertical resolution comparable to conductivity, temperature, depth (CTD) systems (few meters or less) and temporal resolution comparable to moored current meters (few minutes).

Photosynthetically available radiation (PAR) sensors measure scalar irradiance in the visible waveband (~350–700nm) using a spherical light collector. More sophisticated optical instruments for quantifying the oceanic photoenvironment include multi-wavelength spectroradiometers (Smith et al., 1984). In situ fluorometers are used to obtain nearly continuous records of fluorescence in order to estimate chlorophyll a concentration and to infer phytoplankton pigment biomass. Fluorometers are used both with pumping systems and in situ profiling and towed modes, and recently in moored mode. Beam transmissometers measure an inherent optical property of seawater, the beam attenuation coefficient, which relates to the volume of suspended matter or particle concentration in the water column (Dickey, 1988). These instruments can be used to estimate primary productivity.

To determine speciation and abundances of zooplankton and micronekton, multiple opening and closing net systems (e.g., MOCNESS and BIONESS) are frequently used (Ortner et al., 1981). Concentrations of fish eggs and larvae are quite low (typically < 10 larvae $\cdot$ m^{-3}), thus long net tows are often the only way to obtain appreciable sample numbers and little is presently known about their small-scale variability. Briefly, nets are opened at up to eight depths as the system is towed. Mesh sizes range from ~0.06 mm to over 1 mm, and vertical and horizontal scales as small as a few meters and ~100 m can be sampled, respectively. In addition, other system sensors may include thermistors, conductivity sensors, fluorometers, dissolved oxygen sensors, and light sensors.

Another direct zooplankton observation method is based upon the Hardy Continuous Plankton Recorder, which utilizes a sequentially stepped gauze roll to enable the collection of up to 1000 samples. The method can resolve vertical scales of a few meters and horizontal scales of tens of meters. Disadvantages of the method include sample integrity and difficult analysis. Pump-based systems have also been used to directly sample zooplankton. With this method, samples are pumped from depth to the surface for analysis. Relatively high vertical ($O(m's)$) and horizontal ($O(10's\ m)$) resolution can be achieved using profile and tow modes, respectively. The pump system can also be utilized for various other measurements (e.g., nutrients, fluorescence, particle sizes, concentrations). Disadvantages include difficulty in sampling large, fragile, and motile organisms (see Dickey, 1988).

Light-imaging systems for zooplankton observations have been developed relatively recently. An in situ photographic system consisting of a 35-mm camera and a strobe light has been described by Ortner, Hill, and Edgerton (1981). Spatial resolution of ~1 m may be achieved and organisms as small as ~0.1 mm may be identified. This system also utilizes a variety of environmental sensors, and a high frequency acoustic backscatter sensor is being developed to enable real-time decision making for photographic data collection. Another system developed by Johnson, Lange, and Shulenberger (1983) employed a towed body with a strobe light, a video camera, a CTD, a fluorometer, and an acoustic current meter.

Other promising optical methods, which have potential application for in situ small-scale predator-prey studies, include Schlieren video systems (Strickler, 1985) and holographic systems (O'Hern et al., 1985). The latter has been used to record remarkably clear images of several zooplankton (size scales of ~100 mm to ~800 mm). One advantage of this method is that three-dimensional analyses of the individual organisms and their spatial relations can be conducted.

Electronic particle counters have been used to determine zooplankton abundance and physical dimensions indirectly (Herman and Dauphinee, 1980). Spatial patterns of ~1 m can be resolved for copepods ranging in size from about 0.5 to 3.0 mm. One problem with this technique is net clogging by zooplankton. To circumvent this problem, an optical plankton counter has been developed to measure light attenuation and chlorophyll a fluorescence. An automated image analysis system has been developed to reduce the difficulty of discriminating between taxonomic groups of zooplankton (Jeffries et al., 1984). This system utilizes a microscope, a vidicon camera, and a computer microprocessor. Scanned images are analyzed in terms of geometric parameters, area moments, and Fourier descriptors.

Acoustical methods are attractive because they are nonintrusive and can be used for broad spatial coverage applications (Anderson and Zahuranec, 1977). However, only recently has this principle been exploited. Several different acoustical approaches have been summarized by Farmer and Huston (1988) and Rothschild (1988). The first is based upon variation of target strength. For example, individual targets differing in size have different target strengths for differing frequencies. This principle has been exploited by Holliday and Pieper (1980), who developed the Multifrequency Acoustic Profiling System (MAPS). The system may be deployed in either profiling or towing (sawtooth) mode. It consists of a submersible vehicle with an array of 21 side-mounted acoustic transducers (frequencies range from 0.1 to 10 MHz), a pressure sensor, a thermistor, a conductivity sensor, a light sensor, a submersible pump, and hose for chlorophyll a fluorescence sampling onboard ship and another hose that is connected to a shipboard pumping system for sampling zooplankton. The acoustic transducers sample organisms ranging in equivalent spherical radii from roughly 0.1 to 1.0 mm (or animal lengths of ~0.4 to ~4 mm). The acoustic volume scattering data are transformed into size-class estimates of zooplankton abundance and biomass.

The second method utilizes backscatter strength fluctuations to count the number of targets in a given volume (Denbigh and Weintraub, 1986). This technique was motivated by fish stock assessment and draws upon the fact that a simple relationship exists between number density and the second moment of the intensity. Other investigators (Stanton and Clay, 1986) have pursued different aspects of the statistical distribution, which provide relevant information concerning zooplankton and fish.

The third method utilizes phase statistics (Farmer and Huston, 1988). This technique involves successive transmissions that are processed coherently. The second moment of the phase can be related to the number of targets in the sampling volume. To resolve multiple size classes of organisms, several frequencies must be sampled.

The fourth method uses a coherent Doppler measurement with processing of backscattered signals from several pulses in order to determine the rate of phase change. This principle has been used for remote current profiling measurements; thus shear data relevant to organism distributions are a natural product. Farmer and Huston (1988) suggest the possibility of detecting the statistics of movement of individual plankton relative to their neighbors using the calculation of de-correlation times of echoes. This technique has potential for estimating contact rates of zooplankton with phytoplankton.

The final method entails the acquisition of acoustic images based upon target strength distributions in coherent flow structures (Farmer and Freeland, 1983). This technique could be used to determine physical structures and possibly to investigate plankton behaviors. In addition, ultra-high-frequency acoustics have been employed to examine predator-prey interactions (Orr, 1981), and it has been suggested that fish predation of zooplankton communities could be studied using appropriate acoustic frequencies (Koslow, 1981).

Many of the sensors described herein may be considered as modules, which can be interfaced with submersible packages including data acquisition systems and microprocessors. The primary goals of multidisciplinary in situ measurement devices are to (i) sample with the various complementary, multidisciplinary sensors as closely in space and time as possible and (ii) resolve temporal and spatial scales of the ecosystem so as to avoid aliasing (Figures 7.1 and 7.2). Some of these include (i) the bio-optical profiling system (BOPS; Smith et al., 1984), which is used for determining the vertical distribution of several optical and physical variables; (ii) the Batfish system (see review by Herman, 1985) which is tow-yoed to measure spatial scales of variability in zooplankton and physical variables; (iii) the Kiel Sea Rover, which is similar to the Batfish but is designed for longer scales of operation (Woods et al., 1986); (iv) the undulating oceanographic recorder, which is also comparable to the Batfish but includes a suite of optical sensors (Aiken, 1985); (v) the multi-variable profiler (MVP), which is an unattended profiling system for measuring bio-optical and physical parameters including currents (Dickey, 1988); (vi) the multivariable moored system (MVMS), which is used to do time series measurements of several bio-optical and physical parameters (Dickey et al., 1986); and (vii) the MAPS system. These systems have been reviewed recently (Dickey, 1988).

For several years, drifters and drogues have been utilized by physical oceanographers for current measurements; however, their integration into biological studies is relatively recent. One of the principal attractions of drifters and drogues, which are equipped with bio-optical instrumentation, is that broad geographical extents can be sampled. In order for this approach to be viable for general usage, satellite telemetry of data and the production of sensors of moderate cost will be required. Other ways to deploy bio-optical and physical instrumentation packages include submarines (Osborn and Haury, pers. commun.) and remotely controlled underwater vehicles. There will continue to be a need for time series data, which will necessitate development of moored instrumentation capable of sampling more biological parameters. In addition, manned platforms such as R/P FLIP have been demonstrated to be of great utility for multi-disciplinary studies (Dickey, 1988). In the future, platforms comparable to R/P FLIP and perhaps other types of manned platforms (e.g., fixed or towable semi-submersible drilling platforms) developed for industrial marine applications may become available to researchers.

Remote sensing

The in situ instrumentation and systems described earlier are particularly well-suited for vertical profile and time series data acquisition. However, generally they cannot provide synoptic data with high horizontal resolution over extensive geographical regions. For this reason, satellite-borne and airplane-borne remote sensing systems are especially important for oceanographers. The potential geographical coverage of satellite-borne sensors is virtually global, with spatial resolution dependent upon the area observed or the footprint of the sensor. This point is illustrated in Figure 7.7 (after Esaias, 1981), which indicates the spatial and temporal coverage possible with various research platforms along with the domains of several of the important physical processes and biological groups. Satellite oceanographic observations are limited to near-surface measurements because of their reliance upon electromagnetic radiation. Further, sophisticated algorithms must be used to infer the biological and geophysical parameters, and some remote sensors cannot provide meaningful data under cloudy conditions. As a consequence, complementary in situ observations continue to be important.

Time series data that can be derived from satellites are delimited by the time period required to repeat an orbital pass and the lifetime of the satellite system. Although useful oceanographic data have been obtained since the mid-1960s, the first satellite dedicated to oceanographic data acquisition was SEASAT, which collected data over a three-month period in 1978. SEASAT data are still being analyzed; however, many interesting results have been published (see summaries by Saltzman, 1985, and Stewart, 1985). The results derived from SEASAT data sets have demonstrated the feasibility of determining sea surface temperature, surface winds, wave height, sea surface topography, and surface currents using satellite-borne sensors. The following review emphasizes some of the present and future capabilities of satellite remote sensing systems. The principles of remote sensing and more detailed developments of applications are presented in several publications (Huang, 1979; Brown and Cheney, 1983; Maul, 1985; Saltzman, 1985; Stewart, 1985).

Surface ocean currents, tides, and storm surges can be inferred by measuring departures of the sea surface from the earth's geoid. The variability of the surface topography (exclusive of surface waves) has scales ranging from tens of km to the dimensions of ocean basins. Altimeter data obtained from SEASAT observations have been used to map and determine temporal variability of relatively intense surface currents, including the Antarctic Circumpolar Current and the Gulf Stream and its associated rings. The altimeter to be used for the Ocean Topography Experiment (TOPEX) in the early 1990s will be capable of resolving more subtle currents of the world ocean (Wunsch, 1981). Resolvable spatial and temporal scales will be about 5 km and 10 days, respectively. Altimetric data can also be used to provide surface boundary conditions so that geostrophic currents at depth (based upon CTD data) may be computed without the ambiguity of the definition of the level of no motion.

The transfers of heat and momentum across the air-sea interface are important for surface-driven currents, waves, and mixing of the upper ocean. Meteorological satellite sensors provide radiation and atmospheric data relevant to the determination of heat transfer at the air-sea interface (Bretherton, 1981). Sea surface temperatures are presently determined with an accuracy of less than 1°C over spatial scales of about 4 km. The surface heat budget has recently been estimated over the Indian Ocean using satellite data exclusively (Gautier et al., 1987).

Scatterometers are active side-looking radars whose return signals are controlled primarily by Bragg scattering (see Stewart, 1985). These devices have the capability of measuring wind direction and speed and thus are suitable for many applications such as the determination of wind-driven currents. The estimated accuracies for wind speed and direction for the next generation of scatterometer are less than $2 \text{ m} \cdot \text{s}^{-1}$ and 20°, respectively, for swaths of O(500 km) with resolution of 50 km. Coverage will be completed approximately every two days.

The satellite sensors discussed thus far provide data that are primarily relevant to motion and mixing processes and for the most part indirectly relevant to biological processes. However, satellite-borne (and airplane-borne) radiometer (color) sensors have been used to estimate pigment concentrations (Gower, 1981), and ocean primary productivity has been estimated by using ocean color data (Perry, 1986). The Coastal Zone Color Scanner (CZCS), which was launched in 1978, measured visible and near-infrared ocean radiance through 1986. These data have been utilized to estimate chlorophyll a + phaeopigment concentrations within a factor of two. The correspondence of primary productivity to higher trophic levels remains an important problem. Also, it should be noted that essentially cloud-free conditions are required for color imaging and thus long time series have significant data losses. Nonetheless, color imaging has provided considerable insight into spatial distributions of phytoplankton in coastal upwelling regions, in the vicinity of the Gulf Stream, and at the equator. Quite recently, the first global ocean color image was produced (Feldman et al., 1989). Color imagery can also be used to estimate the diffuse attenuation of light in the upper portion of the water column (Gordon et al., 1985). It is likely that other ocean color imaging systems will be placed in orbit within a few years.

Aircraft-borne color imaging and active laser (Airborne Oceanographic Lidar, AOL) systems may be used to determine pigment concentrations as well (e.g., Kim and van der Piepen, 1986). These systems are most useful in coastal regions and can resolve scales down to < 10 m compared with ~1 km for the CZCS. The expense involved for these measurements prohibits routine operations; however, it may be possible to produce more economical and more portable systems, which could increase their usage.

Fisheries researchers have been utilizing satellite data since the mid-1970s. The direct observation of fish schools remains beyond the resolution capability of present satellite sensing systems; thus indirect observations must be used. Many of the fisheries studies have utilized ocean temperature and color data to locate fronts and wind stress maps to estimate transport. These data have been correlated with ground truth observations from research and fishing vessels. An excellent review of satellite data utilization for both research and fishing operations has been presented by Laurs and Brucks (1985).

Lasker et al. (1981) utilized satellite infrared imagery to analyze the spawning of northern anchovy in the California Current. It was established that distinct temperature conditions were correlated with the region where spawning occurred and that upwelling conditions led to the transport of anchovy larvae seaward from the coastal zone. Albacore tuna distributions have been correlated with ocean temperature and color fronts using satellite data off California (Laurs and Brucks, 1985). The small-scale migration of albacore tunas has also been studied by utilizing these same satellite sensors along with expendable bathythermographs (XBTs) and acoustic telemetering devices attached to the tunas.

Near-surface water transport is important for the movement of plankton and larvae from one location to another. Data collected with the SEASAT-A Satellite Scatterometer (SASS) have been utilized to compute near-surface flow (Ekman and Sverdrup transports along with flow converges and divergences) for the Gulf of Mexico (see Laurs and Brucks, 1985). Analysis indicated that regions of enriched spawning and aggregation of eggs and larvae could be predicted based upon wind-stress maps and the derived surface transports.

The study of large marine ecosystems requires remotely sensed data for several purposes including advection and mixing time scales of organisms and mapping of regions for light availability and primary productivity and perhaps even for higher trophic organisms. Large horizontal scales (greater than a few km) of variability near the ocean surface can be best sampled with airplane or satellite sensors. In addition, relatively long time series (years) with temporal resolution of a few days or less are achievable and particularly important for studies of seasonal and interannual variability.

Modelling

It is likely that sophisticated coupled physical-optical-biological models will be required to answer many of the questions concerning distributions of fish populations and larval recruitment, particularly because of the diversity of scales encompassed by the problem. Fortunately, modelling of ocean processes has advanced at a rapid rate within the past two decades and a variety of modelling approaches has resulted. These may be delineated on the basis of scales including (i) large — basin or global scale, $O(1000's$ km); (ii) mesoscale — eddy, ring, or frontal scales $O(10's$ to $100's$ km); and (iii) small scale — less than $O(10's$ km). Models may also be either prognostic or diagnostic. Those that track the motion of individual water parcels (or particles including organisms) are considered to be Lagrangian, whereas models that utilize a fixed coordinate system are Eulerian. The solution of the governing equations are determined using either analytical or numerical techniques. Simulations of more complicated environmental conditions typically require that the latter techniques be utilized. The majority of ocean modelling efforts have been devoted to physical problems, primarily circulation and mixing (O'Brien, 1986; Holland and McWilliams, 1987). However, a growing number of modelling efforts are being devoted to biological problems (Wroblewski and O'Brien, 1976; Okubo, 1980; Walsh, 1980; Kiefer and Kremer, 1981; Klein and Coste, 1984; Evans and Parslow, 1985; Hoffmann, 1985; Taylor et al., 1986; Franks et al., 1986; Wroblewski and Richman, 1987; Wroblewski and Hoffman, 1989).

Large-scale ocean circulation models have evolved from relatively simple analytical models of wind-driven transport (Pond and Pickard, 1983) to complicated numerical models (O'Brien, 1986). One of the obvious goals of oceanic numerical models is to provide realistic depictions of currents and property distributions. The basic premise is that these can be simulated, provided the appropriate governing equations (e.g., for mass, momentum, heat, salt) along with initial and boundary conditions are known. In practice, several difficulties arise. These include (i) the number of computational grid points is large because of the size of the physical domain to be modelled and the need to resolve scales of $O(10's$ km) in the horizontal and $O(10's$ to $100's$ m) in the vertical; (ii) even with the finest resolution now possible with state-of-the-art supercomputers, horizontal scales less than $O(10's$ km) and vertical scales less than $O(10's$ m) must be parameterized; (iii) the time required to perform a solution (integration of the differential equations) is quite long because processes with scales of $O(days)$ must be resolved; and (iv) the initial and boundary conditions are dependent on relatively high-resolution, synoptic oceanic mesurements, which are difficult to obtain and few in number. Nonetheless, relatively realistic simulations of the circulation patterns on basin (Holland and McWilliams, 1987) and even the global scale (Semtner, 1986) have been performed. Several modelling efforts have successfully depicted major features in particular regions of the world ocean (e.g., Holland and McWilliams, 1987).

Many useful field and laboratory studies have provided insights for modelers of smaller scale processes (Gregg, 1987; Hopfinger, 1987). Interestingly, as is the case for mesoscale and large-scale models, small-scale models also have severe computational requirements as they must resolve scales from a few centimeters or less (where dissipation occurs) up to tens of meters or greater (where internal gravity waves exist). Nonetheless, considerable progress is being made using numerical solutions of the exact time-dependent, three-dimensional Navier-Stokes equation (e.g., Siegel, 1988). Each of the models mentioned here has relevance to the modelling of the oceanic ecosystem. In particular, the large- and mesoscale models are important for the transport of nutrients and plankton from one location to another. These are especially useful with regard to questions concerning community structure, the probability of encounter of large groups of predator and prey, and the general geographic distributions of larvae and fish. One particularly relevant study has been reported by Johnson (1985), who utilized a wind-driven model of the general circulation of the Middle Atlantic Bight to study the distributions and dispersion of blue crab larvae.

As indicated earlier, the scaling of motions for the predator-prey problem leads to the conclusion that under many circumstances, the predator will move relative to the prey. For other applications, the problem of particle motion in a turbulent environment has received considerable attention both in the laboratory and using models (e.g., Maxey and Corrsin, 1986). The complicating factors include particle size, density, and shape, along with diverse flow conditions such as waves and turbulence (Clift et al., 1978). In addition, the description of the detailed particle motion is essential to the determination of encounter probabilities. Particle encounter probabilities (or contact rates) can be modelled using the individual particle sizes and their motions along with their concentrations and their spatial distributions. The contact rates between predator and prey have been estimated on a theoretical basis by incorporating the effects of small-scale turbulence by Rothschild and Osborn (1988). They suggest that the turbulence effects are important for determining plankton food requirements, energy balance for foraging, and patch formation and dissipation.

The problem of plankton patchiness and diffusion (including the critical patch size problem) has been addressed by several investigators (Okubo, 1980). Furthermore, models of sinking rates of phytoplankton in turbulent fluids under varying light conditions have been formulated (Granata, 1987; Lande and Wood, 1987). These studies show that turbulent diffusion and sinking interact in sustaining phytoplankton populations.

In order to retain time history information, Lagrangian (or particle-following) models may be utilized. Fortunately, these models have become the focus of several studies and have been applied quite successfully to several fluid problems. One of the more appropriate Lagrangian techniques involves the direct generation of particle trajectories with a stochastic model (Sawford, 1985). Using this method, information concerning light exposure of an organism in a time varying light field may be retained (Woods and Onken, 1982; Wolf and Woods, 1988). This type of model is applicable to the relatively large-scale turbulent flow (large Re), but not to the relatively small scale (small Re). The small-scale end of the motions requires a special model such as one developed by Dickey and Mellor (1979), who presented a formulation for low Reynolds-number–dependent velocity correlation functions. Ultimately, a match of motions at the larger and smaller scales is required.

The development of models for specific scale ranges and applications is well underway. Yet, for those interested in biological problems such as larval recruitment, a unified model will be essential. In other words, the large-scale and mesoscale trajectories of larvae may be simulated satisfactorily, but the local environment and the predator-prey interactions must be adequately modelled as well. It is conceivable that a successful grand model may consist of a system of nested models, each of which has its own function and interfaces to a model for an adjacent scale range.

Another important development being pursued by physical oceanographers is data assimilation. It has been suggested that general circulation ocean models can extrapolate satellite surface information throughout the ocean interior if subsurface data are assimilated (Halpern, 1987). Further, data assimilation techniques can be used to update results of prognostic models in nearly real time. The synergistic utilization of oceanic data and models is probably best achieved through the assimilation approach.

Conclusions

The recruitment problem requires the consideration of physical, optical, and biological processes that can span scales up to nine orders of magnitude. However, significant strides have been taken in multidisciplinary observational technology and computer modelling within the past decade. Furthermore, careful consideration of time and space scales associated with the more energetic processes can be used to optimize both sampling and modelling. An important step for modelers and observers is to be able to accurately parameterize the smaller scale processes such as turbulence, particle motion, and animal feeding activities. The development of a substantially improved understanding of recruitment and large marine ecosystems seems a realistic goal.

Acknowledgments

The author would like to thank Dr. Tim Granata and Dr. David Siegel for their useful comments on this chapter, and Dr. Kenneth Sherman for inviting this contribution. Many of the ideas presented here are based upon several years of work supported by the Office of Naval Research programs in Oceanic Biology and Optics. The present work was supported under ONR contract N00014-87-K-0084. The figures were drafted by Mrs. Janet Dodds.

References

Aiken, J. 1985. The undulating oceanographic recorder Mark 2: A multirole oceanographic sampler for mapping and modeling the biophysical marine environment. In Mapping strategies in chemical oceanography. pp. 315–332. Ed. by A. Zirino. American Chemical Society, Washington, DC.

Anderson, R., and Zahuranec, B. J. 1977. Oceanic sound scattering predictions. Plenum, New York. 859 pp.

Angel, M. V. 1976. Windows into a sea of confusion: Sampling limitations to the measurement of ecological parameters in oceanic midwater environments. In Oceanic sound scattering prediction. pp. 2171–2248. Ed. by R. Anderson and B. J. Zahuranec. Plenum, New York.

Bakun, A. 1986. Definitions of environmental variability affecting biological processes in large marine ecosystems. In Variability and management of large marine ecosystems. pp. 89–108. Ed. by K. Sherman and L. M. Alexander. AAAS Selected Symposium 99, Westview Press, Boulder, CO. 319 pp.

Boucher, J. 1988. Space-time aspects in the dynamics of planktonic stages. In Toward a theory on biological-physical interactions in the world ocean. pp. 203–214. Ed by B. J. Rothschild. NATO ASI Series. Series C: Mathematical and Physical Sciences, Vol. 239. Kluwer, Dordrecht, The Netherlands. 650 pp.

Bretherton, F. P. 1981. Climate, the oceans, and remote sensing. Oceanus 24:48–55.

Brown, O. B., and Cheney, R. E. 1983. Advances in satellite oceanography. Rev. Geophys. Space Phys. 21:1216–1230.

Clift, R., Grace, J. R., and Weber, M. E. 1978. Bubbles, drops, and particles. Academic Press, New York. 380 pp.

Denbigh, P. N., and Weintraub, J. 1986. A statistical approach to fish stock assessment. ICA associated symposium on underwater acoustics. Tech. Univ. of Nova Scotia. pp. 41–42.

Denman, K. L., and Gargett, A. E. 1983. Time and space scales of vertical mixing and advection of phytoplankton in the upper ocean. Limnol. Oceanogr. 28:801–815.

Denman, K. L., and Powell, T. M. 1984. Effects of physical processes on planktonic ecosystems in the coastal ocean. Ocean. Mar. Biol. Ann. Rev. 22:125–168.

Dickey, T. D. 1988. Recent advances and future directions in multi-disciplinary in situ oceanographic measurement systems. In Toward a theory on biological-physical interactions in the world ocean. pp. 555–598. Ed by B. J. Rothschild. NATO ASI Series. Series C: Mathematical and Physical Sciences, Vol. 239. Kluwer, Dordrecht, The Netherlands. 650 pp.

Dickey, T. D., and Mellor, G. L. 1979. The Kolmogoroff $r^{2/3}$ law. Phys. of Fluids 22:1029–1032.

Dickey, T., Hartwig, E., and Marra, J. 1986. The Biowatt bio-optical and physical moored measurement program. Trans. Am. Geophys. Union EOS 67:650.

Esaias, W. E. 1981. Remote sensing in biological oceanography. Oceanus 24:32–38.

Evans, G. T., and Parslow, J. S. 1985. A model of annual plankton cycles. Biol. Oceanogr. 3:327–347.

Farmer, D. M., and Freeland, H. J. 1983. The physical oceanography of fjords. Prog. in Ocean. 12:147–219.

Farmer, D. M., and Huston, R. D. 1988. Novel applications of acoustic backscatter to biological measurements. In Toward a theory on biological-physical interactions in the world ocean. pp. 599–614. Ed by B. J. Rothschild. NATO ASI Series. Series C: Mathematical and Physical Sciences, Vol. 239. Kluwer, Dordrecht, The Netherlands. 650 pp.

Feldman, G., et al. 1989. Ocean color: Availability of data. Trans. Am. Geophys. Union EOS 70:634–641.

Franks, P. J. S., Wroblewski, J. S., and Flierl, G. R. 1986. Prediction of phytoplankton growth response to the frictional decay of a warm core ring. J. Geophys. Res. 91:7603–7610.

Gautier, C., Frouin, R., Simonot, J.-Y., and Iacobelleis, S. 1987. Surface heat budget over the Indian Ocean during the 1979 Monsoon (abstr.). Trans. Am. Geophys. Union EOS 68:1326.

Gordon, H. R., Austin, R. W., Clark, D. K., Hovis, W. A., and Yentsch, C. S. 1985. Ocean color measurements. In Satellite oceanic remote sensing. pp. 297–335. Ed. by B. Saltzman. Academic Press, Orlando, FL.

Gower, J. F. R. 1981. Oceanography from space. Plenum, New York. 980 pp.

Granata, T. C. 1987. Measurements of phytoplankton sinking and growth under varying light intensity and mixing regimes. Ph.D. Thesis. Univ. California, Berkeley. 175 pp.

Grant, H. L., Stewart, R.W., and Moillet, A. 1962. Turbulence spectra from a tidal channel. J. Fluid Mech. 12:241–263.

Gregg, M. C. 1987. Diapycnal mixing in the thermocline: A review. J. Geophys. Res. 92:5249–5286.

Halpern, D. 1987. Data assimilation and ocean general circulation models. Trans. Am. Geophys. Union EOS 68:731–733.

Haury, L. R. 1982. Mesoscale processes: Some biological and physical connections. Trans. Am. Geophys. Union EOS 63:267–275.

Haury, L. R., and Pieper, R. E. 1987. Zooplankton: Scales of biological and physical events. In Marine organisms as indicators. pp. 35–72. Ed. by D. F. Soule and G. S. Kleppel. Springer-Verlag, New York.

Haury, L. R., McGowan, J. A., and Wiebe, P. H. 1978. Patterns and processes in the time-space scales of plankton distributions. In Spatial patterns in plankton communities. pp. 277–327. Ed. by J. H. Steele. Plenum, New York.

Herman, A. W. 1985. Biological profiling in the upper oceanic layers with a Batfish vehicle: A review of applications. In Mapping strategies in chemical oceanography. pp. 293–314. Ed. by A. Zirino. American Chemical Society, Washington, DC.

Herman, A. W., and T. M. Dauphinee. 1980. Continuous and rapid profiling of zooplankton with an electronic counter mounted on a "Batfish" vehicle. Deep-Sea Res. 27A:79–96.

Hinze, J. O. 1959. Turbulence. McGraw-Hill, New York. 586pp.

Hoffmann, E. E. 1985. A physical-biological model of plankton dynamics on the southeastern United States continental shelf. Trans. Am. Geophys. Union EOS 66:1304.

Holland, W. R., and McWilliams, J. C. 1987. Computer modeling in physical oceanography from the global circulation to turbulence. Physics Today 40:51–57.

Holliday, D. V., and Pieper, R. E. 1980. Volume scattering strengths and zooplankton distributions at acoustic frequencies between 0.5 and 3 MHz. J. Acous. Soc. Am. 67:135–146.

Hopfinger, E. J. 1987. Turbulence in stratified fluids. J. Geophys. Res. 92:5287–5304.

Houde, E. D. 1982. Micro- and fine-scale biology. In Fish ecology III. pp. 96–102. Ed. by B. J. Rothschild and C. G. H. Rooth. Univ. Miami Tech. Rep. No. 82008, Miami.

Huang, N. E. 1979. New developments in satellite oceanography and current measurements. Rev. Geophys. Space Phys. 17:1558–1568.

Jeffries, H. P., et al. 1984. Automated sizing, counting, and identification of zooplankton by pattern recognition. Mar. Biol. 78:329–334.

Johnson, D. R. 1985. Wind-forced dispersion of blue crab larvae in the Middle Atlantic Bight. Cont. Shelf Res. 4:733–745.

Johnson, W. P., Lange, R. E., and Shulenberger, E. 1983. TOPBS—A towed oceanographic physical and biological sampler. IEEE Proc. Third Working Symp. on Ocean. Data Systems:141–145.

Kiefer, D. A., and Kremer, J. N. 1981. Origins of vertical patterns of phytoplankton and nutrients in the temperate, open ocean: A stratigraphic hypothesis. Deep-Sea Res. 33:1087–1105.

Kim, H. H., and van der Piepen, H. 1986. Sunlight induced 685nm fluorescence imagery. Ocean Optics VIII, SPIE 637:358–363.

Klein, P., and Coste, B. 1984. Effects of wind-stress variability on nutrient transport into the mixed layer. Deep-Sea Res. 31:21–37.

Klein, P., and Steele, J. H. 1985. Some physical factors affecting ecosystems. J. Mar. Res. 43:337–350.

Koslow, J. A. 1981. Feeding selectivity of schools of northern anchovy (*Engraulis mordax*) in the Southern California Bight. Fish. Bull., U.S. 79:131–142.

Lande, R., and Wood, M. 1987. Suspension times of particles in the upper ocean. Deep-Sea Res. 34:61–72.

Lasker, R., Pelaez, J., and Laurs, R. M. 1981. The use of satellite infrared imagery for describing ocean processes in relation to spawning of the northern anchovy (*Engraulis mordax*). Remote Sens. Environ. 11:439–453.

Laurs, R. M., and Brucks, J. T. 1985. Living marine resources applications. In Satellite oceanic remote sensing. pp. 419–452. Ed. by B. Saltzman. Academic Press, Orlando, FL.

Legendre, L., and Demers, S. 1984. Towards dynamic biological oceanography and limnology. Can. J. Fish. Aquat. Sci. 41:2–19.

Lighthill, J. 1975. Mathematical biofluiddynamics. Society for Industrial and Applied Mathematics, Philadelphia. 281 pp.

Lueck, R. G., and Osborn, T. R. 1982. Dissipation measurements from the FRONTS-80 expedition. Rep. 80. Dept. of Oceanography, Univ. British Columbia, Vancouver.

Lueck, R. G., and Reid, R. 1984. On the production and dissipation of mechanical energy in the ocean. J. Geophys. Res. 89:3439–3445.

Mackas, D., Denman, K., and Abbott, M. 1985. Plankton patchiness: Biology in the physical vernacular. Bull. Mar. Sci. 37:652–674.

Maul, G. A. 1985. Introduction to satellite oceanography. Nijhoff, Dordrecht, The Netherlands. 606 pp.

Maxey, M. R., and Corrsin, S. 1986. Gravitational settling of aerosol particles in randomly oriented cellular flow fields. J. Atm. Sci. 43:1112–1134.

Munk, W., and Wunsch, C. 1979. Ocean acoustic tomography: A scheme for large scale monitoring. Deep-Sea Res. 26A:123–156.

Oakey, N. S., and Elliot, J. A. 1982. Dissipation within the surface mixed layer. J. Phys. Oceanogr. 12:171–185.

O'Brien, J. J. 1986. Advanced physical oceanographic numerical modelling. D. Reidel, Dordrecht, The Netherlands. 608 pp.

O'Hern, T. J., Katz, J., and Acosta, A. J. 1985. Holographic measurements of cavitation nuclei in the sea. ASME Cavitation and Multiphase Flow Forum, Albuquerque, NM.

Okubo, A. 1971. Oceanic diffusion diagrams. Deep-Sea Res. 18:789–802.

Okubo, A. 1980. Diffusion and ecological problems: Mathematical models. Springer-Verlag, New York. 254 pp.

Okubo, A. 1987. The fantastic voyage into the deep: Marine biofluidmechanics, Lecture Notes in Biomathematics 71. Ed. by E. Teramoto and M. Yamaguti.

Orr, M. H. 1981. Remote acoustic detection of zooplankton response to fluid processes, oceanographic instrumentation, and predators. Can. J. Fish. Aquat. Sci. 38:1096–1105.

Ortner, P. B., Hill, L. C., and Edgerton, H. E. 1981. In-situ silhouette photography of Gulf Stream zooplankton. Deep-Sea Res. 28:1569–1576.

Ortner, P. B., Pieper, R. E. and Mackus, D. L. 1983. Advances in zooplankton sampling. In Fish ecology III. pp. 355–379. Ed. by B. J. Rothschild and C. G. H. Rooth. Univ. Miami Tech. Report No. 82008, Miami.

Perry, M. J. 1986. Assessing marine primary productivity from space. Bio. Sci. 36:461–467.

Pond, S., and Pickard, G. 1983. Introductory dynamical oceanography. Pergamon, Oxford, UK. 329 pp.

Rothschild, B. J., 1986. Dynamics of marine fish populations. Harvard Univ. Press, Cambridge, MA. 277 pp.

Rothschild, B. J., 1988. Toward a theory on biological-physical interactions in the world ocean. Kluwer, Dordrecht, The Netherlands, 650 pp.

Rothschild, B. J., and Osborn, T. R. 1988. The effects of turbulance on planktonic contact rates. J. Plank. Res. 10(3):465–474.

Rothschild, B. J., Osborn, T. R., Dickey, T. D., and Farmer, D. M. 1989. The physical basis for recruitment variability in fish populations. J. Cons. int. Explor. Mer 45:136–145.

Rothschild, B. J., and Rooth C. G. H. 1982. Fish ecology III. Univ. Miami Tech. Rep. No. 82008, Miami.

Saltzman, B. 1985. Satellite oceanic remote sensing. Advances in geophysics, 27, Academic Press, Orlando, FL. 511 pp.

Sawford, B. L. 1985. Lagrangian statistical simulation of concentration mean and fluctuation fields. J. Clim. and Appl. Met. 24:1152–1166.

Semtner, A. J., Jr. 1986. Finite-difference formulation of a world ocean model. In Advanced physical oceanographic numerical modelling. pp. 187–202. Ed. by J. J. O'Brien. D. Reidel, Dordrecht, The Netherlands.

Sheldon, R. W., Prakash, A., and Sutcliffe, W. H. 1972. The size distribution of particles in the ocean. Limnol. and Oceanogr. 17:327–340.

Sherman, K. 1988. Large marine ecosystems as global units for recruitment experiments. In Toward a theory on biological-physical interactions in the world ocean. pp. 459–476. Ed by B. J. Rothschild. NATO ASI Series. Series C: Mathematical and Physical Sciences, Vol. 239. Kluwer, The Netherlands. 650 pp.

Sherman, K., et al. 1984. Spawning strategies of fishes in relation to circulation, phytoplankton production, and pulses in zooplankton off the northeastern United States. Mar. Ecol. 18:1–19.

Sherman, K., and Alexander, L. M. 1986. Variability and Management of Large Marine Ecosystems. AAAS Selected Symposium 99, Westview Press, Boulder, CO. 319 pp.

Siegel, D. A. 1988. Large-eddy simulation of the decay of a small-scale oceanic internal gravity wave field. Ph.D. thesis. Univ. of Southern California. 187 pp.

Smith, R. C., Booth, C. R., and Star, J. L. 1984. Oceanographic biooptical profiling system. Appl. Optics 23:2791–2797.

Stanton, T. K., and Clay, C. S. 1986. Sonar echo statistics as a remote sensing tool: Volume and seafloor. IEEE J. of Ocean. Eng. OE-11:79–96.

Steele, J. 1978. Spatial pattern in plankton communities. Plenum, New York. 470 pp.

Steele, J. 1988. Scale selection for biodynamic theories. In Toward a theory on biological-physical interactions in the world ocean. pp. 513–526. Ed by B. J. Rothschild. NATO ASI Series. Series C: Mathematical and Physical Sciences, Vol. 239. Kluwer, Dordrecht, The Netherlands. 650 pp.

Stewart, R. H. 1985. Methods of satellite oceanography. Univ. California Press, Berkeley. 360 pp.

Strickler, J. R. 1985. Feeding currents in calanoid copepods: Two new hypotheses. In Physiological adaptations of marine animals. pp. 459–485. Ed. by M. S. Laverack. Symposia of the Society for Experimental Biology, 23, Pinder Group of Companies, Scarborough, North Yorkshire, UK.

Taylor, A. H., Harris, J. R. W., and Aiken, J. 1986. The interaction of physical and biological processes in a model of the vertical distribution of phytoplankton under stratification. In Marine interfaces ecohydrodynamics. pp. 313–330. Ed. by J. C. J. Nihoul. Elsevier, Amsterdam.

Tett, P., and Edwards, A. 1984. Mixing and plankton: An interdisciplinary theme in oceanography. Ocean. Mar. Biol. Ann. Rev. 22:99–123.

Walsh, J. J. 1980. Shelf-sea ecosystems. In Analysis of marine ecosystems. pp. 159–196. Ed. by A. R. Longhurst. Academic Press, London.

Weller, R. A., et al. 1985. Three-dimensional flow in the upper ocean. Science 227:1552–1556.

Wolf, K. U., and Woods, J. D. 1988. Lagrangian simulation of primary production in the physical environment — the deep chlorophyll maximum and nutricline. In Toward a theory on biological-physical interactions in the World Ocean. pp. 51–70. Ed. by B. J. Rothschild. Kluwer, Dordrecht, The Netherlands.

Woods, J. D., and Onken, R. 1982. Diurnal variation and primary production in the ocean — preliminary results of a Lagrangian ensemble model. J. Plank. Res. 4:735–756.

Woods, J. D., Onken, R., and Fischer, J. 1986. Thermohaline intrusions created isopycnically at oceanic fronts are inclined to isopycnals. Nature 322:446–449.

Wroblewski, J. S., and Hoffman, E. E. 1989. U.S. interdisciplinary modeling studies of coastal-offshore exchange processes: Past and future. Prog. Oceanogr. (in press).

Wroblewski, J. S., and O'Brien, J. J. 1976. A spatial model of phytoplankton patchiness. Mar. Biol. 35:161–175.

Wroblewski, J. S., and Richman, J. G. 1987. The nonlinear response of plankton to wind mixing events — implications for survival of larval northern anchovy. J. Plank. Res. 9:103–123.

Wunsch, C. 1981. The promise of satellite altimetry. Oceanus 24:17–26.

Chapter 8

Direct Simulation of the Effect of Turbulence on Planktonic Contact Rates

Thomas Osborn, Hidekatsu Yamazaki, Kyle Squires

Abstract

Recent work has shown that small-scale turbulence dramatically affects planktonic contact rates. In fact, it is likely that for some plankton, the temporal and spatial variations in turbulence are a major cause of variability in population. This chapter discusses an approach to numerically modelling planktonic encounters in order to adequately represent the temporal and spatial coherence of the turbulent velocity field. Individual particle trajectories are followed in time. As the programming tools are assembled, this Lagrangian modelling can include more and more details of the behavior and response of the individual creatures.

Introduction

The smallest scales of turbulent motion in the ocean are comparable to the scales at which plankton operate and interact. Interactions are affected by turbulence (Rothschild and Osborn, 1988). The Kolmogorov scales can be estimated from the local rate of turbulent energy dissipation, ε, and the kinematic viscosity, ν. The time scale, $\tau_k = (\nu/\varepsilon)^{1/2}$, velocity scale $v_k = (\nu\varepsilon)^{1/4}$, and spatial scale $l_k = (\nu^3/\varepsilon)^{1/4}$ are the scales at which viscosity changes the turbulent motion to what appears at smaller scales as a uniform straining motion. Viscosity is relatively constant for seawater; we will use a nominal value of 10^{-2} cm$^2 \cdot$ s^{-1} and look at the effect of the variation over several decades of the dissipation rate, ε.

ε	10^{-6} cm$^2 \cdot$ s^{-3}	10^{-2} cm$^2 \cdot$ s^{-3}
τ_k	100 s	1 s
v_k	0.01 cm $\cdot$ s^{-1}	0.1 cm $\cdot$ s^{-1}
l_k	1 cm	0.1 cm

Since the temporal, spatial, and velocity scales of oceanic turbulence and planktonic motion are comparable, it is reasonable that turbulence increases the contact rate between predator and prey plankton. The problem is not simple to analyze. A turbulent velocity field has spatial and temporal coherence in addition to satisfying the continuity equation. The relative motion of parcels of water depends on their separation. We cannot treat the effect of the turbulence in a simple fashion — such as a random walk.

The only feasible way to correctly determine the effect of turbulence on contact rates is to have a dynamically correct turbulent velocity field and follow many particles. Averaging can then be performed over the sample population. Behavioral patterns can be imposed on individuals. This detailed treatment of individuals is a valuable characteristic of the Lagrangian particle-following technique. It avoids the problem of Eulerian modelling, where effects and characteristics must be assigned parameters (usually proportional to particle density or some power or the derivative of particle density). The particle-following calculations are computer intensive, as are the calculations of a dynamically correct turbulent velocity field, but they are feasible, using some of the numerical tools available at the NASA/Stanford Center for Turbulence Research.

Calculations and Results

The turbulence model (Lee and Reynolds, 1985) operates on a cubic grid of 64 or 128 points. The calculation is extremely intensive and requires a large amount of storage. The 128 point cube has over 2 million grid points, with three velocity components at each grid point. A program has been developed to track 4096 individual fluid particles to study the Lagrangian autocorrelation. For our work, this program was modified to allow the particles to move relative to the fluid.

The turbulence calculations are performed in a nondimensional format. The only input parameter is the viscosity. One must additionally decide whether to model decaying turbulence or continuously forced turbulence. We prefer the continuously forced case, since the turbulent intensity is constant in time. Having set the viscosity and the forcing scheme, the program is run 600 time steps to achieve steady state, whereupon the volume is "seeded" with the 4,096 particles and their motion is tracked. The particle-tracking program has to include the turbulent calculation because, at present, there is no way to store all of the velocities at each time step. After the trajectories are calculated for 600 time steps, the program is halted. It can be restarted later from that point. A second program is then used to determine encounter between the particles as a function of time. The stored data represent the three coordinates of location for each of the 4,096 particles at each time step, as well as the data from the last time steps required to restart the program.

The parameters of the calculation can be related to the real world by calculating the time step and the grid spacing of the model, and the kinetic energy of the turbulent field in terms of the appropriate Kolmogorov scales. Remember, we input the viscosity to the program, and the dissipation rate, the turbulent energy, the time step, and the energetic length scale are some of the returned parameters. Thus, we can express the output of the model in terms of Kolmogorov units just as we can express real time, distance, and velocity in the ocean in terms of the local oceanic Kolmogorov scales.

Although the turbulent calculation can be nondimensionalized, when the motion of the particles is added, it must be expressed in terms of the available scales, and total nondimensionalization of the calculation is no longer possible. Thus, many cases must be run to cover a wide range of ratios between the turbulent velocities and the particle-swimming velocities.

Direct simulation of Navier-Stokes equation enables us to combine the effects of a turbulent velocity field with a random walk used to model planktonic motion relative to the water. We made use of pseudo-spectral code for turbulent simulations developed by NASA and the Stanford University group. An isotropic turbulent field was generated on 64^3 grid points using a forcing scheme. The velocity calculation started from a field that had reached a steady state. The kinematic viscosity was 0.1. The simulation showed that the Reynolds number was 23 and the rate of dissipation was 202. The time step interval was 0.00046. We incorporated eight different sizes of random walks, in groups of 512 particles. The standard deviations of the random walks were the following:

0, namely no random walk,
v_k
$5v_k$
$10v_k$
$25v_k$
$50v_k$
$100v_k$
$500v_k$

where v_k is Kolomogorov velocity scale

$(\varepsilon v)^{1/4}$ for a given rate of dissipation ε and kinematic viscosity v.

We traced the eight groups of particles over 1,200 steps. Since we are interested in a difference in the contact rate with and without a turbulent field, we also simulated the motion of particles by a pure random walk with no turbulence. The particles were released from randomly distributed locations (by a uniform distribution) inside the simulation domain (box).

The calculation of contact was done with a second program. Particles that exited the box under consideration were moved back in through the opposite face so that the densities of predator or prey were not changed by particles escaping. When the distance between two particles becomes less than a threshold, the particles are considered to be in contact. We treated two cases for the prey particle after the contact. In the "without replacement" case, the particle (prey) was removed from the box. For the "with replacement" case, we kept the particle "hidden" (unavailable for contact but still moving under the effects of the fluid and the random walk) for a certain interval and then allowed contact again. Hiding the particle was done to avoid artificially large contact rates due to repeated contact between the same two particles. We present the result from the "without replacement" case in this chapter. It was found that the "with replacement" case had a source code problem, and a corrected calculation is required. The "with replacement" calculation is the desirable result, as it is the easiest to interpret.

Since we have eight groups of particles, many combinations of "prey-predator" conditions are possible. We calculated 28 cases of "prey-predator" contacts (Table 8.1). Predators with a large random walk contact all of the prey before the 600th time step, so it is necessary to normalize the number of contacts. Let n be the number of prey contacted, n is less than 512 for the small random walks, and N the total number of predators (512). The time step t, when the last prey is contacted, can be used to average the time interval of searching for a prey. The normalized contact rate γ is defined as follows,

$$\gamma = nT/(Nt)$$

where T is the total of time steps. The upper value of each cell is without turbulence and the lower value is with turbulence (Table 8.1). The six cases for the smallest random

Table 8.1. Contact rates between particles. The heading in each column and row refers to the random walk group. The upper number in each pair of data values is a random walk without turbulence, and the lower number is for random walk plus the turbulence.

Prey	Predator						
	2	3	4	5	6	7	8
1	0.09	0.17	0.33	0.77	0.98	1.39	5.08
	0.25	0.34	0.46	0.78	0.99	2.23	6.67
2		0.16	0.32	0.73	0.99	2.21	6.12
		0.31	0.47	0.79	0.99	1.57	5.22
3			0.32	0.76	0.98	1.44	5.88
			0.46	0.78	0.98	2.46	6.74
4				0.76	0.98	1.73	6.32
				0.79	0.98	1.83	5.45
5					0.99	2.13	4.48
					0.99	1.97	8.11
6						2.33	4.29
						2.05	5.41
7							5.17
							5.61

walks reveal the result we expect—that turbulence should increase the contact rates between prey and predator. The rest of the values indicate that the turbulence field is an insignificant factor for the contact rates. Considerably more work is needed; the numbers show large fluctuations, suggesting the need for more averaging. The preliminary results can be put into a biological context (scaled with ε and ν and related to the analytical results in Rothschild and Osborn (Figure 8.1)).

It is apparent that our results, while encouraging, are still primitive. There are both short-term and long-term aspects to the problem. In the short term, we need to modify and correct the computer code that makes the contact calculations. Some of this has been done already. We can then fully analyze the calculations from the 64^3 model that were performed and saved this in fall 1988. Those results can then be rationalized with the analytical calculations and form the basis for a first paper on the approach and the results.

In the longer term, we need to perform some calculations at a higher Reynolds number, presumably on a 128^3 grid. The planktonic motion needs to be expanded from a simple random walk to a series of flights that last over several time steps of the calculation. This change will increase the amount of calculation, because another time scale is introduced (length of flight of the predator along the same path). It will also become important to incorporate behavioral responses to ingestion in the model. After eating, or when well fed, the predators do not hunt as hard; when starving, searching for food and the ability to escape predators is reduced.

Essentially, the short-term projects are improvements of the technique and verification of the results. The long-term projects involve expanded development and utilization of numerical calculation for studying the interaction of the biological system with the turbulent fluid.

Future Plans

In the future, we hope to accomplish several things with and for this model: (i) produce and develop a contact model to which behavioral and environmental characteristics can be added and their effects quantified; (ii) develop some understanding of the effects

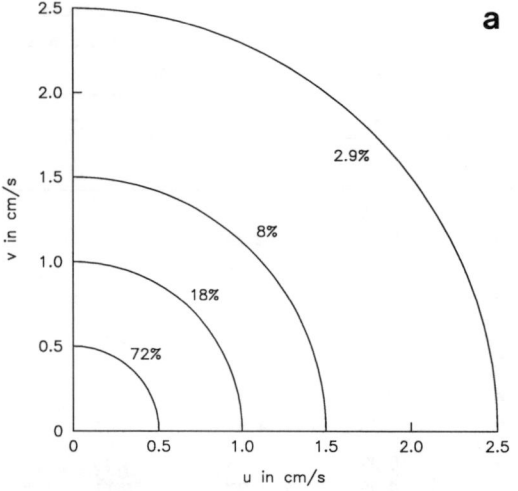
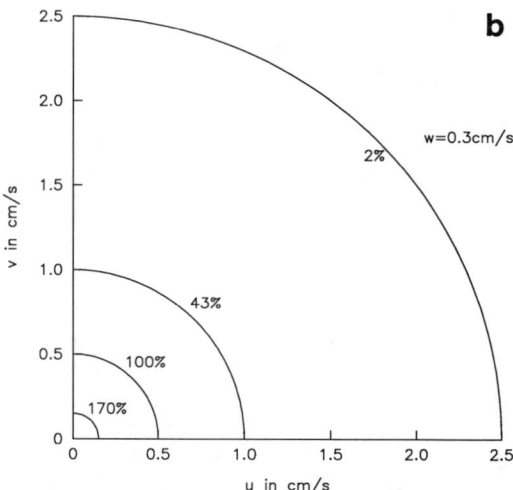

Figure 8.1. (a) Modified Figure 3c from Rothschild and Osborn (1988) showing the increase in contact rate between plankton in random walk with a $0.3 \text{ cm} \cdot \text{s}^{-1}$ turbulent motion. This figure extends to velocities of $2.5 \text{ cm} \cdot \text{s}^{-1}$ where the figure in Rothschild and Osborn only went to $1.5 \text{ cm} \cdot \text{s}^{-1}$. (b) The results for the numerical calculations scaled appropriately to compare with the predictions in (a).

on contact rate of turbulent intensity, particle distributions, capture radius, and behavioral characteristics, among other things; and (iii) look for simplifications of the model based on temporal and spatial correlations. It is the spatial and temporal coherence of the velocity field that requires us to use a "correct" velocity field. If the particles do not exactly follow the water due to inertia, size, and their inherent motility, perhaps the importance of having the correct temporal and spatial coherences for the velocity field is lessened. We might be so fortunate as to only need to track an rms turbulent velocity. These types of questions can be studied with the aim of producing simple estimates of contact rates from species information and physical parameters of the flow.

References

Lee, M. J., and Reynolds, W. C. 1985. Numerical experiments on the structure of homogeneous turbulence. Report No. TF-24 Dept. of Mechanical Engineering, Stanford Univ.

Rothschild, B. J., and Osborn, T. R. 1988. Small-scale turbulence and plankton contact rates. J. Plankton Res. 10(3):465–474.

Chapter 9

Application of Molecular Techniques to the Study of Marine Recruitment Problems

Dennis A. Powers, Fred W. Allendorf, Thomas Chen

Abstract

Two main obstacles to recruitment studies have been the detection of genetic variation within and between species and the assessment of the physiological status of individuals at different life history stages. Molecular techniques can be particularly useful in resolving these problems. The analyses of 5S and 16S RNA, mtDNA, and isozymes have already proven their usefulness in differentiating between species. The latter two methods (mtDNA and isozymes) have also been extremely useful in detecting variation within species. While the development of molecular techniques for the assessment of physiological status is only in its infancy, initial efforts appear promising. The use of RNA-DNA ratios, immunochemical methods, gene specific RNA-RNA and RNA-DNA hybridization, and other molecular methods represent exciting possibilities for future application to recruitment problems in marine ecosystems.

Introduction

Biological oceanographers have recorded the rise and fall of population densities and species compositions for well over a century. Since many of these species impact directly or indirectly on the food supply of man, the causes and predictability of such changes are of societal and scientific concern. In fact, the potential scientific and economic impact is so great that predicting and explaining the mechanisms of secondary production is considered one of the most important problems in biological oceanography.

Most marine organisms produce orders of magnitude more offspring than are needed to replace the parents. Thus, the vast majority of these organisms must die before they become reproductive adults. The individuals that do survive are referred to as recruits. A recruit is a member of any new group and recruitment is defined as the process, action, or state of bringing new individuals into a particular group. Biological oceanographers consider recruitment the process or number of offspring brought into an adult population each year but in a general sense, recruitment applies to the transition between any life history stage and the next (e.g., transitions from eggs to larvae, larvae to juveniles, and juveniles to adults).

Transitions between life history stages entail a degree of mortality. Often the magnitude of this mortality can be staggering. In fact, the probability of survival between egg and adult is often less than a fraction of one percent. It is usually assumed that death is random, especially in relation to genetic background, and that survival is a stochastic process driven primarily by physical, biological, and chemical forces. The ultimate goal is to delineate the detailed interactions between ocean physics, chemistry, and population dynamics, so as to develop a workable model that will allow oceanographers to predict the observed biological variability.

There are a number of different approaches to the study of recruitment. Some biologists focus on the density-dependent spatial and temporal variation of marine organisms, others on the role of environmental parameters, and some biological and physical oceanographers have combined their talents to address the impact of ocean turbulence and other factors on recruitment. Recently, a few scientists have become interested in the importance of genetic variation in the recruitment process and its evolutionary significance.

The importance of genetic variation in critical traits such as survival, growth, development, and age at first reproduction has been demonstrated in a variety of marine organisms. The classic work by Ricker (1981) has shown changes in several important life history characteristics in five species of Pacific salmon (*Oncorhynchus*) that are at least partially due to changes in the genetic composition in these species because of commercial harvesting. Nevo and his co-workers (1985, 1986) have shown that resistance to pollution in marine organisms is associated with genetic variation. It has also been demonstrated that rate of development and hatching of fish is directly related to genetic background (e.g., DiMichele and Powers, 1982a, 1984; DiMichele *et al.*, 1986). In addition, a positive association between the amount of genetic variation and growth rate has been found in many marine species of mollusks (reviewed in Zouros, 1987).

In order to develop a predictive model, we must understand the mechanisms that regulate reproductive success, competitive interactions, predator-prey relationships, and differential survival, including the role of genetic variation. Elucidation of these mechanisms will require (i) a deeper understanding of the genetic architecture of marine species and populations, and (ii) the details of competition and predator-prey interactions.

These issues can be addressed if certain technical difficulties are resolved, such as (i) the identification of species and the detection of genetic variation within a species, and (ii) the assessment of the physiological,

nutritional, developmental, and reproductive status of the marine organisms under study. In many cases, conventional methods can resolve these problems, but sometimes current technology is inadequate or no effective means of resolving these issues is available. In such cases, the application of molecular techniques such as isozyme and immunological methodologies or RNA and DNA techniques may be particularly useful in resolving difficult recruitment problems that were previously unapproachable. However, such approaches are not meant to replace perfectly adequate conventional methods unless they are less expensive and/or more efficient.

This chapter is intended to generate interest among oceanographers regarding the potential for applying molecular techniques toward resolving marine recruitment problems. We provide a few examples of current approaches from the literature, highlight areas of greatest potential for future application, and provide some examples where current biotechnology in the medical field might be applied to critical oceanographic problems.

Identification of Species and the Detection of Genetic Variation Within Species

The identification of species in biological samples is one of the most important and time-consuming tasks faced by biological oceanographers. In fact, it is often the major rate-limiting step in such studies. The time differential between collecting physical and biological oceanographic data is often several orders of magnitude. This difference in sampling and data analysis is a major roadblock to an integrated recruitment study. Acoustic and automatic image analysis techniques for identifying species have made significant advances in recent years, but the early expectation that they would rapidly replace the trained scientific eye has faded. On the other hand, these approaches are in their infancy and, presumably, significant strides will be made during the next decade. The main advantage of such approaches is the potential application in situ and the simultaneous measurement of physical variables. Although, except for a few selected species in rather simple communities, identification to the species level will probably not be achieved for many years, if ever.

Molecular techniques can be particularly useful in the identification and description of taxonomic units, including species, subspecies, populations, and demes. These techniques can sometimes provide species identification with greater accuracy and speed than conventional approaches. Moreover, some methods can be applied to partially digested gut contents in predator-prey studies and the identification of early life history stages of organisms that cannot be determined by the most skilled oceanographer. Finally, these methods can be applied to delineate the genetic architecture within and between species. Such genetic data can contribute to an understanding of the evolutionary, ecological, and developmental processes that structure populations, and the scale(s) at which these processes work.

There are a number of molecular approaches that can be applied to the identification of species and/or the detection of genetic variation within species. All such methods rely on the ability to discriminate between variants of specific genes. Historically, detection has been directed toward specific gene products (e.g., proteins), but recent developments in biotechnology have focused attention at the DNA and RNA level.

Isozymes

During the last two decades, the introduction of electrophoresis coupled with histochemical staining for the identification of protein electromorphs has uncovered a wealth of genetic variation. This genetic variation is reflected at two fundamental levels: (i) multilocus isozymes that indicate one or more evolutionary ancient gene duplications followed by the divergence of the duplicated protein coding loci; and (ii) allelic isozymes (often referred to as allozymes) that are genetic alternatives at a specific locus.

When there is no genetic variation at a particular locus within a species but substantial variation between species, these multilocus isozymes can be used to detect species differences between larvae or other life stages that may be ambiguous from a morphological perspective. For example, the white perch, *Morone americana*, is extremely difficult to distinguish from the striped bass, *Morone saxatilis* in early larval stages and thus studies of larval recruitment of these species are difficult. Biochemical methods have been developed to assess the relative portion of each species in a given collection of larval fish (Morgan, 1975; Sidell et al., 1978). These methods provide relatively fast and unambiguous species identifications, whereas morphological identification has a significant associated error.

Perhaps even more importantly, isozymes have also revealed cryptic species that were not morphologically distinguishable. The most striking example of this is the work of Shaklee and Tamaru (1981) with Hawaiian bonefish (*Albula*). They identified two sympatric sibling species that were isozymically distinct at 58 out of 84 loci. Other examples of isozymically detected cryptic species have been described in oysters, commercially exploited squids, clupeids in the southwestern Pacific, lizardfish, and mackerels (reviewed in Shaklee, 1983).

Hybridization is another example of the increased power of molecular methods to identify species. Natural hybridization is much more common in fish than in other vertebrates (reviewed in Campton, 1987). The detection of hybridization by morphological criteria suffers from many shortcomings. The basic problem is that interspecific hybrids of fishes are often not morphologically intermediate to their parental taxa. The analysis of isozymes encoded by multiple diagnostic loci is the most sensitive and reliable method to identify hybridization. Hybridization is apparently not as common in marine as it is in freshwater species. However, the use of these and other molecular techniques (e.g., mtDNA) in marine species is likely to reveal more hybridization than has previously been thought to exist (e.g., She et al., 1987).

Fisheries biologists use the "stock" concept as a basis for managing commercially imported marine organisms. Identification of these stocks and the impact of harvest-induced mortality (Fetterolf, 1981) are critical elements in any management regime. While morphological life-history characteristics and other variables have been useful in such studies, the analysis of isozymes and their allelic variants have also been widely employed in the study of fish population structure and stock identification. Moreover, they have often helped resolve problems that could not be resolved by conventional approaches.

The advantage of using isozymes over morphological and other classical variables is that (i) the biochemical phenotype is essentially unaffected by the environment, (ii) the biochemical phenotype of each individual is stable through time, and (iii) the observed genetic variation is usually due to a single gene whose alleles are co-dominantly expressed (Ayala, 1975).

Allozymic description of stock composition of marine fishes has become routine in the last 20 years since the pioneering review of de Ligny (1969). A recent book on population genetics and fisheries management (Ryman and Utter, 1987) extensively reviews the voluminous literature on this subject. The paper — "Which witch is which?" — by Fairbairn (1981) exemplifies the value of these studies. Isozyme analysis of samples from three management areas of witch flounder revealed a total of six genetically distinct stocks that differed with respect to age structure, time of spawning, individual growth rate, and temperature and depth of capture. Existing practices had managed these fish as large homogeneous stock units.

An extension of this approach also allows estimating the geographical origin of fish caught in mixed-stock fisheries. It is essential to be able to identify the origin and proportional contribution of different stocks to a mixed fishery. However, this approach has been very difficult to put into practice (Larkin, 1981). Statistical techniques recently have been developed that allow the use of allelic isozyme variants to estimate the contribution of different stocks to mixed-stock fisheries (reviewed in Pella and Milner, 1987).

Recent papers have also shown the utility of using these techniques to describe the transport of organisms by physical features of the ocean environment. Bucklin et al. (1989) have used genetic markers to describe the transport of a calanoid copepod in coastal filaments off the West Coast of the United States. Similarly, Heath and Walker (1987) have used allozymes to describe the pattern of drift and to identify the geographical spawning origin of larval herring (Clupea harengus) caught in the North Sea.

The potential for genetic differentiation between populations of a species depends upon a number of variables including migration rate, the number of individuals within a population, and natural selection at different loci. Determination of genetic exchange between populations has been estimated by a number of experimental and theoretical approaches. Release/recapture studies have been used to examine the migration of individuals, but there are a number of practical restrictions, questionable assumptions, and theoretical constraints that have limited the usefulness of this approach.

In addition to describing the genetic architecture of marine species, allelic isozymes can be used to estimate the amount of gene flow between populations. Allendorf and Phelps (1981) used population genetic theory (Wright, 1969) and computer simulation to show that the amount of allelic divergence, as measured by Wright's F_{st}, between subpopulations is a function of the absolute number, rather than the fraction, of migrants exchanged. This is an important finding, because it emphasizes the need to know the population size in order to estimate reproductive isolation from allozyme data. Allendorf and Phelps also cautioned that the use of larval data to draw conclusions about divergence of reproducing adults could be greatly misleading.

Slatkin (1985) has developed a powerful method for using the frequency of rare alleles to estimate gene flow between populations by determining the average number of migrants exchanged. This approach is both useful and relatively insensitive to changes in any parameter except the number of migrants exchanged and the number of individuals sampled per population.

Waples (1987) recently has used Wright's F_{st} and Slatkin's (1985) method to estimate gene flow in 10 species of marine shore fishes. Estimates of gene flow were highly correlated with the dispersal ability of the species. Waples concluded that genetic differentiation among these fishes was primarily determined by gene flow and genetic drift, rather than by natural selection.

Sometimes the gene frequencies of certain enzyme-synthesizing loci are correlated with directional changes in specific environmental variables like temperature, salinity, and oceanic circulation. In such cases, extensive studies have been launched to elucidate whether one or more of these environmental parameters has acted as a selective force to favor one allelic alternative over the other. Two examples of this approach are provided by the works of Koehn and his colleagues on the Leucine amino pepidase (*Lap*) locus in the mussel, *Mytilus edulis*, and Powers and his colleagues (1988) on the heart type lactate dehydrogenase locus (*Ldh-B*) from the fish, *Fundulus heteroclitus*.

In both examples, a combination of genetics, physiology, and biochemistry was used to demonstrate that selection was acting on these loci. In the former case, Koehn's group showed that salinity and/or temperature were differentially affecting the allelic isozymes of the *Lap-1* locus (e.g., Koehn *et al.*, 1976; Koehn and Hilbish, 1987). In the latter case, Powers and his collaborators showed that the LDH-B allelic isozymes were functionally different (Place and Powers, 1979, 1984), and that these differences allowed them to make predictions concerning differential cell metabolism, developmental rates (DiMichele and Powers, 1982a, 1984), and swimming performance (DiMichelle and Powers, 1982b). Those predictions were substantiated by experimentation and shown to be of selective value in field hatching studies (DiMichele *et al.*, 1986).

Mitochondrial DNA (mtDNA) for intra- and interspecies studies

While the study of allelic isozymes has been useful for the analysis of many species, lack of observable genetic variation in a host of other important species has hindered its wider application to recruitment problems. This has been partly due to the fact that methods such as this grossly underestimate the extent of genetic variability as a result of isopolar amino acid changes and changes in the nucleotides that are not reflected in the protein sequence.

The application of molecular tools by population biologists has revealed a wealth of previously hidden genetic variabilities. Not only do these methods permit the study of a broader array of genetic diversity, but their sensitivity makes it possible to study egg and juvenile stages as well as tissue biopsies of adults.

The most common application by population biologists has been the use of endonuclease restriction digests of mitochondrial DNA (mtDNA) followed by electrophoresis and the construction of mtDNA restriction maps. Since mtDNA is usually between 16 and 19 kilobases, it is small enough to map with an array of 4 and 6 base sequence-specific endonuclease restriction enzymes. The most parsimonious assumption for a change in a restriction site is a single nucleotide change (i.e., a point mutation).

Because mtDNA is maternally inherited, individuals within and between populations can be studied and various matriarchal lineages can be followed geographically, including contributions to new lineages and magnitude of contributions by a particular population to a fishery.

While mtDNA (or any "single locus") could theoretically be used to study gene flow, data derived from natural populations could provide misleading patterns of differentiation because of historical "accidents." The added resolution of mtDNA analysis over allozymes cannot overcome the fact that the entire mtDNA genome is inherited as a single unit.

Nevertheless, mtDNA may be an extremely valuable tool in regard to identifying the geographical origin of individuals caught in a mixed fishery. In this case, we are not trying to understand the evolutionary forces responsible for such a pattern (e.g., differential gene flow, selection). Therefore, the increased resolving power to elucidate genetic divergence between populations makes mtDNA valuable for identifying the geographical origin of individuals. Such information could prove valuable as a tool for stock identification and fisheries management.

There are a number of examples where mtDNA has already provided insight concerning the population structure of marine species that could not be resolved by isozyme studies. For example, eels (*Anguilla* spp.) are known to spawn in the Sargasso Sea and the leptocephalus larvae migrate thousands of kilometers to metamorphose in estuarine waters. Given this life history, it would seem a foregone conclusion that genetic uniformity would be expected over vast regions of the Americas and Europe. However, this has been a debatable point for many years.

The evidence from isozyme studies has been equivocal. Williams *et al.* (1973) found little interlocality variation among elvers but some interlocality differences among adults in North American eels. In more extensive studies (Koehn and Williams, 1978; Williams and Koehn, 1984), it was argued that the small gene frequency differences among populations were due to natural selection. European scientists argued to the contrary.

Comparini and Rodino (1980) have concluded on the basis of allozyme frequency differences that North American and European eels should be considered separate species. However, Williams and Koehn (1984) have argued that their work with North American eels suggests that the amount of allele frequency divergence between European and North American eels indicates only partial reproductive isolation between eels from the two continents.

Avise and his colleagues (1986) resolved this conflict with their classic paper on mtDNA differentiation in North Atlantic eels. Their restriction site polymorphism study of mtDNA showed no genetic divergence among the eels (*Anguilla rostrata*) along the coast of North America and suggested they were all members of a single panmictic population. However, they found that samples of the European eel (*Anguilla anguilla*) were significantly different from those studied along the North American coast.

Bermingham and Avise (1986) used mtDNA restriction length polymorphisms to study the zoogeography of four species of fresh water fish: *Amia calva, Lepomis punctatus, Lepomis gulosus,* and *Lepomis microlophus.* They found that within each species, major mtDNA phylogenetic discontinuities distinguished populations from different geographical areas. From these data, they concluded that dispersal and gene flow were inadequate to override the historically driven geographical changes in sea level and level morphology. Similarly, Gonzalez and Powers (1990), used mtDNA restriction fragment data to show that a previous barrier to gene flow existed between two groups of the teleost, *Fundulus heteroclitus.* This finding allowed them to distinguish between two alternative models, primary and secondary intergradation (Powers and Place, 1978; Powers *et al.,* 1986; Ropson *et al.,* 1990), that helped explain the existence of a series of gene frequency clines along the east coast of North America (reviewed by Powers *et al.,* 1986; Ropson *et al.,* 1990).

There have been many attempts to identify discrete populations of striped bass, (*Morone saxatilis*). Some studies have focused on morphological features (e.g., Setzler *et al.,* 1980), while others have used allelic isozyme variants (Morgan et al., 1973; Grove et al., 1976; Sidell et al., 1980; Rogier et al., 1985). While these studies have been able to delineate populations from the Hudson River, Chesapeake Bay, and Albemarle Sound, attempts to discriminate between spawning grounds of the Chesapeake Bay have been less compelling and have tended to generate opposing conclusions. A major problem with these studies was that striped bass are among the most homozygous vertebrates known and, thus, limit one's ability to discriminate between stocks. Chapman (1987, 1989) has used mtDNA restriction length polymorphism studies to resolve this problem.

Those mtDNA studies provided evidence for discrete stocks of striped bass within the Chesapeake Bay and between the Bay and the Dan River in North Carolina that generally agrees with morphological and some isozyme studies.

Saunders et al. (1986) used mtDNA analysis to show that the horseshoe crab (*Limulus polyphemus*) could be separated into at least two different groups and that the area of divergence was associated with a zoogeographic boundary between warm temperate and tropical marine faunas. Thomas et al. (1986) used mtDNA to study intra- and interspecific variation for rainbow trout and five species of salmon. While an inadequate sampling of populations limited the intraspecies studies, interesting interspecific divergences were observed. Similar intra- and interspecific mtDNA studies have been performed on three trout species. Wilson et al. (1985) demonstrated significant divergence between all populations and the expected evolutionary divergence between trout species.

On the other hand, Kornfield and Bogdanowicz (1987) studied the mtDNA of Atlantic herring (*Clupea harengus*) from the Gulf of Maine and the Gulf of St. Lawrence. They concluded that their data did not support the notion that these were separate genetic stocks.

In addition to the analysis of restriction length polymorphisms of mtDNA, the use of DNA sequence analysis, including mtDNA, is being increasingly employed to study the phylogenetic relationships among marine taxa. While this approach has been slow and rather tedious in the past, the recent introduction of the polymerase chain reaction method (see following section and review by Marx, 1988) to increase the amount of a specific piece of DNA promises to dramatically increase the applicability of this approach in the near future.

DNA "fingerprinting"

Individual-specific fingerprints of human DNA have been very useful in parenthood testing. Tandem-repetitive regions of DNA, called minisatellites, are dispersed throughout the genome of a number of organisms. Jeffreys et al. (1985) showed that a subset of human minisatellites shared a common 10-15-base pair core that had hypervariable regions. Later they demonstrated that a hybridization probe could detect highly polymorphic minisatellites that could be used as DNA fingerprints specific to an individual. A number of laboratories are currently examining marine organisms to see if they have similar minisatellites or other DNA regions that could be used in population studies. The results are very promising (e.g., see Fields et al., 1989).

Nucleic acid hybridization

As stated previously, in recruitment studies, a common problem is species or stock identification at larval stages. Nucleic acid hybridization may be the method of choice provided suitable signature gene probes are available. These "signature gene probes" could be portions of the organism-specific or species-specific DNA or RNA.

DNA probes are often applied to the resolution of phylogenetic questions, but their greatest immediate potential remains as tools for species identification. It is possible to identify single organisms, even single cells, using DNA or RNA oligonucleotide probes that target unique, signature sequences. This technique will be particularly useful for discriminating among organisms that lack distinguishing morphological features. Included in this category are most microorganisms and many larval metazoa. If appropriate sequences are identified, the same approach can be used for distinguishing between members of subpopulations.

A variety of chemical markers can be used to label nucleic acid probes — the most useful are fluorescent, radioactive, or enzymatic (e.g., alkaline phosphatase). For example, single cells have been effectively tagged and microscopically counted using DNA probes complementary to ribosomal RNA signature sequences. Because rRNA is present in multiple copies, they are much easier to detect than single-copy genes. Nonetheless, the detection of a single copy of a gene in a single cell is clearly possible, especially when it is amplified, as will be described in the next section.

A novel approach for counting and identifying species in a community containing a variety of species is the application of in situ hybridization. Radio-labeled nucleic acid probes with complementary sequences can hybridize with DNA inside fixed cells or in fixed thin sections of tissues. These probes can then be visualized by autoradiography, enzyme-linked probes, or another type of reporter group. This approach is particularly attractive because it is easily applicable to field situations.

DNA Amplification via the Polymerase Chain Reaction (PCR)

In order to do restriction length polymorphism studies of mtDNA, repetitive elements, or sequence analysis, an adequate amount of DNA is required. Often, material is not limiting, but when small organisms or only a few cells are available, obtaining enough DNA can become a stumbling block. Under such conditions, the desired DNA is usually cloned and analyzed directly or used as a probe.

Normally, when one wants a substantial amount of a specific piece of DNA, it is cloned into an active replicon, identified by some specific method, and then produced in large quantities in bacteria. However, the initial cloning and identification methods can be very time consuming, especially when the target sequence is not abundant and the starting sample is a complex mixture.

A new and revolutionary approach to the detection and characterization of specific DNA sequences, termed the Polymerase Chain Reaction (PCR), was developed by the Cetus Corporation a few years ago as a simple alternative to the cloning of specific genes. Recently, an important modification of this method has expanded the technique's usefulness (Saiki et al., 1988; Stoflet et al., 1988).

The current procedure employs a thermostable bacterial DNA polymerase and specific oligonucleotide primers to replicate a target DNA sequence in vitro. Amazingly, from as little as a single molecule of the target sequence, enough material for standard analytical procedures (e.g., restriction mapping or DNA sequencing) is produced in about three hours. A microprocessor-controlled device, commercially available from PerkinElmer, automatically takes the reaction through multiple cycles of DNA denaturation/primer annealing/DNA synthesis: The target sequence may be amplified from many samples in parallel, and the entire process is fully automated. This method — and its application to medicine, forensic science, and fundamental molecular biology — was highlighted in the "News and Comments" section of Science (Marx, 1988).

Compared to molecular cloning, the PCR technique is far more rapid (hours vs. weeks) and much more sensitive. Problems that can be addressed include (i) rapid detection and identification of rare microorganisms and minute individual fish larvae invertebrates, (ii) rapid analysis of individual genomes for population studies, (iii) detection and analysis of "rare events" (e.g., gene rearrangements) that occur in a small fraction of the cells in a tissue sample or field collection. The PCR method can also be used to amplify a rare DNA sequence obtained in a complex sample mixture (e.g., total DNA extracted from a mixed phytoplankton or zooplankton community). Moreover, the PCR method appears to work on preserved samples and even on partially degraded samples obtained from gut contents in predator/prey studies. In some cases, the application of the PCR amplification technique to recruitment studies will require the initial identification of suitable regions of the genome, but for many genes, primers (i.e., universal or kingdom) are already available.

Summarily stated, the PCR extends the sensitivity and the rate of molecular biological data acquisition to the extent that entirely new questions may be addressed.

RNA sequencing to distinguish microorganisms

RNA sequencing is a very useful tool for the identification of different species of microorganisms and also for the analysis of their phylogenetic relationships. In addition, quantitative analysis can provide information on species composition of microbial communities (Olsen et al., 1986; Pace et al., 1986).

With some exceptions, the 5S rRNA from most microbial species contains about 115–120 nucleotides that can be directly sequenced by cleavage with base-specific enzymes (Donis-Keller et al., 1977) or chemicals (Peattie, 1979). Briefly, determination of 5S rRNA sequence involves isolation by gel fractionation, ^{32}P-end labeling, cleavage with base-specific enzymes or reagents, and finally separation of RNA fragments on high-resolution polyacrylamide gels. About 400 (see Kuntzel et al., 1983) 5S rRNA sequences have been collected and the data indicate that 5S rRNA has a highly conserved primary structure. When variation does occur, it is not randomly distributed (i.e., some positions change more freely than others).

Although this method has been used successfully to investigate the phylogenetic relationships of several types of microbial communities (Stahl et al., 1984, 1987), the paucity of independently varying nucleotide positions of 5S rRNA limits its phylogenetic usefulness. Beyond phylogenetic trees, 5S rRNA is a suitable and unique model to study relationships between structural constraints, biological function, and evolution.

The 16S rRNA (about 1600 nucleotides) is an appropriate size for broader phylogenetic analysis. Due to its size, the complete sequence of 16S rRNA cannot be determined easily by the methods described earlier. Before the development of DNA cloning and modern nucleotide-sequencing methodologies, the sequence of 16S rRNA was partially determined by the so-called "oligonucleotide cataloging" method using RNase T1 (Fox et al., 1977). Employing this method, Woese and colleagues (e.g., Fox et al., 1977) have characterized the 16S-like rRNAs from over 300 organisms and organelles.

Recently, by the use of advanced nucleotide-sequencing technologies, the full nucleotide sequences of the 16S rRNAs have been determined from about 25 diverse organisms. This information, coupled with the partial 16S sequences accumulated by Woese and collaborators, provides a reasonably detailed picture of the molecule in terms of its primary and secondary structures. Examination of the 16S rRNA sequences reveals regions of constant or nearly constant nucleotide sequences across taxa. The oligonucleotides of these conserved regions have been synthesized. These oligonucleotides are used as primers to provide a fast and direct approach for the determination of 16S rRNA partial sequences. Application of the PCR method to 16S RNA genes promises an explosion of sequences in the near future.

To analyze the population contents of microbial communities by 16S rRNA sequence comparisons, the strategy of shotgun cloning of 16S rRNA genes is used. In this method, DNA is purified from collected biomass, shotgun cloned into the bacteriophage lambda, and individual rRNA genes isolated as recombinant bacteriophages by hybridization with the "mixed kingdom" 16S rRNA probe. Using one of the three synthetic primers in the dideoxynucleotide-chain-termination reactions, the nucleotide sequences of individual rRNA genes are determined. Alternatively, the PCR method can be used to bypass cloning in some instances, and the amplified piece of DNA can be directly sequenced. By referring to existing collections of complete and partial sequences, it is possible to infer the phylogenetic affinities of the organisms in the original community (Vaughn et al., 1984). In addition, the individually cloned rRNA genes can also be used as in hybridization probes to quantitate the species composition of the mixed community.

Immunochemical methods for identifying intra- and interspecies variation

Immunochemical detection of species or population-specific antigens (e.g., proteins, carbohydrates) on the surface of organisms are very useful in recruitment studies. Polyclonal and/or monoclonal antibodies can often be used in food chain studies, the identification of minute larvae, and the analysis of population structure.

Polyclonal antibodies, which are most useful for interspecies studies, are prepared by immunizing an animal with a protein that is specific for a particular species. The serum of the immunized animal can then be used to detect small amounts of the species in question. This method is useful in studying predator-prey interactions by analyzing the gut contents of predators. Moreover, proteins isolated from large adults can be used as antigens to generate antibodies that are useful in detecting and quantitating species-specific larvae when the antigenetic determinants are shared between adults and larvae. This approach can allow the identification of minute larvae that are undetectable by less sensitive methods.

Since polyclonal antibodies are usually of inadequate specificity for population studies, monoclonal antibodies are often preferred. Preparation of a series of specific monoclonal antibodies generally involves the following steps: (i) several mice are immunized with desired antigens (purified or partially purified), and spleen cells are prepared from these mice several weeks after immunization; (ii) the spleen cells are fused to mouse myeloma cells and the hybrid cells are selected and propagated on HAT-selective medium; (iii) monoclonal antibody-producing hybridoma clones are screened by an immunobinding assay, using purified antigen as a probe. Each monoclonal antibody is then characterized extensively in order to determine its specificity to the respective antigen.

While the antibody provides the taxonomic specificity, detection is usually provided by either a direct or indirect coupling to one of the numerous reporter systems. Sometimes a radiolabeled antibody (e.g., ^{125}I-IgG) is used as the reporter. However, some investigators prefer to use a fluorescent label, an enzyme-linked assay (e.g., alkaline phosphotase), or another method to detect the specificity of the reaction. Whatever the method, this approach is potentially one of the most powerful and practical for application to field studies and particularly for the identification of phytoplankton, zooplankton, eggs, larvae, and in food web studies.

Assessing the Physiological, Nutritional, Developmental, and Reproductive Status of Marine Organisms

The physiological, nutritional, developmental, and reproductive status of marine organisms is dependent on both physical and biological parameters. In order to evaluate these variables, one needs a series of biological indicators and appropriate standards. Some methods are currently available, others are being developed, but in many cases, there is a lack of sensitive, accurate, and practical techniques to address specific problems associated with recruitment. It is therefore important to determine the effectiveness and limits of the present technology, but equally important to develop needed technology. The biomedical field is a fruitful reservoir of technology that can be applied to recruitment problems.

Physiological condition is related to the environment by the regulation of gene products in response to environmental stimuli. Since it is possible to test for the expression of these genes and their products (i.e., proteins and mRNA), the development of molecular techniques to quantitate such changes as indices of physiological status would greatly facilitate recruitment studies. Techniques that would assess growth, reproductive, and nutritional status would be particularly useful. At present, very few attempts have been made to develop these molecular techniques for marine organisms. While we shall address a few of these techniques, this area of research is in its infancy and there is a tremendous need for fundamental research. The potential applicability of molecular techniques to this research area is almost without limits.

Nutritional status

A number of variables can be used to determine nutritional status. Length to weight, length to width, hepato-somatic index, color, relative parasite infestation, gut contents, and other measurements are routinely used to assess relative nutrition status. The types and concentration of digestive enzymes, in conjunction with gut contents, have been used as crude indicators of food preference, diet, and general nutritional status (e.g., Patton et al., 1975; Sargent et al., 1979; Seiderer et al., 1987).

In attempts to understand phytoplankton/zooplankton food webs, a number of useful laboratory studies have been done, but field studies have been less useful. Although field studies on phytoplankton allow one to rapidly measure carbon fixation and other quantitative parameters to assess physiology and growth, such studies on zooplankton have been so variable that they are of questionable use. Recently, the quantitation of digestive enzymes (Cox et al., 1983) has been used to determine long-term changes in copepods' feeding. These methods, coupled with gut fluorescence (Mackas and Bohrer, 1976; Dagg and Wyman, 1983) and conventional methods, have proven useful. However, these and related methods are very time consuming, and the results are often affected by confinement. Hakanson (1984) developed a rapid and accurate method that relies on lipid analysis in copepods (i.e., *Calanus pacificus*) as an index of feeding conditions. He found that wax ester content was an excellent index of feeding over a several day period, and triglyceride content was a good index of recent feeding activity. A recent paper (Hakanson, 1987) showed that, using this approach, it is possible to relate lipid content (presumably feeding conditions) of copepods to field phytoplankton pigment.

Salmonid fishes subjected to long periods of starvation initially catabolize non-essential digestive enzymes (Moon, 1983; Loughna and Goldspink, 1984), after which lipid reserves are mobilized (Love, 1980; Mommsen et al., 1980). During starvation, increased concentrations of lipolytic enzymes are commonly found in a variety of fish species (Zammit and Newsholme, 1979; Black and Skinner, 1986). Although growth hormone activity increases lipolytic enzyme activity, it was still surprising when Wagner and McKeown (1986) found that growth hormone levels of starved fish were greater than those of fed controls.

Therefore, decreases in digestive enzymes reflect early phases of starvation, and increases in lipolytic enzymes and growth hormones reflect longer term starvation. It is interesting that Barrett and McKeown (1988) have shown that swimming amplifies this latter effect in starved fish, making growth hormone levels almost four times greater than in the exercised fed controls. It follows that measurements of these enzymes and the determination of plasma growth hormone levels could be good indicators of a fish's nutritional status as long as proper standards and controls were available. However, it is important to point out that growth hormone levels alone would be very misleading, because enhanced levels of this and other hormones are also associated with differential development and enhanced growth of well-fed organisms. On the other hand, the use of a suite of indicators might define various nutritional states. For example, the coupling of conventional methods for assessing nutrition with the biochemical indicators described previously (digestive enzymes, lipolytic enzymes, RNA/DNA, and growth hormones) could be extremely useful, but the details and applicability of such an approach is yet to be defined.

Growth status

Growth is intrinsically coupled to nutritional status as well as to life history stage. However, we shall consider it separately in the hope that it will help clarify some of the special features associated with assessing growth status.

Growth rate studies are normally done in the laboratory by regulating food intake and measuring size and/or weight on a periodic basis. However, these are time consuming, and impractical for field studies. In recent years, attempts have been made to develop simple biochemical techniques to assess growth that could be applied to field samples. Incorporation of ^{14}C-glycine into fish scales, RNA/DNA ratios, RNA content per individual, and tissue-specific RNA concentration have all been used as indicators of growth, metabolism, and physiological condition. While these techniques have limitations, they are potentially useful if proper controls and standards can be developed. One of the more popular techniques, RNA-DNA ratios, is useful when restricted by species, size, life history stage, environmental temperature, and physical activity of the individuals being tested (reviewed by Bulow, 1987). While this technique has limited potential for interspecies application, it can be useful for intraspecies studies of similar life history stages.

The incorporation of ^{14}C-glycine into fish scales in vitro (Ottoway and Simkiss, 1977) was used by Ottaway (1978) to study rhythmic growth activities of fish. This method has been suggested as a useful technique for determining current growth rate in fish (e.g., Busacker and Adelman, 1987). However, as pointed out by Ottaway (1978), the uptake of ^{14}C-glycine into fish scales depends on a number of variables including the in vitro culture conditions, the part of the body from which the scales are taken, and the day and time of year that the fish were caught. This latter point is particularly important because growth hormone exhibits a circadian rhythm in fish blood (Leatherland et al., 1974) and a seasonal cycle in the pituitary (Swift and Pickford, 1965). Moreover, as pointed out earlier, fish growth hormone levels are also affected by nutritional status. Thus, while the incorporation of ^{14}C-glycine into fish scales in vitro has potential as a method to estimate growth status, the standards and limitations of the technique need to be clearly defined before it is applied to field recruitment studies.

When laboratory growth rate studies are done on fish with different amounts of growth hormone, growth rates are easily differentiated (Sekine et al., 1985; Agellon et al., 1988). Recently, it was shown that increased growth rate is directly related to increased RNA/DNA ratio, growth hormone levels, and serum testosterone (Danzmann et al., in press). However, those studies were done under a defined feeding regime and at constant environmental conditions (e.g., temperature). The obvious questions are: Can one estimate the relative growth rate or growth potential of an organism sampled from the field by quantitating such variables as the growth hormone level, testosterone, RNA/DNA, and other parameters? If so, can such data be related to nutritional status.

Consistent with the previous discussion

on nutritional status, one needs to be able to differentiate between increased growth hormone levels that might be enhancing growth and elevated levels that are the result of long-term starvation and physical activity. Coupling a suite of techniques (e.g., conventional methods, enzyme levels, growth hormone concentrations, incorporation of ^{14}C-glycine into fish scales in vitro, RNA/DNA measurements) may resolve these alternatives, but the standards and detailed protocols are yet to be delineated (see preceding discussion concerning rhythmic nature of growth). Clearly, research to explore the potential of these and other molecular techniques to assess growth and nutrition status could make field recruitment studies more definitive.

Detecting reproductive status

Classically, gametogenic capacity is estimated by counting changes in gamete numbers or gonad weight (i.e., gonadosomatic index). While these are useful for estimating reproductive capacity, gamete counts are excessively labor intensive, and gonad weight indices are often misleading. Moreover, such approaches are limited to a few taxa. A possible alternate method for detecting reproductive status might be the assessment of specific reproductive genes like vitellogenin in vertebrate females.

During oogenesis in fish, the egg-yolk precursor protein (vitellogenin) is synthesized in the liver, secreted into the vascular system, and then deposited in the developing oocytes as lipovitellin and phosvitine (Chen, 1983). Therefore, the reproductive status of female fish in a population might be determined by measuring the levels of vitellogenin in the serum, the rate of vitellogenin synthesis in the liver, or the accumulation of vitellogenin mRNA in the liver. Levels of vitellogenin in serum samples can be easily determined quantitatively by the rocket immunoelectrophoresis. This method can detect vitellogenin levels as low as 0.05 mg $\cdot$ ml^{-1} in serum and give reliable quantitation of vitellogenin in both hormone-induced and reproductively active female fish. The rates of vitellogenin synthesis in livers of reproductively active females can be determined by either a radioimmunoprecipitation method (Chen, 1983) or by RNA-DNA hybridization. It should be emphasized that this approach has not been proven; rather, it remains a possibility that needs the type of documentation necessary for any new methodology.

Physiological stress assessment

When faced with environmental stress, organisms must adapt in order to maintain homeostasis. As a result, animals and plants have devised a host of molecular, physiological, and behavioral strategies in order to accomplish this task. Since aquatic poikilotherms usually respond directly to such environmental parameters as temperature, pH, metals, salinity, and pollutants they have been excellent models for the study of adaptation to environmental stress.

In all organisms studied, acute exposure to high-temperature stress results in the rapid activation of a specific family of genes and the efficient and preferential synthesis of a novel set of gene products referred to as "heat-shock" proteins (hsps) (Atkinson and Walden, 1985). A role for hsps in thermal resistance adaptation stems from the observation that induction of hsps will allow cells to transiently tolerate subsequent exposure to temperatures that would be lethal in the absence of hsp synthesis and accumulation (Atkinson and Walden, 1985). There are positive correlations between the temperatures at which hsps are maximally induced, the upper lethal temperature of an organism, and the range of temperatures normally encountered in the habitat of the organism. For example, salmonid fishes are stenothermal and are adapted to relatively cold temperatures. Cell lines derived from these fish maximally synthesize hsps at temperatures below 30°C (Kothary and Candido, 1982; Gedamu et al., 1983). Hepatocytes of a eurythermal and warm temperature-adapted species such as channel catfish will maximally synthesize hsps at temperatures approaching 40°C (Koban et al., 1987). A variety of tissues of a eurythermal teleost, *Fundulus heteroclitus*, demonstrate hsp synthesis consistent with the upper temperature of its thermal environment, but with regulatory responses that vary between tissues (Koban et al., submitted).

The molecular basis by which hsps provide thermal stress resistance is not known, but the universal expression of hsps and their conserved structures across various taxa from bacteria to humans suggests that these proteins must be fundamental in the maintenance of cellular homeostasis. In addition to heat stress, a variety of other perturbations of homeostasis — such as heavy metal exposure, viral infections, synthesis of defective proteins, and oxygen radical formation — will activate hsp genes (Schlesinger et al., 1982; Atkinson and Walden, 1985).

Since increases in hsp synthesis are a good indication of physiological stress, it is assumed that research focused on quantitative estimates of hsp in target species would be a useful indicator for environmental stress in recruitment studies. Several molecular techniques are available that can be applied to that end, including hsp immunoblotting, immunoprecipitation, rocket immunoelectrophoresis, RNA blotting, and Northern analysis (i.e., electrophoresis followed by RNA/DNA hybridization).

It is important to point out that the strong evolutionary conservation of the hsp70 gene complex has allowed the use of antibodies and cDNA probes from diverse taxa (e.g., fruit flies, fish, and man) to be used to probe each others' proteins and RNA. Therefore, a general method developed for one species should have broad applicability to the detection of these stress proteins in other species. However, appropriate standards and protocols have to be developed for each taxonomic group.

Summary

The study of recruitment is an important scientific pursuit with critical societal overtones. As we strive to explain the population dynamics of marine species, we need to acquire a deeper understanding of the underlying mechanisms that drive these processes. In order to delineate the complex predator/prey and competitive interactions between species, it is necessary to (i) identify species, (ii) detect genetic variation within a species, and (iii) assess the physiological, nutritional, developmental, and reproductive status of marine organisms. In many cases, conventional approaches will suffice, but in some instances, molecular approaches may prove valuable. Isozymes, mtDNA, immunological, and other methods such as DNA and RNA probes have already proven their usefulness in detecting genetic variation within and between species. The challenge for the future is to expand the usefulness of these methods and to develop new techniques to quantitatively assess the physiological, reproductive, and nutritional status of marine organisms. We have presented only a few of the numerous possibilities.

References

Allendorf, F. W., and Phelps, S. R. 1981. Use of allelic frequencies to describe population structure. Can. J. Fish. Aquat. Sci. 38:1507–1514.

Agellon, L. B., et al. 1988. Promotion of rapid fish growth by a recombinant fish growth hormone. Can. J. Fish. Aquat. Sci. 45:146–161.

Atkinson, B. G., and Walden, D. B. 1985. Changes in eukaryotic gene expression in response to environmental stress. Academic Press, Orlando, FL. 379 pp.

Avise, J. C., Helfman, G. S., Saunders, N. C., and Hales, S. 1986. Mitochondrial DNA differentiation in North Atlantic eels: Population genetic consequences of an unusual life history pattern. Proc. Nat. Acad. Sci. 83:4350–4354.

Ayala, F. J. 1975. Molecular evolution. Sinauer Assoc., Inc., Sunderland, MA. 277 pp.

Barrett, B. A., and McKeown, B. A. 1988. Sustained exercise augments long-term starvation increases in plasma growth hormone in the steelhead trout, *Salmo gairdneri*. Can. J. Zool. 66:853–855.

Bermingham, E., and Avise, J. C. 1986. Molecular zoogeography of freshwater fishes in the southeastern United States. Genetics 113:939–965.

Black, D., and Skinner, E. R. 1986. Features of the lipid transport system of fish as demonstrated by studies on starvation in the rainbow trout. J. Comp. Physiol. B. 156:497–502.

Bucklin, A., Rienecker, M. M., and Mooers, C. N. K. 1989. Genetic tracers of zooplankton transport in coastal filaments off northern California. J. Geophys. Res. 94:8277–8288.

Bulow, F. J. 1987. RNA-DNA indicators of growth in fish: A review. *In* The age and growth of fish. pp. 45–64. Ed. by R. C. Summerfelt and G. E. Hall. Iowa State Univ. Press, Ames.

Busacker, G. P., and Adelman, I. R. 1987. Uptake of ^{14}C-glycine by fish scales (in vitro) as an index of current growth rate. *In* The age of growth of fish. Ed. by R. C. Summerfelt and G. E. Hall. Iowa State Univ. Press, Ames.

Campton, D. E. 1987. Natural hybridization and introgression in fishes: Methods of detection and genetic interpretations. *In* Population genetics and fisheries management. pp. 161–192. Ed. by N. Ryman and F. M. Utter. Univ. Washington Press, Seattle.

Chapman, R. W. 1987. Changes in the population structure of male striped bass, *Morone saxatilis*, spawning in three areas of the Chesapeake Bay from 1984–1986. Fish. Bull. U.S. 85:167–170.

Chapman, R. W., 1989. Spatial and temperal variations of mtDNA haplotypes frequencies in the striped bass 1982 year class. Copeia 1989:344–348.

Chen, T. T. 1983. Identification and characterization of estrogen-responsive gene products in the liver of the rainbow trout. Can. J. Biochem. Cell Biol. 61:802–810.

Comparini, A., and Rodino, E. 1980. Electrophoretic evidence for two species of Anguilla leptocephali in the Sargasso Sea. Nature 287:435–437.

Cox, J. L., Willason, S., and Harding, L. 1983. Consequences of distributional heterogeneity of *Calanus pacificus* grazing. Bull. Mar. Sci. 33:213–226.

Dagg, M. J., and Wyman, K. D. 1983. Natural ingestion rates of the copepods *Neocalanus plumchrus* and *N. cristatus* calculated from gut contents. Mar. Ecol. Prog. Ser. 13:37–46.

Danzmann, R. G., Van der Kraak, G. J., Chen T. T., and Powers, D. A. In press. The metabolic effects of bovine growth hormone and biosynthetic rainbow trout growth hormone and rainbow trout reared at high temperature. Can. J. Fish. Aquat. Sci.

de Ligny, W. 1969. Serological and biochemical studies on fish populations. Oceanogr. Mar. Biol. 7:411–513.

DiMichele, L., and Powers, D. A. 1982a. LDH-B genotype- specific hatching times of *Fundulus heteroclitus* embryos. Nature 296:563–564.

DiMichele, L., and Powers, D. A. 1982b. Physiological basis for swimming endurance difference between LDH-B genotypes of *Fundulus heteroclitus*. Science 216:1014–1016.

DiMichele, L., and Powers, D. A. 1984. Developmental and oxygen consumption rate differences between *Ldh-B* genotypes of *Fundulus heteroclitus* and their effect on hatching time. Physiol. Zool. 57:52–56.

DiMichele, L., Powers, D. A., and DiMichele, J. 1986. Developmental and physiological consequences of genetic variation at enzyme synthesizing loci in *Fundulus heteroclitus*. Am. Zool. 26:201–208.

Donis-Keller, H., Maxam, A., and Gilbert, W. 1977. Mapping adenines, guanines and pyrimidines in R or A. Nucleic Acids Res. 4:2527–2538.

Fairbairn, D. J. 1981. Which witch is which? A study of the stock structure of witch flounder (*Glyptocephalus cynoglossus*) in the Newfoundland region. Can. J. Fish. Aquat. Sci. 38:782–794.

Fetterolf, C. M., Jr. 1981. Foreword to the Stock Concept Symposium. Can. J. Fish. Aquat. Sci. 38:iv-v.

Fields, R. D., Johnson, K. R., and Thorgaard, G. H. 1989. DNA fingerprints in rainbow trout detected by hybridization with DNA of bacteriophage M13. Trans. Am. Fish. Soc. 118:78–81.

Fox, G. E., Pechman, K. R., and Woese, C. R. 1977. Comparative cataloging of 16 S ribosomal ribonucleic acid: Molecular approach to prokaryotic systematics. Lit. J. Syst. Bacteriol. 27:44–57.

Gedamu, L., Culham, B., and Heikkila, J. J. 1983. Analysis of the temperature-dependent temporal pattern of heat-shock-protein synthesis in fish cells. Biosci. Rep. 3:647–658.

Gonzalez-Villasenor, L. I., and Powers, D. A. 1990. Mitochondrial DNA restriction site polymorphisms in the teleost *Fundulus heteroclitus* supports secondary intergradation. Evolution 44:27–37.

Grove, T. L., Berggren, T. S., and Powers, D. A. 1976. The use of innate genetic tags to segregate spawning stocks of striped bass (*Morone saxatilis*). Estuar. Process. 1:166–176.

Hakanson, J. L. 1984. The long and short term

feeding condition in field-caught *Calanus pacificus*, as determined from the lipid content. Limnol. Oceanogr. 29:794–804.

Hakanson, J. L. 1987. The feeding condition of *Calanus pacificus* and other zooplankton in relation to phytoplankton pigments in the California Current. Limnol. Oceanogr. 32:881–884.

Heath, M. R., and Walker, J. 1987. A preliminary study of the drift of larval herring (*Clupea harengus* L.) using gene-frequency data. J. Cons. int. Explor. Mer 43:139–145.

Jeffreys, A. J., Wilson, V., and Thein, S. L. 1985. Individual-specific "fingerprints" of human DNA. Nature 316:76–79.

Koban, M., Graham, G., and Prosser, C. L. 1987. Induction of heat-shock protein synthesis in teleost hepatocytes: Effects of acclimation temperature. Physiol. Zool. 60:290–296.

Koban, M., Yup, A. A., Agellon, L. B., and Powers, D. A. Submitted. Molecular adaptation to the thermal environment. Heat-shock response of the eurythermal teleost *Fundulus heteroclitus*. J. Comp. Physiol.

Koehn, R. K., and Hilbish, T. J. 1987. The adaptive importance of genetic variation. Am. Sci. 75:134–141.

Koehn, R. K., and Williams, G. C. 1978. Genetic differentiation without isolation in the American eel, *Anguilla rostrata*. II. Temporal stability of geographic patterns. Evolution 32:624–637.

Koehn, R. K., Milkman, R., and Mitton, J. B. 1976. Population genetics of marine pelecypods. IV. Selection, migration and genetic differentiation in the blue mussel *Mytilus edulis*. Evolution 30:2–32.

Kornfield, I., and Bogdanowicz, S. M. 1987. Differentiation of mitochondrial DNA in Atlantic herring, *Clupea harengus*. Fish. Bull. U.S. 85:561–568.

Kothary, R. K., and Candido, E. P. M. 1982. Induction of a novel set of polypeptides by heat shock or sodium arsenite in cultured cells of rainbow trout, *Salmo gairdneri*. Can. J. Biochem. 60:347–355.

Kuntzel, H., Piechulla, B., and Hahn, U. 1983. Consensus structure and evolution of 5S rRNR. Nucl. Acids Res. 11:893–900.

Larkin, P. A. 1981. A perspective on population genetics and salmon management. Can. J. Fish. Aquat. Sci. 38:1469–1475.

Leatherland, J. F., McKeown, B. A., and John, T. M. 1974. Circadian rhythm of plasma prolactin, growth hormone, glucose and free fatty acid in juvenile kokanee salmon *Oncorhynchus nerka*. Comp. Biochem. Physiol. 47:821–828.

Loughna, P. T., and Goldspink, G. 1984. The effects of starvation upon protein turnover in red and white myotomal muscle of rainbow trout, *Salmo gairdneri*. J. Fish. Biol. 25:223–230.

Love, R. M. 1980. The chemical biology of fishes. Vol. 2. Academic Press, New York.

Mackas, D. L., and Bohrer, R. 1976. Fluorescence analysis of zooplankton gut contents and an investigation of diet feeding patterns. J. Exp. Mar. Biol. Ecol. 25:77–85.

Marx, Jean L. 1988. Multiplying genes by leaps and bounds. Science 240:1408–1410.

Mommsen, T. P., French, C. J., and Hochachka, P. W. 1980. Sites and patterns of protein and amino acid utilization during the spawning migration of salmon. Can. J. Zool. 58:1785–1799.

Moon, T. W. 1983. Metabolic reserves and enzyme activities with food deprivation in immature American eels, *Anguilla rostrata* (LeSueur). Can. J. Zool. 61:802–811.

Morgan, R. P. II. 1975. Distinguishing larval white perch and striped bass by electrophoresis. Chesapeake Sci. 16:68–70.

Morgan, R. P. II, Koo, T. S. Y., and Krantz, G. E. 1973. Electrophoretic determination of populations of striped bass, *Morone saxatilis*, in the Chesapeake Bay. Trans. Am. Fish. Soc. 102:21–32.

Nevo, E., Noy, R., Lavie, B., Beiles, A., and Muchtar, S. 1986. Genetic diversity and resistance to marine pollution. Biol. J. Linnean Soc. 29:139–144.

Nevo, E., Noy, R., Lavie, B., and Muchtar, S. 1985. Levels of genetic diversity and resistance to pollution in marine organisms. FAO Fish. Rep. No. 352 Suppl., pp. 175–182.

Olsen, G. J., Lane, D. J., Giovannoni, S. J., and Pace, N. R. 1986. Microbial ecology and evolution: A ribosomal RNA approach. Am. Rev. Microbiol. 40:337–65.

Ottoway, E. M. 1978. Rhythmic growth activity in fish scales. J. Fish. Biol. 12:615–623.

Ottoway, E. M. and Simkiss, K. 1977. "Instantaneous" growth rates of fish scales and their use in studies of fish populations. J. Zool., Lond. 181:407–419.

Pace, N. R., Stahl, D. A., Lane, D. J., and Olsen, G. T. 1986. The analysis of natural microbial populations by ribosomal RNA sequences. Adv. Microbiol. Ecol. 9:1–55.

Patton, J. S., Nevenzel, J. C., and Benson, A. A. 1975. Specificity of digestive lipases in hydrolysis of wax esters and triglycerides studied in Anchovy and other selected fish. Lipids 10(10):575–583.

Peattie, D. A. 1979. Direct chemical method for

sequencing RNA. Proc. Natl. Acad. Sci. USA 76:1760–17645.

Pella, J. J., and Milner, G. B. 1987. Use of genetic marks in stock composition analysis. In Population genetics and fisheries management. pp. 247–276. Ed. by N. Ryman and F. M. Utter. Univ. Washington Press, Seattle.

Place, A. R., and Powers, D. A. 1979. Genetic variation and relative catalytic efficiencies: Lactate dehydrogenase B allozymes of Fundulus heteroclitus. Proc. Nat. Acad. Sci. USA 76(5):2354–2358.

Place, A. R., and Powers, D. A. 1984. Kinetic characterization of the lactate dehydrogenase (LDH-B$_4$) allozymes of Fundulus heteroclitus. J. Biol. Chem. 259(20):1309–1318.

Powers, D. A., and Place, A. R. 1978. Biochemical genetics of Fundulus heteroclitus (L.). I. Temporal and spatial variation in gene frequencies of Ldh-B, Mdh-A, Gpi-B, and Pgm-A. Biochem. Gen. 16(5,6):593–607.

Powers, D. A., Chen, T. T., DiMichele, L., Chapman, R. W., and Gonzalez-Villasenor, L. I. 1988. A molecular approach to recruitment problems: genetics and physiology. In Toward a theory on biological-physical interactions in the world ocean. pp. 411–440. Ed. by B. J. Rothschild. Kluwer, Dordrecht, The Netherlands.

Powers, D. A., et al. 1986. Genetic variation in Fundulus heteroclitus: Geographic distribution. Am. Zool. 26:131–144.

Ricker, W. E. 1981. Changes in the average size and average age of Pacific salmon. Can. J. Fish. Aquat. Sci. 38:1636–1656.

Rogier, C. G., Ney, J. J., and Turner, B. J. 1985. Electrophoretic analysis of genetic variability in a landlocked striped bass population. Trans. Am. Fish. Soc. 114:244–249.

Ropson, I., Brown, B., and Powers, D. A. 1990. Biochemical genetics of Fundulus heteroclitus VI Geographic variation and the gene frequencies of 15 loci. Evolution 44:16–26.

Ryman, N., and Utter, F. M. (Editors.) 1987. Population genetics and fisheries management. Univ. Washington Press, Seattle.

Saiki, R. K., et al. 1988. Primer-directed enzymatic amplification of DNA with a thermostable DNA polymerase. Science 239:487–491.

Sargent, J. R., McIntosh, R., Bauermeister, A., and Blaxter, J. H. S. 1979. Assimilation of the wax esters of marine zooplankton by herring (Clupea harengus) and rainbow trout (Salmo gairdneri). Mar. Biol. 51:203–207.

Saunders, N. C., Kessler, L. G., and Avise, J. C. 1986. Genetic variation and geographic differentiation in mitochondrial DNA of the horseshoe crab, Limulus polyphemus. Genetics 112:613–627.

Schlesinger, M. J., Ashburner, M., and Tissieres, A. 1982. Heat shock, from bacteria to man. Cold Spring Harbor Laboratory Press. Cold Spring Harbor, NY. 440 pp.

Seiderer, L. J., Davis, C. L., Robb, F. T., and Newell, R. C. 1987. Digestive enzymes of the anchovy Engraulis capensis in relation to diet. 35:15–23.

Sekine, S. T., et al. 1985. Cloning and expression of cDNA for salmon growth hormone in E. coli. Proc. Nat. Acad. Sci. USA 82:4306–4310.

Setzler, E. M., et al. 1980. Synopsis of biological data on striped bass, Morone saxatilis. FAO Synopsis 121:69 pp.

Shaklee, J. B. 1983. The utilization of isozymes as gene markers in fisheries management and conservation. In Isozymes: Current topics in biological and medical research, Vol 11. pp. 213–247. Alan R. Liss Publ. Co., New York.

Shaklee, J. B., and Tamaru. 1981. Biochemical and morphological evolution of Hawaiian bone fishes. Systematic Zool. 30:125–146.

She, J. X., Autem, M., Kotulas, G., Pasteur, N., and Bonhomme, F. 1987. Multivariate analysis of genetic exchanges between Sole aegyptiaca and Solea senegensis (Teleosts, Soleodae). Biol. J. Linnean Soc. 32:357–371.

Sidell, B. D., Otto, R. G., and Powers, D. A. 1978. A biochemical method for distinction of striped bass and white perch larvae. Copeia 1978(2):340–343.

Sidell, B. D., Otto, R. G., Powers, D. A., Karweit, M., and Smith, J. 1980. A reevaluation of the occurrence of subpopulations of striped bass (Morone saxatilis, Walbaum) in the upper Chesapeake Bay. Trans. Am. Fish. Soc. 109:99–107.

Slatkin, M. 1985. Rare alleles as indicators of gene flow. Evolution 39:53–65.

Stahl, D. A., Lane, D. J., Olsen, G. J., and Pace, N. R. 1984. Analyses of hydrothermal vent-associated symbionts by ribosomal RNA sequences. Science 224:409–411.

Stahl, D. A., et al. 1987. Phylogenetic analysis of certain sulfide-oxidizing and related morphologically conspicuous bacteria by SS ribosomal ribonucleic acid sequences. Int. J. Syst. Bacteriol. 37:116–133.

Stoflet, E. S., Koeberi, D. D., Sarkar, G., and Sommer, S. S. 1988. Genomic amplification with transcript sequencing. Science 239:491–494.

Swift, D. R., and Pickford, G. E. 1965. Seasonal variations in hormone content of the pituitary gland of perch (Perca fluviatilis). Gen. Endroc. 5:5354–5365.

Thomas, W. K., Withler, R. E., and Beckenbach, A. T. 1986. Mitochondrial DNA analysis of Pacific salmonid evolution. Can. J. Zool. 64:1058–1064.

Vaughn, J. C., Sperbeck, S. J., and Hughes, J. M. 1984. Molecular cloning and characterization of ribosomal RNA genes from the brine shrimp. Biochem. Biophys. Acta. 783:144–151.

Wagner, G. F., and McKeown, B. A. 1986. Development of a salmon growth hormone radioimmunoassay. Gen. Comp. Endocrinol. 62:452–458.

Waples, R. S. 1987. A multispecies approach to the analysis of gene flow in marine shore fishes. Evolution 41:385–400.

Williams, G. C., and Koehn, R. K. 1984. Population genetics of North Atlantic catadromous eels (*Anguilla*). *In* Evolutionary genetics of fishes. pp. 529–560. Ed. by B. J. Turner. Plenum Press, New York.

Williams, G. C., Koehn, R. K., and Mitton J. B. 1973. Genetic differentiation without isolation in the American eel, *Anguilla rostrata*. Evolution 27:192–204.

Wilson, G. M., Thomas, W. K., and Beckenbach, A. T. 1985. Intra- and inter-specific mitochondrial DNA sequence divergence in Salmo: Rainbow, steelhead, and cutthroat trouts. Can. J. Zool. 63:2088–2094.

Wright, S. 1969. Evolution and the genetics of populations. Vol. II. The theory of gene frequencies. Univ. Chicago Press, Chicago.

Zammit, V. A., and Newsholme, E. A. 1979. Activities of enzymes of fat and ketone-body metabolism and effects of starvation on blood concentrations of glucose and fat fuels in teleost and elasmobranch fish. Biochem. J. 184:313–322.

Zouros, E. 1987. On the relation between heterozygosity and heterosis: An evaluation of the evidence from marine mollusks. *In* Isozymes: Current topics in biological and medical research, Vol. 15. pp. 255–270. Ed. by N. C. Rattazi, J. G. Scandalios, and G. S. Whitt. Alan R. Liss Publ. Co., New York.

Chapter 10

Application of Image Analysis in Demographic Studies of Marine Zooplankton in Large Marine Ecosystems

Mark S. Berman

Abstract

Changes in large marine ecosystems (LMEs) can be observed by measuring key biological, chemical, and physical components over the entire range of the LME at frequent intervals. Parameters such as primary and secondary production, zooplankton abundance, and species dominance are basic to our understanding of the structure and function of the LME. Advances in modern technology can provide these sorts of data with greatly increased efficiency. Computerized image analysis is proving to be a powerful tool for the analysis of zooplankton community structure. A prototype image analyzer, designed jointly by the University of Rhode Island and National Marine Fisheries Service (NMFS), counts, measures, and identifies planktonic animals to major taxonomic groups about eight times faster than conventional analysis. A high-resolution television camera captures images of preserved zooplankton and sends them to a parallel-processing computer network. The computers measure a suite of 40 size-and-shape parameters from each animal, and use discriminant analysis to identify the taxon of the animal. Examples of the application of this device to oceanographic research will be discussed in this chapter.

Introduction

As the study of marine ecosystems has become more sophisticated, it has become clear that techniques developed early in the century when aquatic ecology was basically descriptive will no longer suffice. Dragging nets through the water and laboriously identifying and counting the phyto- and zooplankton caught therein provides valuable information and is still a standard technique used throughout the world. It does, however, have serious limitations. It requires skilled, dedicated technicians, trained to recognize the large number of species and developmental stages found in typical plankton samples and able to look into microscopes day after day, fighting eye strain and boredom to develop the basic data needed to assess aquatic ecosystems. The costs of training and salaries for these people is high, and turnover of personnel is great. The time needed to process each sample can range up to eight hours.

The net result is that marine ecosystem research has been costly and limited by the difficulty in obtaining the basic data needed for analysis. It is not uncommon for plankton samples gathered during research cruises to be archived uncounted. As we start to use LMEs as basic management units, such problems become more serious. The areas of interest are much larger than those previously studied, and the number of plankton samples needed to be collected and examined has grown proportionally. New technologies are needed to supply the data necessary to characterize the ecosystems and the key trophic interactions controlling them in a timely and efficient manner. Two promising approaches are multispectral acoustic samplers (Pieper and Holliday, 1984) and computerized image analysis, discussed in this chapter. Both techniques speed the counting and measurement of plankton, but with a consequence of reduced information about the identification of individual animals. In the case of the acoustic sampler, no specific data are collected; advanced image analyzers are capable of group-level taxonomic identifications. Both techniques promise to speed progress in our understanding of the structure and function of LMEs. However, both also require the development of new ways to view and assess the data.

When research became focused on the functioning of the aquatic ecosystem in the 1970s, new technologies were developed to facilitate the needed analysis. Usually, these produced large amounts of data quickly, but acceptance of these devices was slow, often because the data produced were not the same as those traditionally collected. A prime example of this was the use of the electronic gate (Coulter) counter for analysis of the phytoplankton community. This device measures the volume of particles passing through a small aperture by recording the changes in electrical resistance as the particles displace an electrolyte solution. The gate counter is able to count and size thousands of cells or particles per minute, and reports the data as a size frequency spectrum. Phytoplankton specialists trained to think about the phytoplankton community in terms of numbers of individual species made limited applications of the new technology, often using it only to measure concentrations of monospecific cultures. Zooplankton ecologists, however, used the gate counters to measure the size-frequency spectra of the phytoplankton before and after copepods had grazed on it. Comparison of these data allowed conclusions to be drawn relating to the filtering rates of the copepods and their abilities to feed selectively on certain sized particles.

The application of a new technology allowed rapid progress in our understanding of the trophic interactions between phytoplankton and zooplankton, despite the fact that specific information about the makeup of the phytoplankton community was not obtained. Results of some of these studies have been called into question because of the possibility that the measurements of particle volume can be affected by the shape of the particle (Harbison and McAlister, 1980), and the possibility that fragments resulting from copepods breaking up large cells or chains lowered the apparent feeding rates on small particles (Deason, 1980). Such issues have limited use of electronic gate counters in copepod feeding studies to very specific applications, but the conclusions drawn from the original work have, for the most part, been confirmed by other techniques. While the electronic gate counter was of limited value to those who wanted to study the composition of the phytoplankton community, it proved to be a valuable tool for researchers willing to develop new techniques for exploring the trophic interactions on the lower end of the marine food chain.

Image analysis as an oceanographic research tool is now at a stage analogous to that of the gate counter in the early 1970s. New and better technology is being developed, and for the first time, off-the-shelf image analyzers, designed specifically for marine biological applications, are becoming available commercially. Early work in this field was accomplished using devices intended for work in other fields, often metallurgical analysis. One such system, the Quantimet 720, is discussed in detail by Fawell (1976). Although these devices had basically the same components as modern image analyzers, image quality was variable, speed was much less than the present standard, and complex image tailoring and measurement schemes were absent. The result was a system capable of rapidly counting the number of objects in the field of view of a television camera and making measurements of area somewhat more slowly. A major problem with many of these systems was that they were poorly suited to imaging the semitransparent specimens in a zooplankton sample. Thus, they required a great deal of operator intervention, including manually setting threshold and contrast levels for each screen, and often pointing to specific objects of interest with a light pen. This slowed the time of analysis, resulting in a questionable gain over traditional techniques.

Modern oceanographic image analyzers were designed to alleviate these problems, but in most cases, the data are limited to size-frequency distributions. Some prototype image analyzers have limited ability to recognize taxonomic groups (Jeffries et al., 1984; Chehdi et al., 1986); even in these cases, the usefulness of the image analyzer depends on the willingness of researchers to go beyond the traditional analysis of the ecosystem based on species count to more dynamic models based largely on particle size and distribution (Sheldon et al., 1972; Gerritsen and Strickler, 1977; Steele and Frost, 1977; Beyer and Laurence, 1980).

In following sections of this chapter, I will describe the machine designed by a joint NMFS–University of Rhode Island (URI) research project. Later sections describe results of pilot studies using this device, and a final section will cover the direction in which image analysis should progress to meet its full potential as a tool for analysis of LMEs.

Technology

An image analyzer is composed of three basic sections: image acquisition, frame grabber, and computer. In the NMFS/URI analyzer, the image acquisition section is built around a high-contrast Newvicon television scanner and a three-axis mechanical stage. The sample is placed on the stage and viewed by the scanner. The system is constructed so that a dissecting microscope (for most zooplankton), macro lenses (for large fish scales, otoliths, and krill), or an inverted compound microscope (for phyto- and bacterioplankton) can be mounted on the camera to increase the size range of objects studied. Dark-field illumination is used with the dissecting microscope, and diffuse substage transmitted light with the lens systems. A black-and-white image is sent from the camera to the frame grabber, where it is digitized and stored on a 256×256 pixel matrix. The image is then thresholded, with each point set to black or white, instead of its original gray level. This binary image is then sent to the computer for analysis. The computer system first tailors the image, then extracts features. Image tailoring includes defining the edges of all objects in the field, checking to see that each object is completely within the field of view (a closed curve), and within a previously set size range, and, finally, smoothing the image. Smoothing is accomplished through three serial erosions and expansions. Its effect is to eliminate narrow parts of the image, typically swimming appendages and antennae. These body parts fall in random orientation in preserved samples and thus add a large element of uncertainty to measurements and computerized taxonomic identification; the elimination of these elements from the image serves to increase the accuracy of the automated analysis.

After the image is tailored it is subjected to a series of measurements. In the NMFS/URI prototype, 40 parameters, including simple morphometrics, moment-invariant functions, and Fourier descriptors are recorded in a computer disk file. These parameters can be used subsequently to classify the animals to taxonomic group with discriminant analysis. At the magnifications we typically employ for continental shelf zooplankton samples, the machine can identify nine to twelve taxonomic groups, including copepods, chaetognaths, fish larvae, fish eggs, pteropods, amphipods, and euphausiids.

Because the amount of analysis involved in thresholding, tailoring, and feature extraction of an image consisting of over 65,000 pixels is extensive, the engineers who designed this system decided to increase its speed by using a parallel-processing computer network, instead of one central processor. The network was implemented with a series of custom designed satellite computers, each complete on a single board, consisting of a Motorola 68000 CPU, a floating point processor, static and dynamic memories, switching, control, and I/O circuitry. Each satellite works independently, completing all processing on a specific image. Thus, when three satellites are in the system, three separate images are being processed simultaneously, effectively tripling speed of the device. When a satellite has finished processing an image (usually containing three to five planktonic animals) it dumps the data file into disk memory, orders the automated stage to move to the next sampling point, and signals the frame grabber to sample the video stream, restarting the cycle.

Speed of this system is dependent on several factors including the size and complexity of the animals being studied. With three satellites, it usually requires about an hour to process a sample of 300–500 animals. Accuracy of taxonomic identification varies with the types of animals in the sample and magnification. Typical classification results based on test sets not used to create the discriminant functions average better than 90% correct (Table 10.1).

Results

We have conducted several studies with this system to meet the dual purposes of furthering our understanding of marine ecosystems and assessing the capabilities of the image analyzer so that improvements can be made in this and future designs. I will describe two of these studies to demonstrate the flexibility of the system and the diverse applications for which it can be used.

A suite of zooplankton samples comprising a transect from the Providence River, through Narragansett Bay and Rhode Island Sound to the edge of the continental shelf south of New England (Figure 10.1), was assembled from NMFS's MARMAP program, and an estuarine sampling program run by Dr. Perry Jeffries of URI. Each sample consisted of an oblique tow using a 253 μm mesh net. Samples were analyzed both through conventional techniques and the NMFS/URI prototype image analyzer. Temperature, salinity, and depth show a classic estuarine pattern along the transect (Figure 10.2), with an increase in bottom temperature indicative of slope water at the outermost station. Analysis of the plankton community is based on size-frequency distributions of animals identified by the image analyzer as copepods (Figure 10.3). The results demonstrate two characteristics of the community: (i) increasing diversity along the transect from estuarine to open water conditions, and (ii) the tendency for the size of the animals to increase offshore. The change in diversity is evident in the increasing number of peaks, and increasing range exhibited in the size frequency spectra as the transect progresses offshore. The increasing size along the transect is shown clearly in Figure 10.4. Mean copepod size at each station is significantly different at the $p = 0.05$ confidence level except for the adjacent Rhode Island Sound and Block Island stations, which showed no significant difference. Two samples departed from the relationship between copepod size and station location, the midshelf and midbay stations. Data logs indicate that the net had clogged with salps at the midshelf station, which

Table 10.1. Accuracy of classification into major taxonomic groups. Based on nine parameters (% correct).

	Classified to:				
From:	Chaetognath	Copepod	Fish egg	Fish larva	Euphausiid
Chaetognath	<u>100</u>	0	0	0	0
Copepod	0	<u>98</u>	0	0	2
Fish egg	0	7	<u>93</u>	0	0
Fish larva	0	8	0	<u>78</u>	14
Euphausiid	0	2	0	15	<u>82</u>

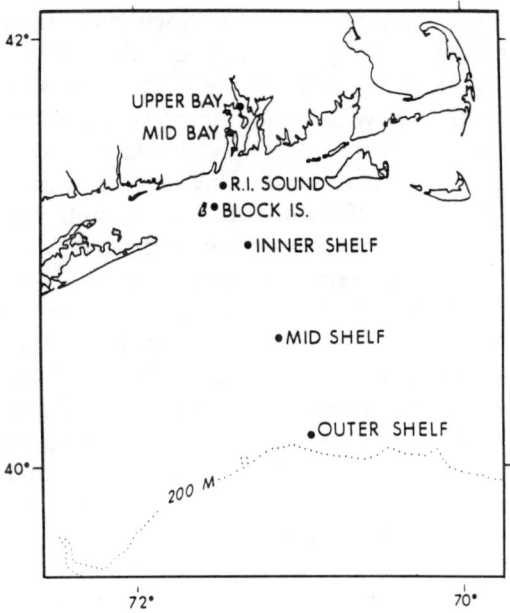

Figure 10.1. Station locations of the transect from Narragansett Bay (Rhode Island) to the edge of the continental shelf.

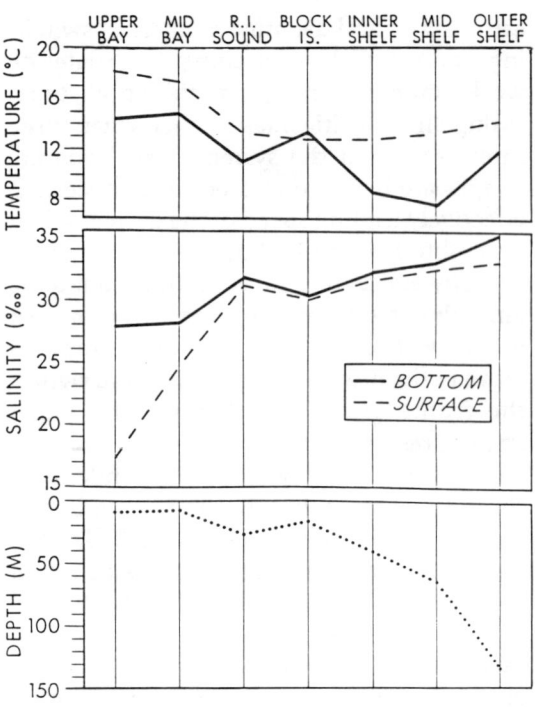

Figure 10.2. Depth, salinity, and temperature profiles (surface and bottom water) at each station along the transect.

likely caused undersampling of the large *Calanus finmarchicus,* thus reducing mean copepod size in the sample. The anomalous mean size in the midbay sample was due to the analysis rather than the sampling. The discriminant function used to classify the animals in these samples forced identifications into one of five groups shown in Table 10.1. These groups were selected with the expectation that they would represent all but a few of the plankters caught. This was not the case at the midbay station, however. Figure 10.5 shows that approximately 60% of the midbay sample was made up of cladocerans, a group not represented in the training set. The cladocerans were misidentified as copepods, thus incorrectly reducing the calculated mean size of the copepods. This problem has since been remedied by adding cladocerans to the training sets used to analyze estuarine samples.

While it is true that use of this sort of automated analysis results in a loss of specific taxonomic identification, the loss need not be too much of a setback in many cases.

A detailed knowledge of the system being studied combined with manual identification of the animals in occasional samples will generally allow the peaks on size-frequency spectra to be identified to genus or species. For example, in the outer-shelf sample, almost all copepods belonged to two species: 75% were *C. finmarchicus* and 24% were *Metridia lucens* (Table 10.2). The size-frequency spectrum for this sample was made up of two peaks, one centered at 1.2 mm and a second, broader peak with a mode at 2.1 mm (Figure 10.3). Integrating the area under each curve indicates that about 30% of the copepods in the sample belong with the smaller peak, and the remaining 70% with the larger. Thus, a plankton ecologist familiar with the area could assume, based on the position of the station, time of year, size of the modes, and perhaps manual counts at another station in the area, that the copepod community at this station was dominated by *C. finmarchicus* and *M. lucens* at about a 2.5:1 ratio.

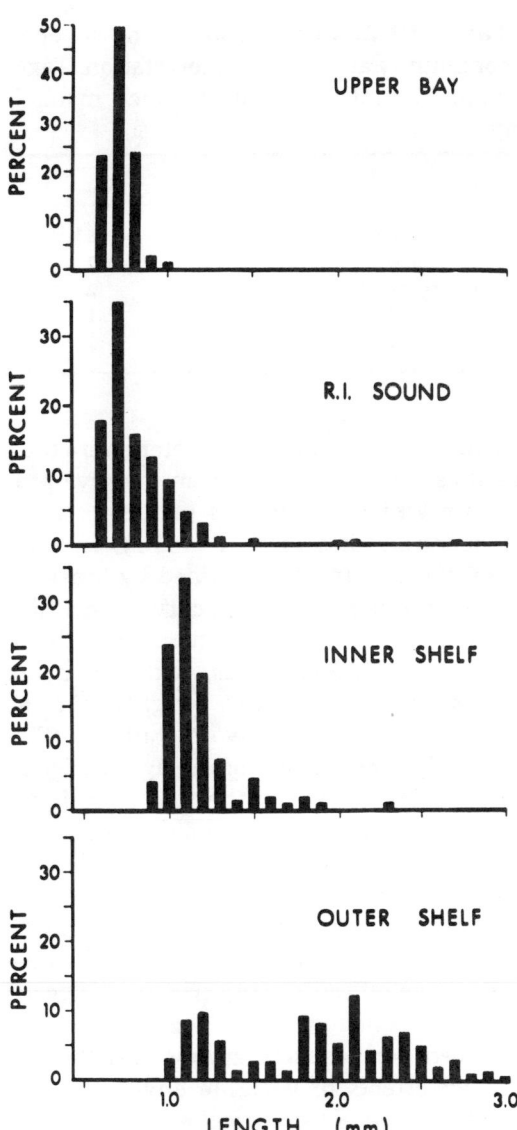

Figure 10.3. Size-frequency spectra of the animals identified by the image analyzer as copepods. Alternate stations were left off this figure for clarity.

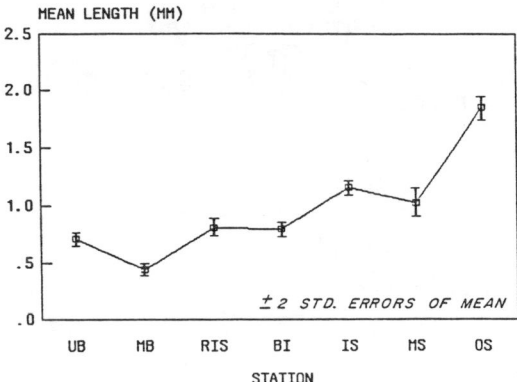

Figure 10.4. Mean length of copepods at each station along the transect.

In this example, the image analyzer was used in its fully automated mode, with the sample being moved from one field to the next by the mechanical stage operating under computer control. There is also a manual mode for this device, used for larger objects that can be handled efficiently one at a time. In this mode, an object is placed on the stage and scanned, and a false color image is displayed on the monitor. The operator at this point must judge whether the image is acceptable (i.e., illumination, contrast, and positioning all set correctly to yield an image suitable for analysis) and either reset the image or command the computer to continue. The manual mode was used in the next example, in which the image analyzer was used to help deduce the age structure of a sample of Antarctic krill.

Determining the age of krill has proved to be a contentious problem. Differing interpretation of length-frequency data has led to longevity estimates ranging from two to four years (Ruud, 1932; Marr, 1962; Ivanov, 1970; Macintosh, 1972). More recent evidence that krill may shrink during nonfeeding winter periods (Ikeda and Dixon, 1982) calls into question all conclusions based solely on size-frequency. Ettershank (1983) suggested that fluorescent age pigment (FAP), a metabolic breakdown product that collects in the brain and correlates with physiological age, might be an alternative parameter for determining age of krill. The NMFS/URI image analyzer was used to relate morphometric characteristics with the age of krill as determined through both length-frequency and FAP analysis in order to find a rapid technique for age determination (Berman *et al.*, 1989).

Krill were obtained from one haul taken in Prydz Bay, Antarctica, during the SIBEX II

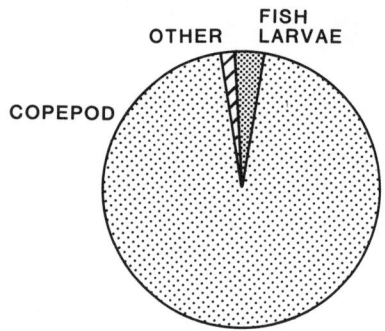

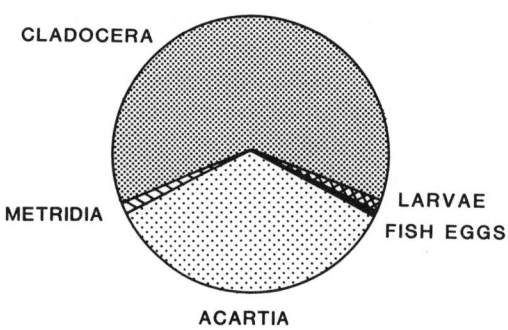

Figure 10.5. Composition of the plankton community at the midbay station as identified by the image analyzer (**top**) and manual identification (**bottom**).

Table 10.2. Composition of the copepod community at the outer shelf station. Taxonomic identifications determined manually.

Species	No.
Calanus finmarchicus	196
Metridia lucens	62
Centropages typicus	2
Oithona similis	1
Pseudocalanus minutus	1

cruise, January 1985, and were fixed and stored according to Steedman (1976). The animals were sexed, weighed, and measured manually, their images analyzed, and FAP extracted using the technique of Ettershank (1984a). Data shown here are based on the 252 adult female krill analyzed.

Smoothed size frequency and fluorescence curves were similar in shape (Figure 10.6). A two-stage technique was used to break each curve up into its component overlapping normals (each normal presumably representing one year class). First, Cassie's (1954) graphical technique was used to determine the number of normals and estimate the mean, standard deviation, and proportion of each. These parameters were then used as starting points for an iterative procedure which maximizes goodness of fit (Macdonald and Pitcher, 1979). In each case, the best fit was obtained by breaking the distribution up into six curves. The third age group was the most frequent, in both the length and FAP analyses, with other age groups accounting for roughly similar proportions using the two techniques (Figure 10.7). Comparison of the ages assigned to each individual by the two techniques show that physiological age did not correlate exactly with chronological age (as defined by length-frequency analysis), but there was 75% agreement between them to plus or minus one year. This lack of precise correspondence reflects the uncertainty of age assignments due to the high degree of overlap between adjacent age groups and the unknown effects of variable environmental conditions on physiological age.

Stepwise discriminant analysis of the image analysis data selected eight morphological parameters, including five Fourier descriptors, to represent chronological age, and four parameters, including two Fourier descriptors, to represent physiological age. Discriminant functions based on these parameters had accuracies of 87% for chronological age and 52% for physiological age.

This study, although preliminary, supports recent findings that Antarctic krill have a longer life span than previously estimated (Ettershank, 1983; Ikeda, 1985; Rosenberg et al., 1986). Assuming two years to reach maturity (Ettershank, 1984b) both FAP and

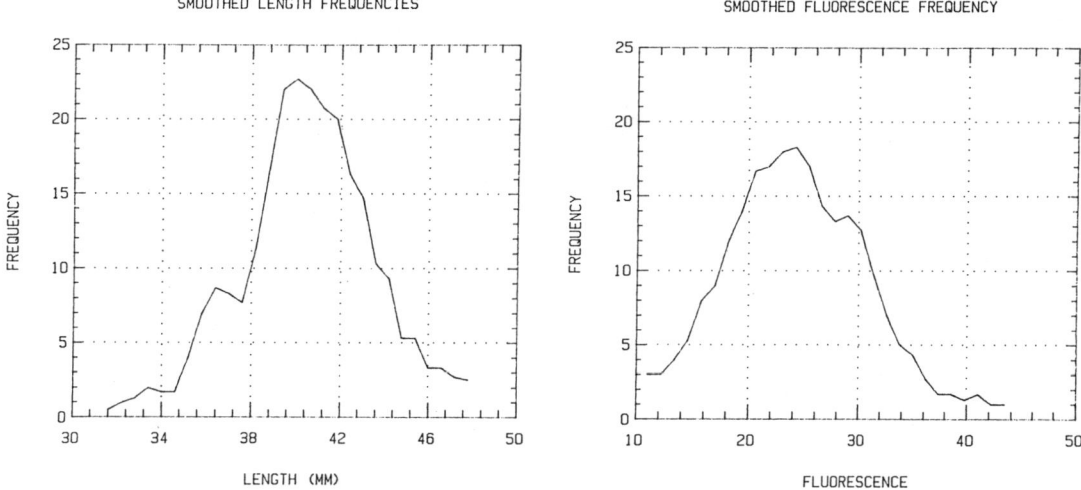

Figure 10.6. Length (**left**) and FAP (**right**) frequency spectra for the adult female krill smoothed with a three-point running average.

length data suggest a life of eight years. Further, it demonstrates that image analysis can be used as a technique for rapid determination of chronological age in krill, making future studies of the life histories of these animals more practical.

Future Directions

Image analysis has the potential to play an increasingly important role in marine science, especially in the analysis of lower trophic levels of large marine ecosystems. However, if this potential is to be realized, it is important to recognize the strengths and weaknesses of the technology, and design future analyzers and applications with these in mind. The most important advantage that an image analyzer can offer is increased speed of data acquisition. All future development of systems for ecosystem applications must maximize this speed. This means ignoring the temptation to develop techniques for species-level identification, if this requires higher magnification and more time-consuming analysis than present devices use. Perhaps if there are major breakthroughs in computer design and technology, this will become a realistic goal but, for the present, we should concentrate our efforts on speeding group-level identifications as well as developing new ways to analyze communities and intertrophic relationships through size-frequency spectra.

Interestingly enough, it appears that even with present technology, it is not the speed of the computer that limits the productivity of the image analyzer, but rather the sample-handling technique. After a known aliquot is extracted from a plankton sample, it is necessary that each animal be counted only once. Thus we have developed techniques for scanning aliquots that ensure this. The simplest is dividing the aliquot into a large number of discrete portions in a spot plate. The magnification is selected so that each spot exactly fits one field of view, thus eliminating the problem of animals divided between two frames. This technique is effective, but requires a great deal of operator intervention, filling each spot and changing the plate every few minutes. A better solution is one that involves the scanning of a continuous sample. We have tried several such schemes, but all involve dealing with animals divided between two fields of view. At best, this requires time-consuming techniques to rectify, and use of these techniques is often confounded by the motion of the ani-

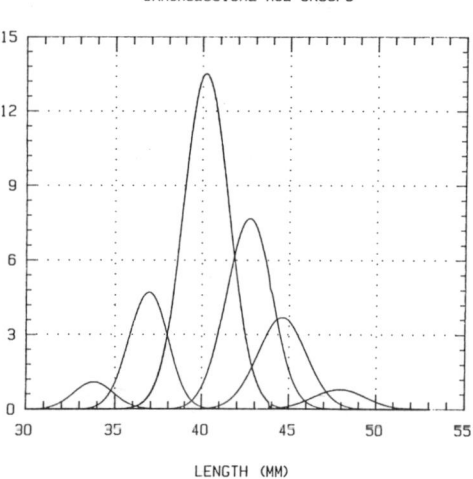
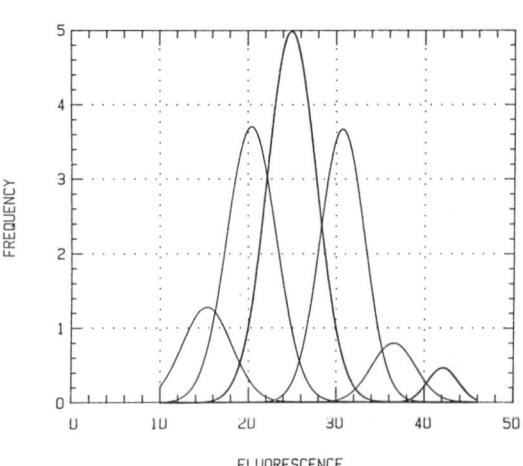

Figure 10.7. Chronological (**left**) and physiological (**right**) age groups derived from the length and FAP spectra, respectively.

mals in the sample caused by stage movements. This last problem can be avoided by using silhouette photographs of the samples rather than the actual animals suspended in a preservative solution. The disadvantage of silhouette photography is that it is an additional time-consuming step that reduces the speed advantage of the automated analyzer. A better solution seems to be a flow-through system currently being developed at URI, in which a dilute sample is pumped through a small imaging chamber. The pump, under computer control, stops when an animal is in the field of view, allowing the frame grabber to capture the image. This promises to be fast and accurate, and requires little operator attention.

The design of this flow-through system suggests the next important development that should be undertaken. That is to build the image acquisition section into a submersible body that can be towed behind a ship to give real-time readouts of plankton size, taxonomic group, and abundance. Other sensors built into this device could measure temperature, salinity, depth, and chlorophyll fluorescence. An instrument of this sort would allow the rapid characterization of the water column and plankton community needed for the routine monitoring of large marine ecosystems. Far fewer net samples would be needed; those for calibration or study of specific problems where identification of planktonic animals to species and developmental stage are truly necessary.

The coupling of a real-time, in-situ image analyzer with one of the new models of ecosystem structure and function based on particle size and distribution such as those developed by Sheldon *et al.* (1972), and Steele and Frost (1977) will make analysis of LMEs faster and more practical, helping scientists to maximize the potential use of LMEs as a management concept.

References

Berman, M. S., McVey, A. L., and Ettershank, G. 1989. Age determination of Antarctic krill using fluorescence and image analysis of size. Polar Biol. 9:267–271.

Beyer, J. E., and Laurence, G. C. 1980. A stochastic model of larval growth. Ecol. Modelling 8:109–132.

Cassie, R. M. 1954. Some uses of probability paper in the analysis of size frequency distributions. Aust. J. Mar. Freshw. Res. 5:513–552.

Chehdi, K., Boucher, J. M., and Hillion, A. 1986. Pattern recognition of zooplankton. *In* Eighth International Conference on Pattern Recognition. CH2342/86 IEEE. pp. 789–791.

Deason, E. E. 1980. Potential effect of phytoplankton colony breakage on the calculation of zooplankton filtration rates. Mar. Biol. 57:279–286.

Ettershank, G. 1983. Age structure and cyclical annual size change in the Antarctic krill, *Euphausia superba* Dana. Polar Biol. 2:139–193.

Ettershank, G. 1984a. Methodology for age determination of Antarctic krill using the age pigment lipofuscin. BIOMASS Handbook 26. 14 pp.

Ettershank, G. 1984b. A new approach for assessing longevity in the Antarctic krill *Euphausia superba* Dana. J. Crust. Biol. 4 (Spec. No. 1):295–305.

Fawell, J. K. 1976. Electronic measuring devices in the sorting of marine zooplankton. In Zooplankton fixation and preservation. Ed. by H. F. Steedman. UNESCO Press, Paris. 350 pp.

Gerritsen, J., and Strickler, J. R. 1977. Encounter probabilities and community structure in zooplankton: A mathematical model. J. Fish. Res. Board Can. 34:73–82.

Harbison, G. R., and McAlister, V. L. 1980. Fact and artifact in copepod feeding experiments. Limnol. Oceanogr. 25:971–981.

Ikeda, T. 1985. Life history of Antarctic krill *Euphausia superba*: A new look from an experimental approach. Bull. Mar. Sci. 37(2):313–318

Ikeda, T., and Dixon, P. 1982. Body shrinkage: A possible overwintering strategy of the Antarctic krill *(Euphausia superba* Dana). J. Exp. Mar. Biol. Ecol. 62:143–151.

Ivanov, B. G. 1970. On the biology of the Antarctic krill *Euphausia superba*. Mar. Biol. 7:340–351.

Jeffries, P., et al. 1984. Automated sizing, counting and identification of zooplankton by pattern recognition. Mar. Biol. 78:329–334.

Macdonald, P. D. M., and Pitcher, T. J. 1979. Age groups from size-frequency data: A versatile and efficient method of analyzing distribution mixtures. J. Fish. Res. Board Can. 36:987–1001.

Macintosh, N. A. 1972. Life cycle of Antarctic krill in relation to ice and water conditions. Discovery Rep. 36:1–94.

Marr, J. W. S. 1962. The natural history and geography of the Antarctic krill *(Euphausia superba* Dana). Discovery Rep. 32:33–464.

Pieper, R. E., and Holliday, D. V. 1984. Acoustic measurements of zooplankton distributions in the sea. J. Cons. int. Explor. Mer 41:226–238.

Rosenberg, A. A., Beddington, J. R., and Basson, M. 1986. Growth and longevity of krill during the first decade of pelagic whaling. Nature 342:152–153.

Ruud, J. T. 1932. On the biology of Southern Euphausiidae. Hvalrad. Skr. 2:1–105.

Sheldon, R. W., Prakash, A., and Sutcliffe, W. H., Jr. 1972. The size distributions of particles in the ocean. Limnol. Oceanogr. 17:327–340.

Steedman, H. F. 1976. Zooplankton fixation and preservation. UNESCO Press, Paris. 350 pp.

Steele, J. H., and Frost, B. W. 1977. The structure of plankton communities. Royal Soc. London Phil. Trans. Series B. 280:486–534.

Chapter 11

Growth, Survival, and Recruitment in Large Marine Ecosystems

Geoffrey C. Laurence

Abstract

Marine fish populations, which are highly fecund, suffer an enormous mortality between the egg stage and the age of recruitment to the fisheries. Most of the mortality is thought to occur in the early life stages during the first year. It has become an axiom that this reduction in numbers must be caused by a forced density-dependent, or biological, function in one of the early stages. Otherwise, fish population numbers would either expand in an unlimited way, or decline to extinction. Some fisheries scientists, however, postulate that density-independent, or physical, factors are controlling mechanisms. Still others argue it is a combination. Whatever the cause(s) and their modes of action, it can be demonstrated easily that small changes in growth and mortality during the early stages can have a profound effect on the variability of numbers eventually recruited. A survey of LMEs shows some systems where favored hypotheses control recruitment. However, none are unequivocal on a reliably predictive basis. Cod and haddock stocks in the Georges Bank system are presented as a case in point that has been particularly resistive to explanatory theory. Extensive hypotheses formulations, experimentation in the laboratory and mesocosms, and experimental surveys coupled with intensive site studies in the field conducted during the present decade have demonstrated that there is probably no single factor responsible for year-to-year recruitment variability. The efficacy of considering recruitment variability in a probabilistic manner because of the uncertainty of causative factors is discussed.

Introduction

Although the term recruitment is firmly established in the lexicon of fisheries science, it has remained an enigmatic concept. Its most common usage is in reference to the size of and timing associated with a year class entering a fishery. This is a discrete event where there is a passing of biomass from the unexploited to the exploited segment of a resource. It is sort of a reference point for population dynamics. A more holistic view is one where cause and effect are considered. The idea here is that fishery production includes the processes of fecundity, spawning, growth, survival/mortality, and interactions with and effects of other trophic levels in the prerecruit stages as well as effects of the physical environment — a very complicated situation.

No matter what the overview of recruitment, its single most important factor in a pragmatic sense is its variability. Estimation and prediction of this variability on an annual basis would be an enormous aid to fishery managers in allocating fishery production. The managers could deal with such questions as how much exploitation can a stock stand before failure? An understanding of the causal mechanisms and processes associated with the variability could answer important scientific and ecological questions regarding the regulation of populations.

Two major parameters that can be measured, although with difficulty, during the prerecruit stage of fishes are growth and survival/mortality. They are important because they represent the degree or expression of population success. They are the end result of the synergy of all the regulatory factors. This chapter will consider growth and survival/mortality of fish populations in a recruitment context within the concept of large marine ecosystems (LMEs). Particular emphasis will be given to the cod and haddock stocks of Georges Bank.

Recruitment Processes

Most fisheries scientists agree that factors affecting the early life stages (egg, larva, and juvenile) during the first year of life set the level of recruitment. The reasons for this are twofold: first, this is the time of greatest variability in growth and mortality rates, thus producing the potential for the widest range of population numbers and individual size; and second, density-dependent factors must be operating for animals as highly fecund as fishes to keep them from expanding without limit in good times or falling to extinction in bad times. All evidence points to this happening during the early life stages for fishes (Gulland, 1965).

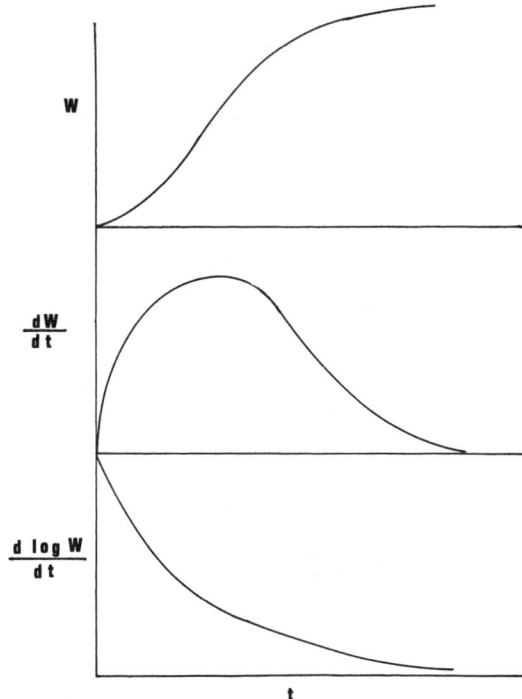

Figure 11.1. Representative curves of fish growth: **(top)** absolute growth, **(middle)** growth rate, **(bottom)** specific growth rate.

Growth

The growth of fishes is remarkable due to the high change in rates during the life cycle and the variability at a given time (Figure 11.1). Rates during the first year of life are extremely high — as much as 10–20% · d^{-1} for weight and a total periodic change of as much as 5 orders of magnitude during the larval stage. However, the overall tendency is for specific growth rate (% per unit time) to decrease with increasing age. The factors normally considered as directly affecting growth during the first year of life are prey availability, including density and distribution; prey type or quality; encounter with prey temporally and spatially, or searching ability; prey capturing rate; temperature, because fishes are cold-blooded, and light, because most early life stage fishes are visual feeders (Table 11.1).

Variable growth in early life stage fishes exists at all levels, from the individual to the population. It is axiomatic that individuals vary in growth, and the literature is full of examples of annual differences in growth for particular populations as well as intra-annual differences for cohorts of the same species (Leak and Houde, 1987). Figure 11.2 is a generalization of the difference growth rate effect can have on size at age. The impact of this is likely the variability in time or duration at a particular stage or size that may be vulnerable to certain mortality factors. In fact, some fishery scientists think that

Table 11.1. Factors affecting feeding and growth of early life stage fishes.

Factor	Effect
Temperature	Physiological rates
Light	Visual perception of prey
Prey type-quality	Preference and nutrition
Prey density-distribution	Prey encounter
Search ability	Prey encounter
Capture rate	Prey intake success

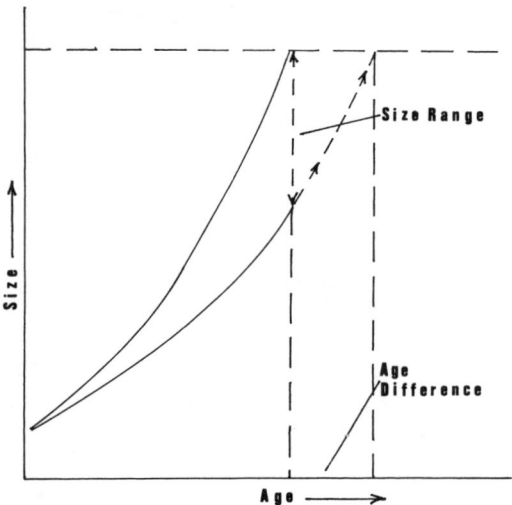

Figure 11.2. Growth rate effect on size at age. Two different growth curves demonstrate the differences in size at a given age or age at a given size.

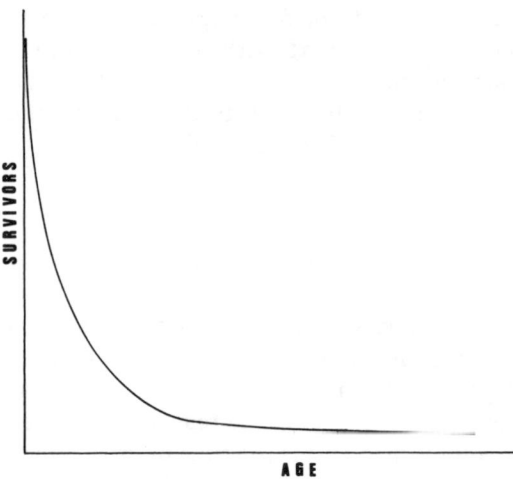

Figure 11.3. Generalized curve of early life stage fish mortality. There is typically a heavy, earlier mortality with a subsequent reduction.

growth rate may be the major factor affecting recruitment levels (Houde, 1986).

Mortality/survival

The general pattern for mortality of marine organisms is for a decrease with increasing weight (McGurk, 1986) and/or age (Cushing, 1974). Marine fishes follow this trend; however, the decrease is not normally proportional or constant. The earliest stages usually suffer a much higher death rate than the older stages (Figure 11.3). In fact, 50 to 99% of the total mortality usually occurs in the first year of life. The most important aspect of this first year mortality in a "critical" sense is the timing and magnitude (Gulland, 1965). The term "critical" is used in reference to differences in good and bad year-class formation. A relatively short period of critical mortality can drastically alter a mortality curve (Figure 11.4).

Growth and mortality interaction

Although growth and mortality are important factors affecting recruitment levels, it is difficult to consider them separately. Their interactive relationship revolves around the triotrophic principle (Laurence, 1981; Figure 11.5) of "eat or be eaten." This is commonly thought of in terms of starvation or predation, the two dominant factors of mortality for the early life stages. The basis for this relationship is the Ricker-Foerster (1948) hypothesis that differences in food level generate differences in growth, which results in a differential vulnerability to predation mortality. Poor feeding conditions result in slower growth, which in turn results in a longer time to be subject to a particular predatory field and, thus, a larger stage specific mortality.

Density dependence

The major question concerning growth and mortality does not revolve around their identification, mensuration, or interrelationship, but how they function in the density-dependent system of population regulation. The term density dependent has been an integral part of fisheries population dynamics studies for close to half a century; however, it has been used ambiguously. Its strict definition, and the one used in this chapter, implies the

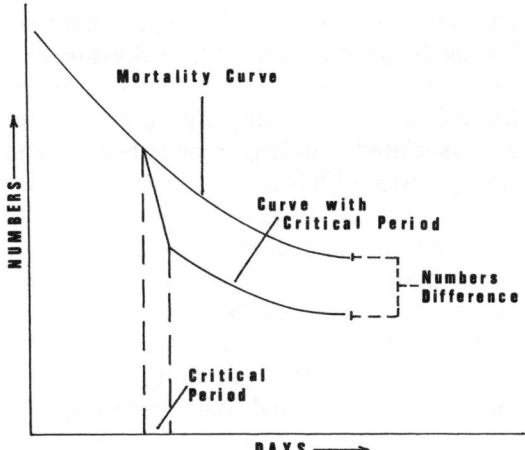

Figure 11.4. The concept of a critical period in the early life stages of fishes. A short critical period can drastically reduce numbers surviving.

lessening of the effect of a factor at a constant level on lower population numbers compared to higher, and/or a competition between siblings or members of the same population or species for a limited resource.

The evidence for density dependence in fishes is overwhelming. All populations must have some density-dependent regulators to prevent them from overpopulating or declining to zero. Furthermore, many fish populations are subject to extremely high levels of exploitation, often two to three times the level of natural mortality, without a reduction in population size or recruitment levels. This certainly indicates a density-dependent compensation. The major questions then become (i) What stage(s) in the life cycle is (are) are affected by density dependence? and (ii) How is density dependence operational?

The principle stages or processes of the life cycle that could be affected by density dependence are egg, larval, juvenile, adult, and the reproductive process (fecundity-gametogenesis). There is little evidence for density-dependent effects on adults because of relatively low population numbers in comparison with resources that potentially may be limiting. Fecundity also would not seem to be density dependent (Shepherd and Cushing, 1980). Adult weight is usually proportional to fecundity, and since growth in adults is not normally density dependent there is no reason to suspect a linkage. The egg stage is probably not affected either. Eggs have an endogenous source of nourishment and, consequently, don't need to compete for limited food. Also, the vast majority of eggs in the marine environment are pelagic and, as the tendency for eggs is to be dispersed by the physics of the environment, predation on aggregations (not strictly definitional density dependence) is unlikely. Those few species that do produce demersal eggs, however, might be subjected to increased egg predation. This leaves only the larval and juvenile stages and, indeed, most evidence points to these as bearing the im-

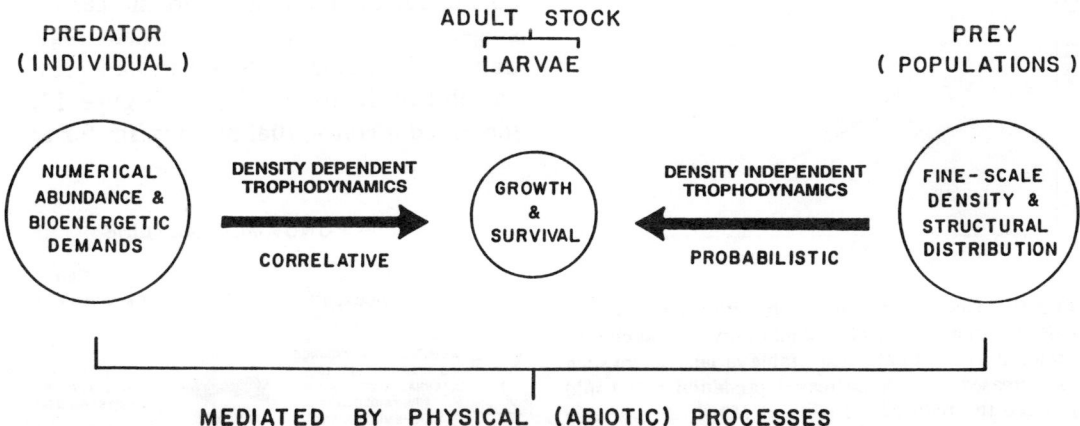

Figure 11.5. Triotrophic relationship showing early life stage fish feeding and predation interactions. (See Laurence, 1981, for detailed discussion.)

pact of density dependence.

A number of density-dependent interactive mechanisms that include feeding, growth, and predation have been proposed. Cushing and co-authors (Cushing and Harris, 1973; Cushing, 1974; Shepherd and Cushing, 1980) have generally favored a growth-predation interaction similar to the original Ricker-Foerster hypothesis occurring during the larval stage. They advocate density-dependent growth in the larval stage controlled by food availability. This growth then controls the rate at which larvae are subjected to limited food or a predatory field causing mortality (Figure 11.6). Faster growing larvae are presumably more successful. Jones (1973) also favors density-dependent larval growth, but with a different interpretation. His "limiting food" concept proposes that larvae and their prey exist and grow together in cohorts. In this context, it is possible to see that food limitation — and, hence, density-dependent growth — could happen due to differential cohort densities. Jones also proposes that larval individual growth rate is variable and predetermined genetically. Therefore, in a food-limited situation, larvae with high innate growth rates and food requirement are less likely to obtain food at the necessary rate to provide for growth than slower growing individuals. As a result, high growth rates are associated with high mortality rates and slow growth with lower mortality.

Recent studies have tended to counter the larval density-dependent growth concepts proposed by Cushing and Jones. Laurence's analyses (1985) of the feeding requirements of cod and haddock larvae on Georges Bank from laboratory, field, and modeling data showed that larvae were too dilute spatially to compete with each other for limited food, or to significantly impact on the food available, for that matter. This would tend to minimize the concept of density-dependent larval growth on Georges Bank. A field study conducted in Biscayne Bay, Florida, with larval bay anchovy (Leak and Houde, 1987) noted no statistical correlation of instantaneous mortality coefficient with growth rate. The implication in these studies is not that variable growth is not influential — because it most certainly is — but, rather, that it may not be under density-dependent control.

If density-dependent growth is not a factor, in some cases at least, and we must still concede a density-dependent influence in the early life stages, then what might be a factor, and how might it work? A link with predation appears to be a candidate. Houde (1986) identified predation as a more important factor than starvation in the early life stages with respect to recruitment regulation in situations of demonstrated growth variability. Laurence (1981, Figure 11.7) identified a conceptual mechanism for cou-

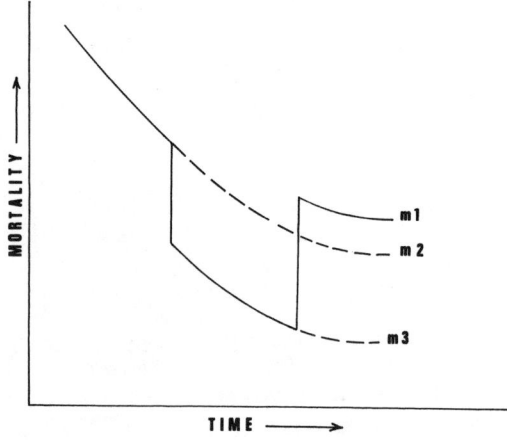

Figure 11.6. Changing mortality with age under variable conditions. An unaltered mortality had no change occurred is given by M2. A favorable variable factor such as increased food or decreased predation rate could produce the reduced mortality of M3. If the variable factor returns to its original level, then the new mortality M1 may be higher than before because the numbers alive are relatively greater than they would have been. (Adapted from Cushing, 1974.)

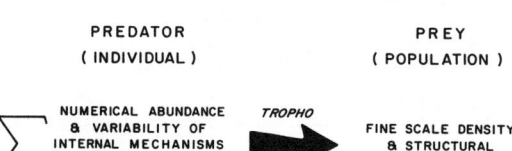

Figure 11.7. Basic concept of a mechanistic trophodynamic link for predator-prey relations of early life stage fishes. (From Laurence, 1981.)

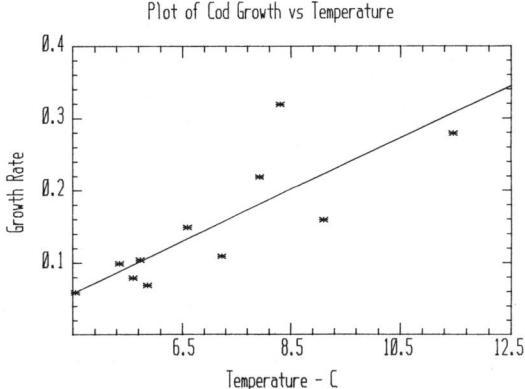

Figure 11.8. Basic linear relationship of temperature and growth rate for North Atlantic adult cod stocks. The relationship is inverse as growth rate is the K parameter of the Von Bertalanffy growth equation. (Adapted from Taylor, 1958.)

pling density dependence with predation during the larval stage in which the predator's individual internal bioenergetic requirements summed over its numerical population abundance acted on or with the prey's fine-scale density and population spatial distribution. This could produce a compensatory predation mortality at high prey population levels and less of an effect at lower levels of prey. McGurk (1986) has recently considered this concept in detail, quantifying it with the use of extensive field data to develop a model of egg and larval mortality (attributed to predation) and their relationship to spatial patchiness.

Density-dependent growth and mortality during the juvenile stage have not been investigated intensively. Since specific growth rate declines with age, direct mortality due to starvation appears unlikely for juveniles because they can respond to a shortage of food with a decreased growth rate (Jones, 1973). This does not, of course, rule out the potential for variable growth rates to affect survival through other interactive factors. Predation on juveniles is insidious. Whether it is density dependent or not in a critical sense relative to recruitment is debatable. It is most certainly important. Cohen et al. (1988) show that relative cumulative mortality for the juvenile stages of Georges Bank fishes is as much as that for the egg and larval stages. Predation is the likely cause for this, and it has the potential to affect recruitment.

Density-Independent Factors

Density-independent factors have traditionally been considered as external physical factors that directly affect individuals or populations by acting on physiology, behavior, or distribution. Examples are temperature's direct influence on the physiology of growth (Figure 11.8) or abnormal currents sweeping a population of larvae from a nursery ground. We have also seen, however, that biological-trophic interactions that might traditionally be considered to be density dependent can function in a density-independent manner.

Intrinsic or Innate Factors

There is a class of factors that act in a density-independent way and have not received the attention they should have. Biological in nature, they affect individuals and/or populations in a genotypic, mutagenic, or evolutionary mode. The process occurs during development in the early life stages. It has its origin in the disruption of genetic control of the initiation, sequencing, and development of basic biochemical, cell, tissue, and organ systems. In animals as highly prolific as fishes, it is not unreasonable to assume less-than-rigid genetic control during early development. In a situation of "loose" genetic-developmental control, it also is easy to see the environment exerting an influence. This could be considered a "compound" density-independent effect. In any case, it is all part of the evolutionary process of speciation and adaptability to environmental change.

Vladimirov (1975) presented a very good concept of this intrinsic influence as it pertains to critical periods in the early development of fishes. He proposed that a critical period was the inherent result of developmental events under specific conditions. Death is the outward manifestation of this critically impaired development. The basic factors are the qualitative state of the organism and environmental influences. Indirect evidence in support comes from laboratory and field studies. Laboratory developmental assays examining physical factor control of mortality in factorial experimental designs have been numerous. It has not been unusual for these studies to note an inherent residual mortality unattributable to any experimental factors (Laurence and Rogers, 1976). In a field study, Leak and Houde (1987) reported high and variable cohort mortality rates for egg and larval bay anchovy. This mortality was statistically unrelated to any causative factors measured in the research. Also, a significant portion of this mortality occurred during the egg stage when complicating feeding effects were not present. In the absence of direct evidence to the contrary, Leak and Houde assumed this egg mortality was caused by predation. There is a tendency, with justification, for most researchers to *a priori* name predation as the cause of unknown natural mortality. However, it is not unreasonable to consider intrinsic factors as well.

A generalized system of recruitment factors

The construction of a coordinated holistic system of the factors and processes affecting recruitment is useful to interpret the concepts associated with its variability. The system is hierarchical (Figure 11.9). Output at the top is an indication of the level of recruitment and/or its variability, preferably in a predictive, quantitative form. Indicators just beneath the top level are the quantitative measurements of growth and mortality. They provide an index of the population's state of well-being. The lower levels consist of an operational segment with density-dependent biological factors, intrinsic developmental factors, and density-independent physical factors at the base, and a

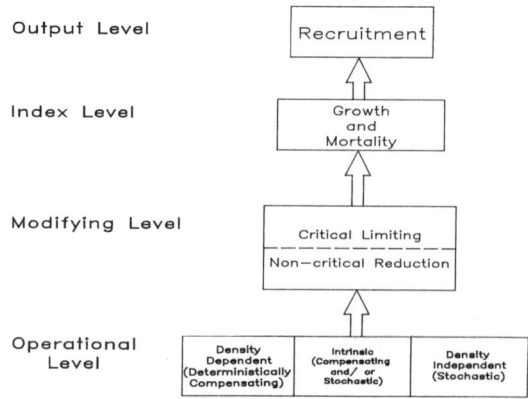

Figure 11.9. A conceptual holistic system of factors and processes affecting recruitment.

modifying level immediately above the base affecting the control of the operational factors in a critical or noncritical manner.

The base-level density-dependent factors operate in a deterministically compensating way at high and low population levels. Intrinsic factors are somewhat probabilistic and, generally, suppress a bumper crop, but not entirely. Density-independent physical factors are highly stochastic and have the potential to severely limit population success.

The concept of criticality in the modifying level of this system is attributed to Gulland (1965). This is a two-phase process. The first phase, always present, consists of a high, density-dependent mortality (perhaps modulated by intrinsic factors) that reduces large numbers of early life stages and is higher for high initial numbers and lower for low. This may be considered a coarse adjustment. The second or subsequent phase, which may or may not occur, is the critical fine-tuning where stochastic density-independent factors set the difference between a good and bad year class (Figure 11.10).

Recruitment Research in LMEs

LMEs as a conceptual unit encompass large geographic areas within which biological populations have evolved and adapted reproductive strategies in response to particular bathymetry, hydrography, and circula-

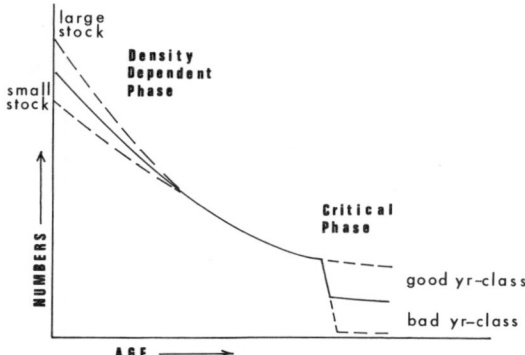

Figure 11.10. The concept of criticality in reducing population numbers of early life stage fishes. It is a two-phase process involving consistent density-dependent affects and stochastic density-independent factors. (Adapted from Gulland, 1965.)

tion (Sherman and Alexander, 1986). The LME concept is modified by geopolitical, legal, and economic implications and is a continually evolving scientific and sociologic process (Sherman, 1987). Coordinating research and synthesizing available information on a comparative basis within the LME concept may prove helpful in understanding recruitment processes.

Research results for population regulation processes have been reported and compiled for 20 defined LMEs (ICES, 1987; Jones, 1987; NOAA Ad Hoc Scientific Committee, 1987; Sherman, 1987). Table 11.2 presents the primary and secondary recruitment-related hypotheses (when available) according to type for each of the LMEs. Only two of the 20 did not have sufficient information (Antarctic and Insular Pacific). There is no overwhelming favorite hypothesis or factor affecting recruitment in a generalized sense; the hypotheses are somewhat specific to LMEs. However, a breakdown of the hypotheses according to whether they are physical or biological in origin and the association with LME category is interesting. A majority of the primary hypotheses are categorized as physical (Table 11.3). Most of the primary hypotheses classified as biological are those associated with fishing exploitation (man's predation). There is good cause to consider this as a special case and not bio-logical in the traditional sense. Associating hypothesis type with LME classification showed that a majority of the continental shelf systems favored physical hypotheses, as did all of the upwelling-frontal-current LMEs (Table 11.4). Although many of the LMEs may be primarily influenced by density-independent physical factors, there is certainly no consensus type of physical factor, and the situation is far from uniform. A thorough understanding of the functional mechanisms controlling recruitment and its prediction on a reliable basis has probably not been achieved yet for any LME.

Georges Bank Recruitment Research

Georges Bank, one of the world's most productive fisheries systems (Backus, 1987), has been particularly resistant to the development of explanatory theory for the mechanistic regulation of variability of many of its fish stocks. NOAA's Northeast Fisheries Center (NEFC) functional recruitment studies of the Bank have focused on cod and haddock early life stages. This multispecies approach within an ecosystem context has combined the results of laboratory experiments, mesocosm techniques, simulation modeling and field investigations (Table 11.5). The overall strategy has been to use the various approaches to generate and analyze data for the formulation of testable hypotheses during field experiments and surveys. The central theme of the hypotheses developed concerns trophodynamics, although some physical factors have been considered.

Laboratory experiments were conducted with embryonic and larval cod and haddock to define the critical ranges of factors affecting development, growth, and survival. The ranges of temperature and salinity combinations for optimal and suboptimal development of eggs was determined (Laurence and Rogers, 1976; Figure 11.11). Controlled feeding experiments linked growth rates with prey concentrations and temperature (Laurence, 1974, 1978; Figure 11.12). Bio-

Table 11.2 Recruitment-related hypotheses for large marine ecosystems.

System	Type	Prime hypothesis	Type	Secondary unrefutable hypothesis	Type	Refutable hypothesis	Type
Iberian Peninsula (pelagic clupeids)	Upwelling	Physical (upwelling production)	Density independent	—	—	—	—
Northeast U.S. Continental Shelf-Georges Bank (cod-haddock, other demersal)	Continental Shelf	Predation (man)	Density dependent	Physical (stratification, advection, circulation)	Density independent	Early life stage starvation	Density dependent
Barents Sea-Norway (cod)	Continental Shelf	Physical (stratification-advection)	Density independent	Prey availability	Density dependent	—	—
Northeast Arctic-Norway (cod)	Continental Shelf	Physical (current transport)	Density independent	—	—	—	—
North Sea (herring)	Continental Shelf	—	—	Physical (advection) Prey availability and predation	Density independent Density dependent	—	—
Yellow Sea (pelagic-demersal mixed)	Continental Shelf	Predation (man)	Density dependent	—	—	—	—
Gulf of Thailand (demersal)	Continental Shelf	Predation (man)	Density dependent	—	—	—	—
Great Barrier Reef (various reef species)	Continental Shelf (reef)	Predation	Density dependent	—	—	—	—
Kuroshio/Oyashio Current (pelagic clupeids, scombrids)	Frontal-current	Physical (mixing-production)	Density independent	Prey availability and predation	Density dependent	—	—
Humboldt/Peru Current (sardine)	Upwelling	Physical (mixing-production)	Density independent	—	—	—	—
California Current (pelagic clupeids)	Upwelling	Physical (stratification)	Density independent	—	—	—	—

Table 11.2 (continued)

System	Type	Prime hypothesis	Type	Secondary unrefutable hypothesis	Type	Refutable hypothesis	Type
Baltic	Continental Shelf	Pollution	Density independent	—	—	—	—
Benguela Current (pelagic-demersal mixed)	Upwelling-frontal	Physical (frontal shifts)	Density independent	—	—	—	—
East Bering Sea	Continental Shelf	Physical (mixing production)	Density independent	—	—	—	—
Gulf of Alaska (walleye pollock)	Continental Shelf	—	—	Predation-cannibalism, prey availability	Density dependent	—	—
				Physical (bathymetry, currents)	Density independent		
Gulf of Mexico (pelagic-demersal mixed)	Continental Shelf	Physical (frontal)	Density independent	Prey availability	Density dependent	—	—
Scotian Shelf (cod-haddock)	Continental Shelf	—	—	Food availability, predation, and parasitism	Density dependent	—	—
				Physical (stratification, circulation)	Density independent		
Antarctic	Continental Shelf	NA	—	—	—	—	—
Insular Pacific (reef species)	Reef	NA	—	—	—	—	—
Southeastern U.S. Shelf	Continental Shelf	Physical (transport, frontal)	Density independent	—	—	—	—

Table 11.3. Summation of physical and biological related recruitment hypotheses for LMEs. Parentheses indicate number associated with fishing (man's) predation.

Type	Primary hypothesis	Secondary hypothesis
Physical	11	4
Biological	4 (3)	6

chemical studies established indices for determining condition and immediate past growth history (Buckley, 1979; Figure 11.13). Overall results from the laboratory research corroborated the expected finding that temperature and prey concentration directly affected growth and development. More important was the fact that empirical values with ranges were recorded for the first time, and these were used to design mesocosm experiments, design field surveys, and construct hypotheses for field experiments.

Mesocosms are very large experimental systems designed to simulate or encapsulate a natural environment. This approach was used as a transitional method to bridge the gap between laboratory results and the design and conduct of field research. The hypothesis that thermal stratification was beneficial to haddock larval growth and survival was tested in a land-based mesocosm (Bergen et al., 1985). Duplicated stratified and well-mixed units were used to rear larvae in controlled experiments. Results clearly indicated that the stratified environment designed to simulate the thermocline region of Georges Bank produced better growth and survival than a well-mixed environment (Table 11.6).

Table 11.4. Association of hypothesis type with LME classification. Parentheses indicate classification total.

Type	Continental Shelf (12)	Upwelling-frontal-current (5)	Reef (1)
Physical	8	5	0
Biological	4	0	1

Ongoing field research on Georges Bank consists of broad-scale survey cruises to determine abundance and distribution and process-oriented cruises concentrating on functional mechanisms of trophodynamics. Large- and small-scale physical processes affecting the biological distributions are monitored on all cruises.

Broad-scale abundance and distribution is surveyed annually (Smith and Morse, 1985). In general, the principal spawning ground occurs over the eastern half of the Bank (Figure 11.14). A large anticyclonic gyre coupled with strong rotary tidal current maintain eggs and larvae on the Bank. Drift is east to west along the southern flank of the Bank within the 50 to 100 m contours. Evidence from finer scale cruises shows that eggs and larvae are concentrated in discrete populations on the order of tens of kms. These populations are transported about 1 to 2.5 km $\cdot$ d^{-1}, which is consistent with the long-term residual current for the area (Lough, 1984; Figure 11.15). It is believed that the concentrations belong to the same spawning populations.

Favored prey organisms of cod and haddock larvae were determined from stomach analyses. Copepods, particularly early life stages, were the most common items consumed (Kane, 1984). Small-scale spatial distribution and variability of prey were measured (Laurence et al., 1984). All prey items within a spatial scale relevant to the daily searching ability of larvae were randomly distributed except for copepod eggs which were contiguous (Figure 11.16).

A study of cod and haddock daily otolith growth increments during the larval and juvenile stages provided information on growth and development in the field and allowed the calculation of growth curves for the first year of life (Bolz and Lough, 1983, 1988). In addition, this research provided an estimate of the direct effect of temperature on field growth (Figure 11.17) as well as back-calculated estimates of spawning date.

Multidisciplinary, small-scale site studies were conducted to test specific hypotheses. Buckley and Lough (1987) used dis-

Table 11.5. Early life stage recruitment research relative to Georges Bank conducted by the Northeast Fisheries Center, National Marine Fisheries Service.

	Laboratory		Experimental mode Mesocosm		Field
Factor	Response	Factor	Response	Factor	Response
			Density dependent		
Food-feeding	Growth (Laurence, 1974, 1978; Buckley, 1979; Laurence et al., 1981)	Food-feeding	Growth-mortality (Bergen et al., 1985)	Food	Prey abundance and distribution (Laurence et al., 1984; Laurence, 1985; Kane, 1984)
Food-feeding	Bioenergetics (Laurence, 1978)			Feeding	Growth (Bolz & Lough, 1983, 1988; Buckley & Lough, 1987; Laurence, 1985)
				Predation	Mortality (Cohen & Grosslein, 1982; Sissenwine, 1984)
			Density independent		
Temperature-salinity	Development-mortality (Laurence & Rogers, 1976)	Thermal stratification	Growth-mortality (Bergen et al., 1985)	Temperature	Growth (Bolz & Lough, 1983)
				Thermal stratification	Growth (Buckley & Lough, 1987)
				Thermal stratification	Distribution (Lough, 1984)
				Advection	Mortality (Cohen et al., 1986; Smith & Morse, 1985; Laurence & Burns, 1982)
				Storms	Distribution-mortality (Lough, 1984; Cohen et al., 1988)
				Circulation	Distribution (Lough, 1984)

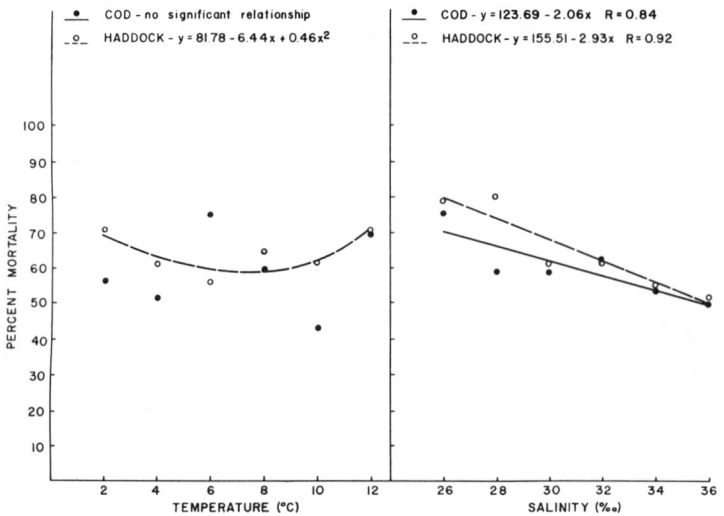

Figure 11.11. The relationships of temperature and salinity to percent embryo mortality of Atlantic cod and haddock from laboratory experiments. Data points for the response at each temperature are the mean of all salinities and for each salinity the mean of all temperatures. (From Laurence and Rogers, 1976.)

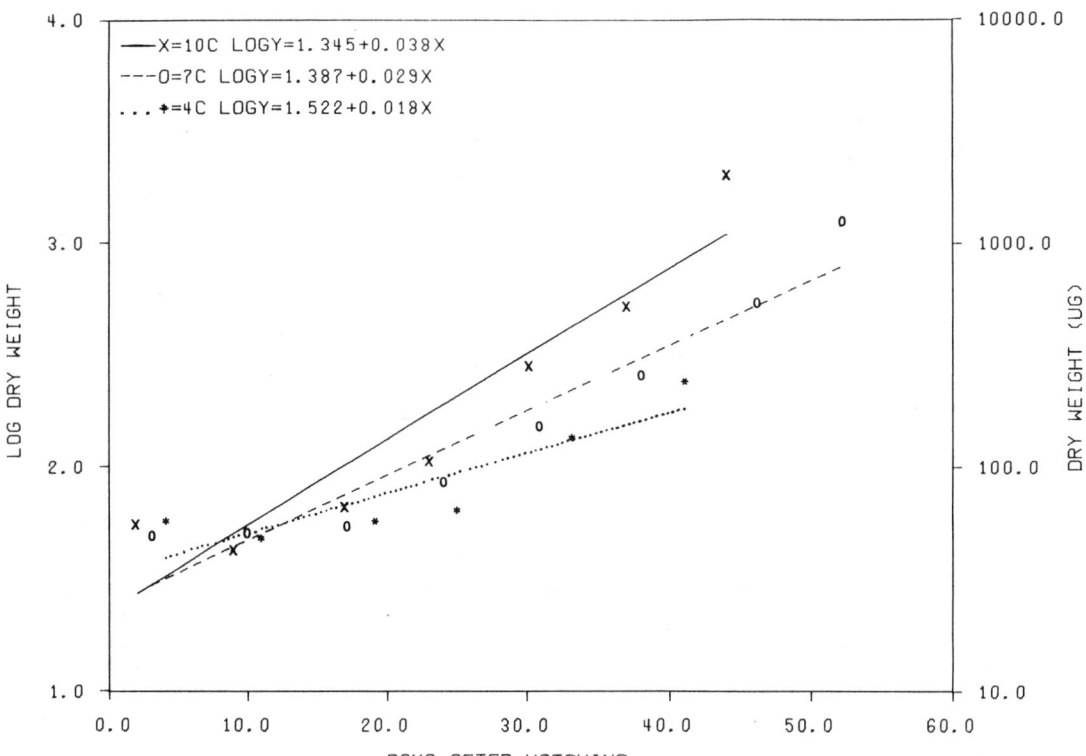

Figure 11.12. Dry weight growth in the laboratory of cod larvae for period of yolk absorption to metamorphosis at three temperatures. [Reprinted from Laurence, Mar. Biol. 50:1–7, with permission. Copyright 1978 by Springer-Verlag.]

crete sampling techniques to measure vertical distribution of larval cod and haddock and prey, and biochemical indices were used to estimate growth and condition at stratified and well-mixed sites on Georges Bank. Results indicated that haddock larvae were in better condition, were more abundant (Figure 11.18), and had better growth rates

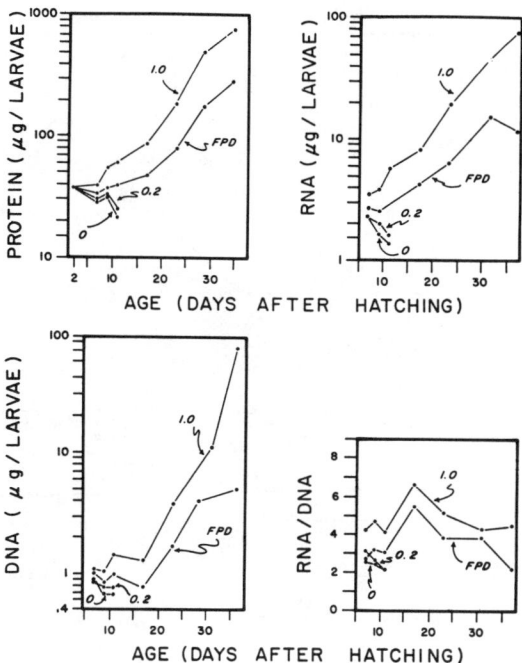

Figure 11.13. Changes in the RNA-DNA ratio, protein, RNA, and DNA content of larval cod during development at four prey densities in laboratory experiments. Data points are means of three replicate samples containing 12–30 larvae each. The four prey densities used were 0, 0.2, and 1.0 plankters · ml^{-1} and a fluctuating plankton density (**FPD**) ranging from less than 0.2 to greater than 4 plankters · ml^{-1}. (From Buckley, 1979.)

at the stratified sites compared to the well-mixed location. Since measured prey levels were considerably higher at the stratified sites, this supported the hypothesis that thermal stratification can promote conditions for favorable growth and survival.

Possible physical advection of early life stages off Georges Bank has been a subject of much debate. Warm core rings entrain water from Georges Bank; this is a potential source of removal of pelagic early life stages of fishes (Flierl and Wroblewski, 1985). This situation is complicated, however, because the timing and position of a ring off Georges Bank is critical, and the actual physical mechanics of the entrainment process relative to the behavior and locomotor abilities of the fishes is not known. To date, the evidence has been debatable, with no direct proof of significant entrainment of cod and haddock (Laurence and Burns, 1982; Cohen et al., 1986, 1988).

Severe storms are also a potential source for water-transport loss of early life stages off the Bank. The effects of a powerful storm were monitored in April 1982 (Ramp, 1982; Cohen et al., 1986). Wind measurements, current measurements, satellite imagery,

Table 11.6. Growth and survival of 500 haddock larvae after 28 days in a stratified and well-mixed mesocosm. (From Bergen et al., 1985.)

Parameter	Trial 1		Trial 2	
	Stratified	Well-mixed	Stratified	Well-mixed
Temperature (°C)	9.0/5.0	6.5	9.0/5.0	6.5
Average plankton density	84	13	38	50
Final number of larvae (# · tank^{-1})	10	0	44	0
Initial length (mm)	4.74	4.74	4.85	4.85
Final length (mm)	9.12	0	9.68	0
Initial weight (mg)	.09	.09	.11	.11
Final weight (mg)	2.88	0	1.32	0
Initial RNA/DNA	3.32	3.32	3.32	3.32
Final RNA/DNA	4.32	0	4.15	0
Survival (%)	2.0	0	8.8	0
Mortality coefficient (z)	0.14	—	0.08	—
Growth in dry weight (% · d^{-1})	12.4	0	8.9	0
Growth in standard length (% · d^{-1})	2.3	0	2.5	0

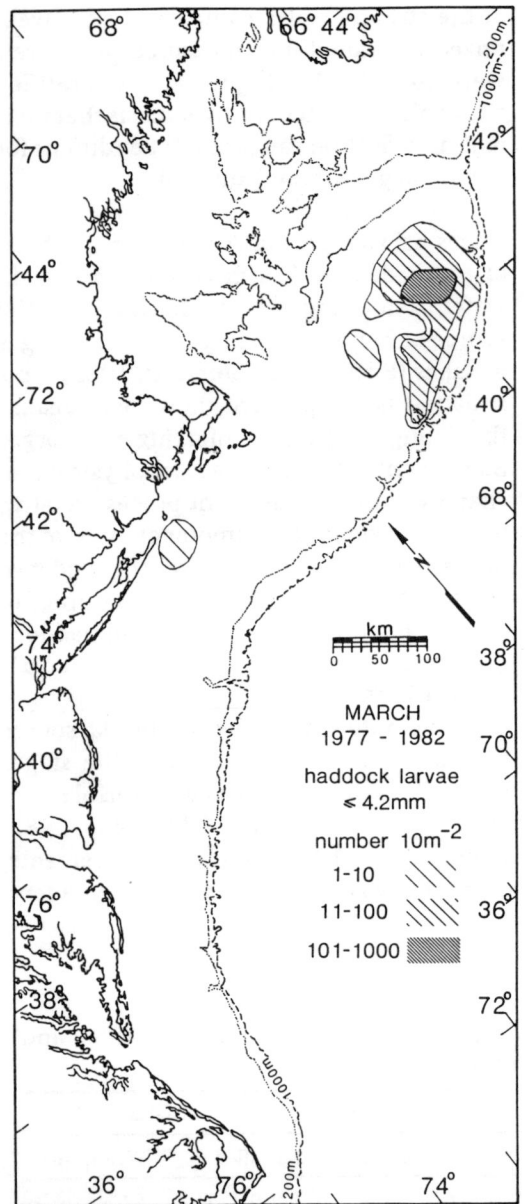

Figure 11.14. Distribution of newly hatched haddock larvae on Georges Bank, 1977–1982. (From Smith and Morse, 1985.)

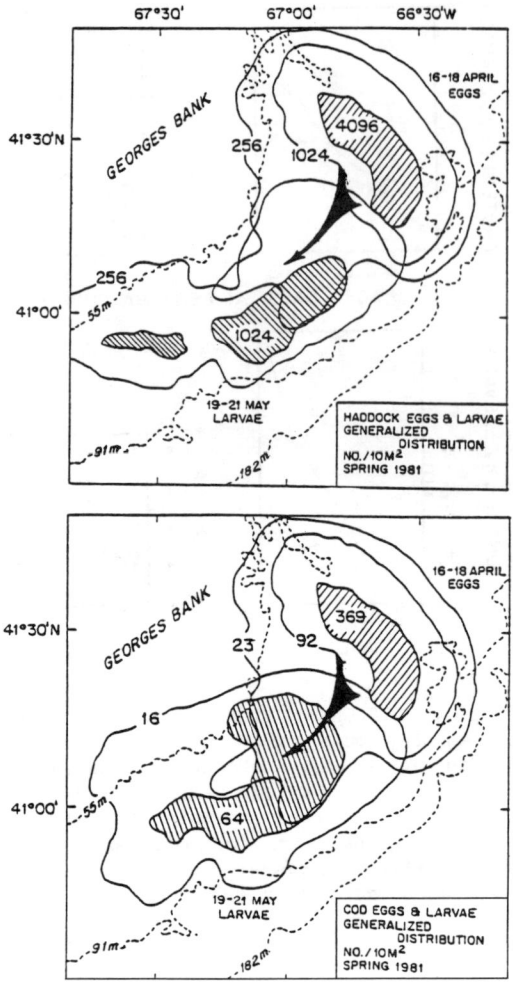

Figure 11.15. Haddock and cod egg and larval distributions on Georges Bank generalized from April and May 1981 grid surveys showing movement pattern. (From Lough, 1984.)

and wind-driven Ekman transport calculations indicated a substantial displacement of the shelf-slope front accompanied by advection to the south-southeast. Although no biological sampling was done at this time, the implications are obvious. A storm was shown to have a strong effect on vertical stratification of the water column on Georges Bank during May 1981. Lough (1984) conducted site studies during this storm that documented the destruction of the thermocline and the dispersal of cod and haddock larvae and their prey (Figure 11.19).

Predation is most certainly an important cause of mortality in prerecruit stages on Georges Bank. However, it is hard to detect and measure, and the evidence developed thus far is mainly indirect. No important predators of eggs and larvae have been identified as yet, and the magnitude of predation

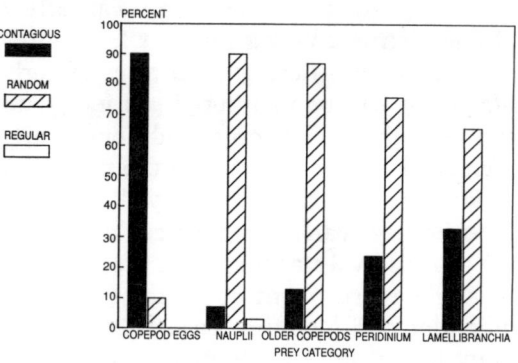

Figure 11.16. Percentage of distributional classification of all samples of the relevant prey organisms of larval cod and haddock on Georges Bank during May 1980. (From Laurence et al., 1984.)

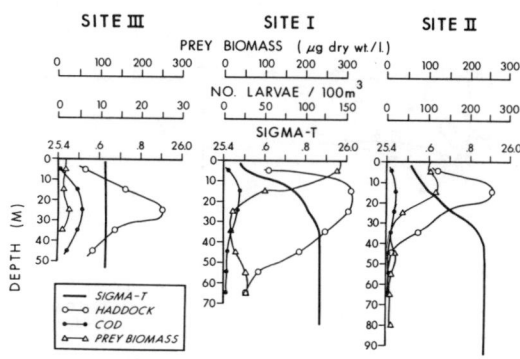

Figure 11.18. Sigma-t profile and vertical distribution of cod and haddock larvae and their prey at three study sites on Georges Bank during May 1983. Note difference in scale for density of larvae for each site. (From Buckley and Lough, 1987.)

mortality for these stages remains unknown. Cumulative mortality for the juvenile stage of cod and haddock has been calculated to be as great or greater than cumulative mortalities for eggs and larvae (Sissenwine, 1984; Table 11.7). The construction of energy budgets for Georges Bank have shown that fish production consists mainly of prerecruits, and that most of it is consumed by the fish themselves (70%, Cohen and Grosslein, 1982, 1987). However, this is contradicted by the fact that data on food habits collected by the NEFC show minor predation on cod and haddock by fish.

The actual measurement or calculation of mortality rates for early-life-stage cod and haddock on Georges Bank has been minimal. Lough (1984) sampled a discrete patch of cod and haddock eggs and larvae during a 35-day period and computed daily mortality rates of 6 to 8%. Laurence (1985) developed a simulation model of food-limited growth and survival of cod and haddock larvae using all data available from NEFC research and other sources. A strong direct influence of prey levels on larval population survival was demonstrated (Figure 11.20).

It seems quite clear that the system controlling prerecruit success of cod and haddock on Georges Bank is probably not dominated by a single causative factor. Research and analyses completed to date implicate a probable synergism of factors that may change from year to year (Cohen et al., 1988). Although we may not know yet the functional mechanics of prerecruit survival variability, we have learned a number of facts and inferences including the following: (i) recruitment can be highly variable, as can mortality during the first year of life or during any discrete early life history stage (i.e., egg, larva, juvenile); (ii) primary production can be seasonally high on Georges Bank; (iii) starvation mortality is a significant factor but probably not a population-limiting or

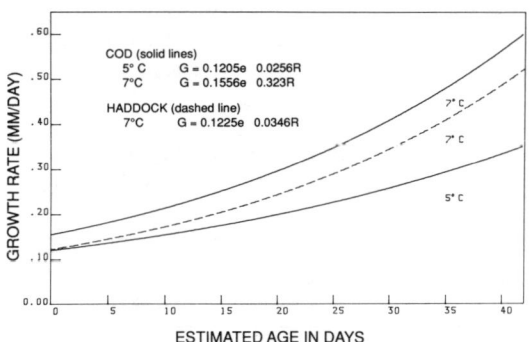

Figure 11.17. Growth comparison of larval Atlantic cod and haddock at different temperatures. The 5°C curve for cod and the 7°C for haddock are based on an analysis of the number of daily increments counted on otoliths from larvae collected during the spring of 1981 on southeastern Georges Bank. The 7°C curve for cod is hypothetical and incorporates the percent per day increase in growth for a 2°C temperature rise found in laboratory rearing experiments (From Bolz and Lough, 1983.)

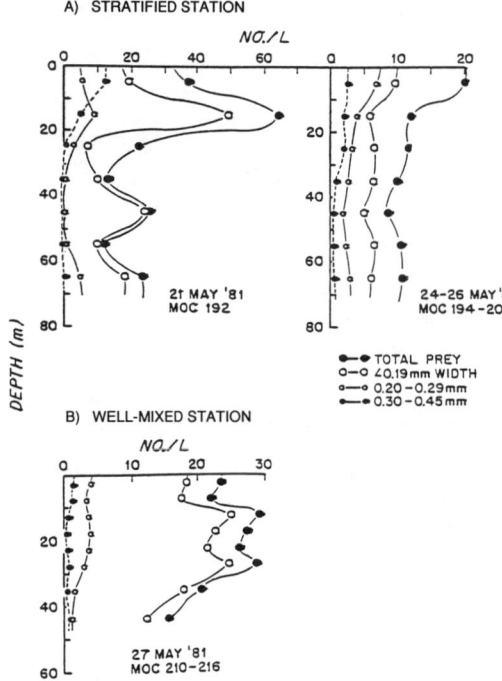

Figure 11.19. Vertical distribution of larval prey field on (**A**) stratified station before and after storm, 22–24 May 1981; and on (**B**) shoal, well-mixed station, 27 May 1981. (From Lough, 1984.)

single controlling factor at the normal prey concentrations found on the Bank; (iv) larvae do co-occur with their prey in relation to physical factors (i.e., thermocline); (v) large-scale advective features have the potential to affect early life stage survival, but this remains to be clearly demonstrated; (vi) probabilistic meteorological events can affect physical structure and relevant larval-prey continuity; (vii) field growth rates of early life stages can be monitored routinely and can be high; (viii) poorly conditioned early life stages can be detected in the field; (ix) relative cumulative mortality during the juvenile stage can be as great as that during the egg or larval stages and has the potential to affect recruitment variability; and (x) predation of fish on fish is a significant mortality factor on Georges Bank and has the potential to affect the level of recruitment.

Conclusions

The complexity of the recruitment situation on Georges Bank where there is apparently no dominating signal is perplexing. A consideration of LMEs in general is much the same. Although the controlling mechanism seems apparent for some LMEs, there is still much uncertainty and variability between systems. The major question then becomes: How do fishery scientists conduct process-oriented research in a subject area of unpredictable variability? We in the NEFC have advocated adoption of a two-phase strategy for cod and haddock studies. The first step will be to conduct temporally and spatially intensive surveys for all life stages during the entire first year of life. This will allow the

Table 11.7. Relative instantaneous mortality for the larval (hatching to 90 days), juvenile to age 1 (90 days to year end), and juvenile age 1–2 periods for Georges Bank haddock, 1974–1977.[1]

	Year			
	1974	1975	1976	1977
Spawning stock ($\times 10^6$)	10.9	12.0	10.0	45.3
Stock fecundity ($\times 10^{12}$)	6.5	7.2	6.0	27.2
Eggs hatched $\times 10^{12}$ ($\times 0.34 \times$ fecundity)	1.2	2.5	2.0	9.3
Larval abundance ($\times 10^9$)	56.9	159.2	76.9	483.7
MT (L)	4.7	2.8	3.3	3.0
1-year abundance ($\times 10^6$)	1.8	94.0	23.0	4.0
MT (L–1)	12.6	7.4	8.1	11.7
2-year abundance ($\times 10^6$)	1.3	77.0	19.0	—
MT (1–2)	0.12	0.20	0.19	—

[1]Northeast Fisheries Center data from larval survey estimates and VPA estimates of stock sizes.

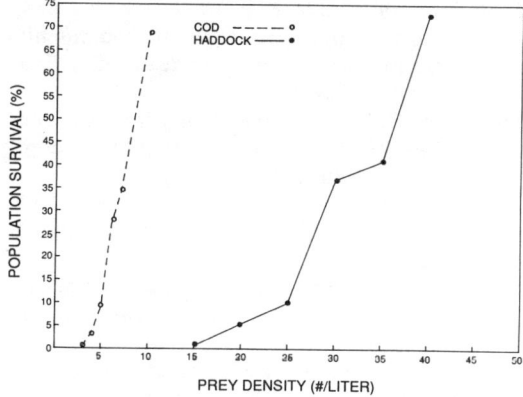

Figure 11.20. Simulated population survival at different constant prey densities for larval cod and haddock. (From Laurence, 1985.)

calculation of mortality rates to determine if there is a critical stage. These surveys will be conducted for several years to see if there are interannual differences. Once the timing and criticality of mortality is ascertained, then specific hypotheses will be formulated to test processes. It is imperative that these hypotheses be formulated within an inductive framework. They must be provable or rejectable and in a logical sequence so that information from the proof or rejection rationally leads to the next experiment. This inductive approach should be the best way to deal with a complex, variable, and unpredictable research subject like recruitment.

References

Backus, R. H. (Editor). 1987. Georges Bank. MIT Press, Cambridge, MA. 593 pp.

Bergen, R. H., Laurence, G. C., Oviatt, C. A., and Buckley, L. J. 1985. Effect of thermal stratification on the growth and survival of haddock larvae *(Melanogrammus aeglefinus)*. ICES C.M. 1985 Mini-Symposium No. 6.

Bolz, G. R., and Lough, R. G. 1983. Growth of larval Atlantic cod, *Gadus morhua*, and haddock, *Melanogrammus aeglefinus*, on Georges Bank, spring 1981. Fish. Bull., U.S. 81:827–836.

Bolz, G. R., and Lough, R. G. 1988. Growth through the first six months of Atlantic cod, *Gadus morhua*, and haddock, *Melanogrammus aeglefinus*, based on daily otolith increments. Fish. Bull., U.S. 86:223–237.

Buckley, L. B. 1979. Relationships between RNA-DNA ratio, prey density, and growth rate in Atlantic cod *(Gadus morhua)* larvae. J. Fish. Res. Board Can. 36:1497–1502.

Buckley, L. J., and Lough, R. G. 1987. Recent growth, biochemical composition, and prey field of larval haddock *(Melanogrammus aeglefinus)*, and Atlantic cod *(Gadus morhua)* on Georges Bank. Can. J. Fish. Aquat. Sci. 44:14–25.

Cohen, E. B., and Grosslein, M. D. 1982. Food consumption by silver hake *(Merluccius bilinearis)* on Georges Bank with implications for recruitment. In Gutshop '81. Fish food habits studies, Proceedings of the Third Pacific Workshop, Wash. Sea Grant. pp. 286–294. Ed. by G. M. Calliet and C. A. Simenstadt. Univ. Wash., Seattle.

Cohen, E. B., and Grosslein, M. D. 1987. Production on Georges Bank compared with other shelf ecosystems. In Georges Bank, pp. 383–391. Ed. by R. H. Backus. MIT Press, Cambridge, MA.

Cohen, E. B., Mountain, D. G., and Lough, R. G. 1986. Possible factors responsible for the variable recruitment of the 1981, 1982, and 1983 year classes of haddock *(Melanogrammus aeglefinus* L.) on Georges Bank. NAFO SCR Doc. 86/110, 27 pp.

Cohen, E. B., Sissenwine, M. P., and Laurence, G. C. 1988. The "recruitment problem" for marine fish populations with emphasis on Georges Bank. In Toward a theory on biological-physical interactions in the world ocean. pp. 373–392. Ed. by B. J. Rothschild. NATO ASI Series C: Mathematical and physical sciences, Vol. 239. Kluwer, Dordrecht, The Netherlands. 650 pp.

Cushing, D. H. 1974. The possible density-dependence of larval mortality and adult mortality in fishes. In The early life history of fish. pp. 103–111. Ed. by J. H. S. Blaxter. Springer-Verlag, Berlin.

Cushing, D. H., and Harris, J. G. K. 1973. Stock and recruitment and the problem of density dependence. Rapp. P.-v. Reun. Cons. Perm. int. Explor. Mer 164:142–155.

Flierl, G. R., and Wroblewski, J. S. 1985. The possible influence of warm core Gulf Stream rings upon shelf water larval fish distribution. Fish. Bull., U.S. 83:313–330.

Gulland, J. A. 1965. Survival of the youngest stage of fish and its relation to year-class strength. ICNAF Spec. Publ. No. 6:363–371.

Houde, E. D. 1986. Potential for growth, duration of early life stages, and regulation of recruitment in marine fish. ICES C.M. 1986/L:28, 11 pp.

International Council for the Exploration of the Sea [ICES]. 1987. Working Group on Larval Fish Ecology Report to the Biological Oceanography Committee. 1987. ICES C.M. 1987/L:28. 80 pp.

Jones, R. 1973. Density dependent regulation of the numbers of cod and haddock. Rapp. P.-v. Reun. Cons. int. Explor. Mer 164:156–173.

Jones, R. 1987. Report of the IREP Steering Committee. ICES C.M. 1987/A:3, 5 pp.

Kane, J. 1984. The feeding habits of co-occurring cod and haddock larvae from Georges Bank. Mar. Ecol. Prog. Ser. 16:9–20.

Laurence, G. C. 1974. Growth and survival of haddock (Melanogrammus aeglefinus) larvae in relation to planktonic prey concentration. J. Fish. Res. Board Can. 31:1415–1419.

Laurence, G. C. 1978. Comparative growth, respiration and delayed feeding abilities of larval cod (Gadus morhua) and haddock (Melanogrammus aeglefinus) as influenced by temperature during Laboratory studies. Mar. Biol. 50:1–7.

Laurence, G. C. 1981. Modelling — an esoteric or potentially utilitarian approach to understand larval fish dynamics? Rapp. P.-v. Reun. Cons. int. Explor. Mer 178:3–6.

Laurence, G. C. 1985. A report on the development of stochastic models of food limited growth and survival of cod and haddock larvae on Georges Bank. In Growth and survival of larval fishes in relation to the trophodynamics of Georges Bank cod and haddock. NOAA Tech. Mem. NMFS-F/NEC-36. pp. 84–159. Ed. by G. C. Laurence and R. G. Lough. 150 pp.

Laurence, G. C., and Burns, B. R. 1982. Ichthyoplankton in shelf water entrained by warm core rings. EOS, Trans. Am. Geophys. Union 63(45):998. Abstract #032A-07.

Laurence, G. C., and Rogers, C. A. 1976. Effects of temperature and salinity on the comparative embryo mortality of Atlantic cod (Gadus morhua) and haddock (Melanogrammus aeglefinus). J. Cons. int. Explor. Mer 36(3):220–228.

Laurence, G. C., Green, J. R., Fofonoff, P. W., and Burns, B. R. 1984. Small-scale spatial variability of plankton on Georges Bank with particular reference to prey organisms of larval cod and haddock. ICES C.M. 1984/L:9. 10 pp.

Leak, J. L., and Houde, E. D. 1987. Cohort growth and survival of bay anchovy Anchoa mitchilli larvae in Biscayne Bay, Florida. Mar. Ecol. Prog. Ser. 37:109–122.

Lough, R. G. 1984. Larval fish trophodynamic studies on Georges Bank: Sampling strategy and initial results. In The propagation of cod (Gadus morhua L.). pp. 395–434. Ed. by E. Dahl, D. S. Danielssen, E. Moksness, and P. Solemdal. Flodevigen rapportser 1.

McGurk, M. D. 1986. Natural mortality of marine pelagic fish eggs and larvae: role of spatial patchiness. Mar. Ecol. Prog. Ser. 34:227–242.

NOAA Ad Hoc Scientific Committee on Recruitment. 1987. Directed research NOAA's Cooperative Recruitment "CORE" Program. A Development Plan. 55 pp.

Ramp, S. 1982. Georges Bank current meter monitoring survey scientific report. Appendix. Prepared for Mobil Research and Development Corp. Tech. Rep. #3, ENDECO, Inc., Marion, MA.

Ricker, W. E., and Foerster, R. E. 1948. Computation of fish production. Bull. Bingham Oceanogr. Collect. 11:173–211.

Shepherd, J. G., and Cushing, D. H. 1980. A mechanism for density-dependent survival of larval fish as the basis of a stock-recruitment relationship. J. Cons. int. Explor. Mer 39:160–167.

Sherman, K. 1987. Large marine ecosystems as global units for recruitment experiments. ICES C.M. 1987/L:38. 16 pp.

Sherman, K., and Alexander, L. M. (Editors). 1986. Variability and management of large marine ecosystems. AAAS Selected Symposium 99, Westview Press, Boulder, CO. 319 pp.

Sissenwine, M. P. 1984. Why do fish populations vary? In Dahlem workshop on exploitation of marine communities. pp. 59–64. Ed. by R. May. Springer-Verlag, Berlin.

Smith, W. G., and Morse, W. W. 1985. Retention of larval haddock Melanogrammus aeglefinus in the Georges Bank region, a gyre-influenced spawning area. Mar. Ecol. Prog. Ser. 24:1–13.

Taylor, C. C. 1958. Cod growth and temperature. J. Cons. int. Explor. Mer 23(3):366–370.

Vladimirov, V. I. 1975. Critical periods in the development of fishes. J. Ichthyol. (Transl.) 15:851–868.

Chapter 12

Perspectives on Larval Fish Ecology and Recruitment Processes
Probing the Scales of Relationships

Christopher T. Taggart and Kenneth T. Frank

Abstract

The scale of the test in larval fish ecology research must match the scale of the hypothesis. This chapter illustrates the importance of resolving significant scales of variation by drawing on published studies, unpublished data, and theoretical modeling. Aliasing a cyclic variation in egg abundance is used to show how various sampling frequencies can result in an apparent increase, decrease, or near-random pattern in abundance estimates. Egg, larval, and postlarval mortality estimates may be biased in a time-dependent manner by assumptions concerning constant production or immigration and emigration, the spatial distribution of the target population, and the exponential decay model. A theoretical model is used to illustrate how misleading interpretations can be drawn from the relationship between larval mortality and instantaneous measures of larval predator and prey abundances that are biased by different rate processes having different scales of operation. The results of an extensive field program are used to illustrate how water-mass dynamics and the characteristic fauna associated with them can produce asynchronous patterns of coupled predator/prey oscillation that have no biological basis. Identifying the actual relationships is shown to be achieved only by resolving the biological processes at and below the scales of the physically driven processes. Finally, we will illustrate how localized small-scale, high-resolution approaches to hypothesis testing can be scaled upward and tested at the scale of large marine ecosystems (LMEs).

Introduction

Robust predictions of recruitment variation among freshwater, anadromous, catadromous, or marine fish stocks inhabiting relatively small "closed" inland lakes and seas or more "open" large marine ecosystems have proven to be unobtainable after more than a century of fairly intensive efforts.

We suggest that two principal reasons exist for the present lack of predictive capability. First, there is no general theoretical framework acceptable to all researchers within which to place the hypothetical-deductive scientific method. Second, and perhaps most important, is our collective failure to test hypotheses (within or outside of what loose theoretical framework that does exist) correctly scaled to the relationships thought to exist. In brief, we emphasize that the scale of the test must match the scale of the hypothesis, an approach championed by Harper (1977) in the field of terrestrial plant ecology.

In this chapter we stress the importance of resolving the significant temporal and spatial scales of variation by drawing on published studies, unpublished data, and some theoretical modeling. We show how localized, small-scale, high-resolution approaches to hypothesis testing can be scaled upward and tested at the scale of LMEs. Some solutions are suggested to improve the predictive capabilities of environmentally based models of recruitment variation.

Scale and the Scientific Method

Logistics often dictate the scale of data acquisition in ichthyoplankton studies (Smith, 1978). Sampling is spaced on the order of tens of km covering areas of hundreds to thousands of km^2 over time scales of weeks to months. However, the greatest rate changes and variations in growth, mortality, dispersion, and reaggregation (schooling) apparently occur during the egg, larval, and early postlarval stages, and at scales below those most commonly employed for sampling. This could explain in part the general failure of hypotheses related to the early life history to be rejected (or supported) simply because of inappropriately scaled sampling. This can lead to a situation where the sampling scale employed distorts our perception of the significant time and space scales.

The scientific method proceeds more or less as follows (Andrewartha, 1957): from existing observations and data, a conceptual model is formulated by deduction. Using mathematically supported logic, the model is made operational and restated as a testable hypothesis. The empirical test requires measurable independent variables (Peters, 1980). The test can result in either (i) rejection or the inability to reject the hypothesis; (ii) suggestions that the operational model is flawed due to unsound logic; (iii) suggestions of a flawed model because of erroneous or insufficient empirical knowledge and/or a faulty interpretation of the knowledge, or (iv) statistically indefensible rejection of the hypothesis because of insufficiently rigorous data collection and analysis. We suggest that the latter two results are the most critical in fisheries research, because scale-related problems easily result in biased data and analyses that can form the basis for new models supported by untested assumptions.

Recruitment Variation and the Environment

Explanations of recruitment variation range from the belief that recruitment is some function of stock size or biomass, to the idea that it is solely dependent on variations in the environment. Neither is really exclusive of the other (Rothschild, 1986), yet it is undeniable that the environment exerts a major influence on recruitment variation in marine finfish and shellfish populations. Many published studies (see, for example, Cushing and Dickson, 1976; Hempel, 1978; Lasker, 1978; Bakun and Parish, 1980; Bardach and Santerre, 1981; Bakun et al., 1982; Shepherd et al., 1984; Sissenwine, 1984; Kawasaki, 1985; Frank et al., 1988) have reviewed the vast body of literature reporting relationships between recruitment (or some proxy thereof) and some physical-environmental factor(s). In spite of this ever-increasing body of correlative evidence, the fisheries scientific community is far from agreement on the significance of these studies for the following reasons: poor predictive power due to statistical problems related to sample size; post publication failure; a posteriori approaches where several variables are tested but their number and the lags employed (in time-series data) are not reported (distorted statistical significance); autocorrelative problems in both the recruitment data and the techniques used to generate them, and in the environmental data (Bradford and Peterman, 1989; Thompson and Page, 1989); and the limitations of the environmental data that are available (see also Gulland, 1952, 1965; Bell and Pruter, 1958; Walters and Ludwig, 1981; Sissenwine, 1984).

Such criticisms are indeed valid. Most environment-recruitment correlation studies are based on the premise that year-class strength is established during the early life history. If correct, then it can be argued that most of these studies have used inappropriate time and space scales for measuring and comparing the biological and physical variables. A fundamental requirement in the analysis of the influence of environmental effects on the growth and survival of the early life stages is measurement at appropriate time and space scales. Failure to meet this requirement can have profound effects on the data and subsequent analysis and interpretation. Even when hypotheses determine the sampling strategy, this does not ensure their proper evaluation. Scaling is a multifaceted problem. It ranges from simple aliasing to the use of sampling gear. We now highlight some scale-related problems and their association with theory and hypothesis testing.

Aliasing

Aliasing is simply the problem that exists when a significant amount of variation in a process is not resolved due to an observational scale (spatial or temporal) that is too large. The biasing effect of the phenomenon is most critical to resolving cyclic processes. We present a hypothetical example in Figure 12.1, which shows that the abundance of eggs in a water column is a function of tidal velocity (eggs are resuspended or advected to the sampling area during each tidal cycle). Egg sampling conducted with a fixed daily periodicity would not reflect the true cyclic nature of abundance variation. Instead, it would present an apparent decay in abundance (if sampling ended after a few days) that could be misinterpreted as mortality or hatching (Figure 12.1, line A). Shifting the fixed daily sampling by approximately six hours would result in an apparent increase in abundance (Figure 12.1, line B). If, in either case, sampling continued for several days, the apparent cyclic behavior would still bear no relationship to reality. If the daily sampling time was not fixed, but varied by a few hours among days, then a more-or-less random pattern would be apparent (Figure 12.1, line C). Although in this example the variation in the biological phenomenon is physically driven (tidal), this fact is not readily apparent from any of the three sampling strategies employed. It is clear that aliasing effects easily distort the estimates of statistical parameters (e.g., mean, variance) that are frequently the basis for hypothesis testing. This phenomenon can occur at all scales due to physical (tides, storms, seasons, etc.) and biological (diel migrations, serial spawning, etc.) processes that have characteristic periodicities.

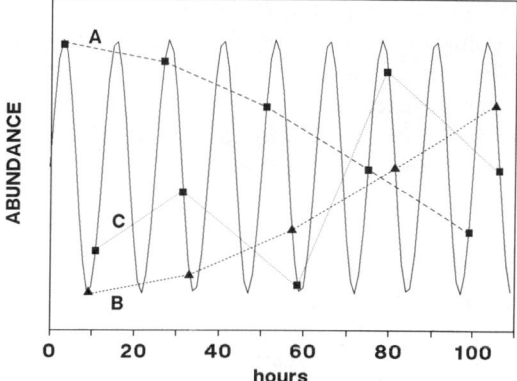

Figure 12.1. An example of aliasing the abundance of eggs (solid line), that vary as a function of tidal velocity, by sampling with different but fixed daily periodicities (lines A and B) and with a slightly varying (± a few hours) daily periodicity (line C).

Scale Effects on Mortality Estimates and Assumptions

High mortality estimates appear typical for marine fish eggs and larvae. However, few estimates are based on replicate samples, and their range and variability reflect both natural variation and sampling biases that stem from the gear type, sampling frequency, and the spatial and temporal resolution employed to assess mortality (Taggart and Leggett, 1987a,b). Sampling frequency typically ranges from once every 4–14 days (d) for yolk-sac larvae (Dahlberg, 1979), to as long as 100 d for post larvae (Graham and Townsend, 1985). In most studies, changes in abundance due to immigration and emigration (diffusion and advection) are not considered. While field sampling is typically large-scale, it is assumed that (i) the target population is adequately sampled, (ii) dispersion does not bias the mortality estimates, and (iii) immigration or production of new individuals is nil, or positive but constant (Taggart and Leggett, 1987b). Furthermore, it is generally assumed that the best estimate of mortality is defined by the exponential decay model fitted to abundance estimates over a series of sampling intervals. The implicit assumption is that deviations from the fitted exponential reflect "sampling error" and not variation due to immigration, emigration, or mortality.

Mortality and sampling intervals: An empirical relationship

In reviewing the literature on the mortality estimates (M) of marine fish eggs and larvae (Table 12.1), we noted that studies reporting high mortality rates (Dragesund and Nakken, 1971; Fortier and Leggett, 1985; Hewitt et al., 1985; Taggart and Leggett, 1987b) were those that were designed to directly measure mortality over short intervals (0.5 to 3 d). They were also studies that corrected for immigration and emigration to and from the target populations. In the data summarized in Table 12.1, we also noted an apparent relationship between the estimates of daily mortality and the length of the interval between sampling surveys that were used to estimate abundance and subsequently derive the mortality estimates using the exponential decay model. Linear regression analyses were used to examine the relationship between estimates of daily mortality (% d^{-1}) and the sampling intervals (d) used to derive the estimates. The data were normalized with the arcsin-square root and logarithmic transformations.

Daily mortality decayed exponentially for all developmental stages with increasing duration of the sampling interval (Figure 12.2). The equation describing the relationship (Table 12.2) was highly significant ($r^2 = 0.68$). The mortality estimates for different stages also showed an exponential decay in mortality with sampling interval (Figure 12.2). Each was statistically significant, though the strength of the relationships weakened with increasing development. It is unlikely that the developmental stage is solely responsible for the phenomenon, as an analysis of covariance showed that the slopes of the egg, larval, and postlarval relationships were homogeneous ($F = 0.31$, $p = 0.731$, $n = 58$). Clearly, our argument would be stronger using a species-specific approach, but insufficient data forced us to use a composite.

Mortality and sampling intervals: Theoretical relationships

Modeling results (Figure 12.3) show the expected relationships between the daily mor-

Table 12.1. Data sources from the published literature giving the number of daily mortality estimates of marine fish eggs, larvae, and postlarvae. These data were used to determined the relationship illustrated in Figure 12.2.

Source	Eggs	Larvae	Postlarvae	Comments
Dahlberg, 1979	11	12	13	Numerous species; numerous references
Graham & Townsend, 1985	—	—	17	Atlantic herring; 7 different cohorts
Frank & Leggett, 1986; and unpublished data	—	18	—	Capelin; 8-11 d enclosure experiments
Fortier & Leggett, 1985	—	1	—	Capelin; 48 h drift study
Taggart & Leggett, 1987b	—	40	—	Capelin; multiple short-term estimates
Ware & Lambert, 1985	—	1	—	Atlantic mackerel; average estimates
Munk et al., 1986	—	2	—	Atlantic herring; 1984 yr-class, 2nd cohort

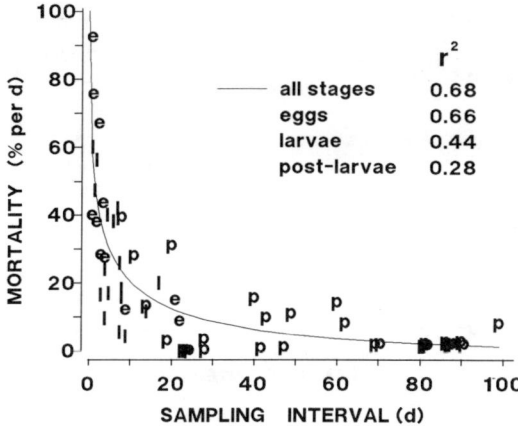

Figure 12.2. The relationship between estimated daily mortality (%/d) and the time interval over which the estimate was derived for marine fish eggs (e), larvae (l), and post larvae (p) from the sources listed in Table 12.1.

tality estimated for a series of increasing sampling intervals (maximum 18 d) using the exponential decay model and the interval between the abundance estimates used to calculate z (instantaneous mortality) and M for various theoretical changes in z over time. Not surprisingly, the relationship is flat and linear when z is constant and therefore independent of the sampling interval. When z declines linearly with time, there is a monotonic decay in apparent mortality with the interval. However, when z declines exponentially with time, so does the relationship between apparent mortality and the in-

Table 12.2. Regression statistics for the relationship between the arcsin square-root of daily mortality in all stages, eggs, larvae, and postlarvae of marine fish species (see Table 12.1 and Figure 12.2.), and the natural logarithm of the interval between sampling surveys used to derive the mortality estimates. All relationships are significant ($p < 0.01$).

Development stage	n	Slope	Intercept	r^2
All stages	58	−9.01066	47.47616	0.678
Eggs	11	−12.32164	54.54774	0.660
Larvae	17	−10.33155	46.63857	0.441
Postlarvae	30	−6.70334	39.28084	0.278

terval, in a manner similar to the empirically defined relationship (Figure 12.2). The addition of an immigration term to the population model (Figure 12.3) results in a more rapid decay in the estimated mortality with increasing sampling intervals, a decay rate that further approaches the relationship in Figure 12.2. However, it can be easily shown that the estimated decay in a population over time using z (estimated at the end of a given interval) is only correct when z is constant. The estimated decay in population size increasingly deviates from the actual decay when z declines either linearly or exponentially with time and as the interval used to estimate z increases.

Mortality and sampling intervals: Possible explanations

Based on the outcome of these simulations, the empirical relationship (Figure 12.2) has at least two possible explanations. First, daily mortality (Table 12.1) was poorly described by a simple exponential decay model, and a two-parameter model is required to account for mortality during the earliest developmental period (e.g., size- or age-dependent mortality; see, for example, Peterson and Wroblewski, 1984). If true, a simple exponential model fitted to the same data would yield accurate estimates only when the data were obtained over very short intervals concentrated near the beginning of the developmental stage affected. Longer intervals would yield biased mortality estimates in the absence of constant mortality. Second, a simple exponential decay model is representative of mortality, but the abundance variation is unrelated to mortality, and the effect on apparent mortality increases with the measurement interval. Variation in immigration and/or emigration rates, size- (age-) dependent changes in catchability (gear-related), and changes in contagiousness would all produce such an effect. McGurk (1986) touched on the potential for bias in mortality estimates caused by changes in larval patchiness with time, but

156 / Taggart & Frank

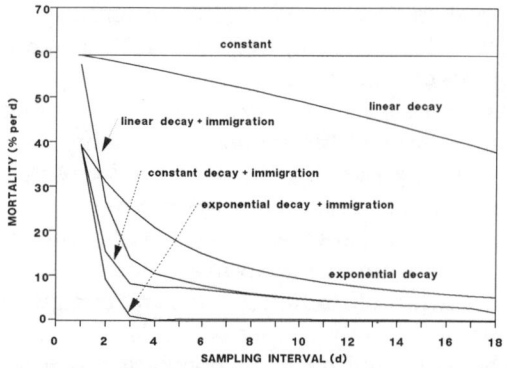

Figure 12.3. The expected relationship between daily mortality estimated using the simple exponential decay model at increasing sampling intervals when actual mortality is constant or decays linearly or exponentially with and without an immigration term.

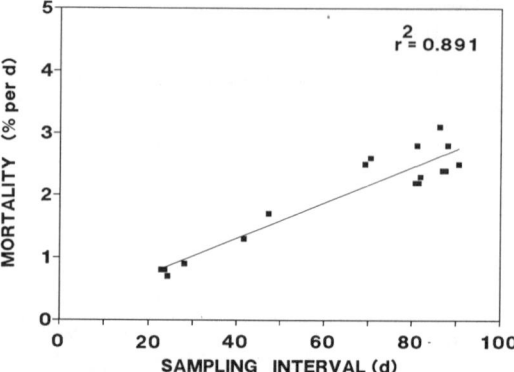

Figure 12.4. The relationship between larval and post larval herring mortality and the length of the sampling interval calculated from Graham and Townsend (1985).

he invoked predation to explain the relationship between patchiness and mortality. Our analysis suggests that temporal scale effects related to sampling design alone will artificially bias the mortality estimates.

The effect of increasing population losses through either increased sampler avoidance or emigration (diffusion and/or advection) would not account for the relationship and may be demonstrated by the relationship between larval and postlarval herring mortality estimates and sampling intervals taken from Graham and Townsend (1985). Their data show that the mortality estimates of several larval herring cohorts increased linearly with increasing survey intervals (Figure 12.4) and indicates that either time- (age- or size-) dependent sampler avoidance or advective losses were responsible. Graham and Townsend (1985) assumed that advective losses were insignificant, whereas the relationship in Figure 12.4 indicates that such losses may have significantly biased the mortality estimates.

Mortality and sampling intervals: Possible implications

The effect of eliminating mortality-independent population losses and gains on the relationship between daily mortality and sampling interval is well illustrated by the daily mortality estimates provided by Frank and Leggett (1986, and unpublished data). In their study of mortality in capelin larvae in large, in situ enclosures, the biasing effects of diffusion, advection, and larval production were eliminated. Their data show that there is a very slight decline in the mortality rate of larvae in different rearing conditions when evaluated against increasing sampling intervals (Figure 12.5). This relationship is consistent with the idea that z is constant (Figure 12.3).

Taggart and Leggett (1987b) determined 40 different field estimates of short-term daily mortality in sequentially emerging populations (here defined as cohorts) of capelin larvae. There was no relationship be-

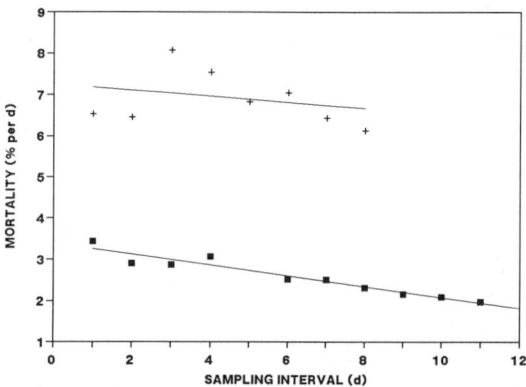

Figure 12.5. The relationship between mortality in enclosed larval capelin populations and the length of sampling intervals calculated from Frank and Leggett (1986, and unpublished data).

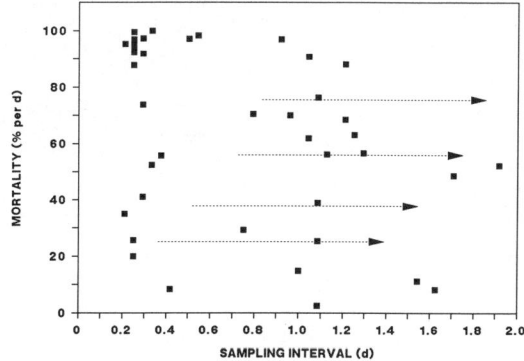

Figure 12.6. The relationship between 40 different field estimates of larval capelin mortality and the length of sampling intervals calculated from Taggart and Leggett (1987c). Arrows depict the hypothetical mortality trajectory among different cohorts if mortality within cohorts was constant.

tween the mortality estimate and the duration of the sampling interval (Figure 12.6). However, immigration (hatching and emergence) and emigration (through advection) were considered when calculating the estimates. Immigration alone contributed considerable bias to the mortality estimates. If mortality was constant within any given cohort, but varied among cohorts as in Figure 12.5, then the hypothesis that mortality was more or less constant within cohorts but varied among cohorts is feasible (Figure 12.6). This implies that those cohorts with the low mortality rates will persist longer relative to those with high mortality rates. Such a suggestion is consistent with Lambert's (1984) hypothesis that relatively few of many cohorts may be responsible for year-class strength.

The relationships presented here, and the demonstrated influence of immigration and advection on mortality estimates (Taggart and Leggett, 1987b), indicate that mortality estimates are easily biased by assumptions concerning (i) constant production or immigration and emigration, (ii) the spatial distribution of the target population, and (iii) the validity of the simple exponential decay model. Failure to consider these constraints can lead to biases in mortality estimates that seemingly accelerate with time. McGurk (1989) has reached a similar conclusion and has called for the development of an analytical advection-diffusion-mortality model that incorporates a time-varying z.

Predator-Prey Relationships: A Simulation Example

Some of the leading hypotheses to explain recruitment variation deal with starvation and predation. Each of these can be physically mediated. Recently, arguments have been made supporting the predation hypothesis over the food limitation hypothesis due to lack of supporting evidence (e.g., Sissenwine, 1984). Therefore, one might expect to find that larval mortality or abundance and predator abundance are correlated in some way. In fact, such arguments can form the basis of support for the predation hypothesis over the starvation hypothesis. The implication is that predation exerts a major effect on mortality, and therefore recruitment. Predation may be affected by the condition of the larvae. However, the scales of predation and starvation processes and the scales of sampling can easily distort the true nature of relationships. This point was made by Taggart and Leggett (1987c) when they warned that misleading inferences can be drawn from data that are collected near instantaneously and used to test competing hypotheses that pertain to near-instantaneous processes (predation) and to those whose effect is cumulative over a relatively longer period (starvation).

To illustrate this phenomenon, we simulated a larval population simultaneously purposely subjected to high daily mortality directly related to food abundance and relatively low predation mortality directly related to predator abundance. In the model, predator and food abundances (Figure 12.7) were drawn randomly from rectangular distributions (predator mean = 0.333; food mean = 0.5). Predator-(P) related larval mortality operated instantaneously at time t(d) and was directly proportional to predator abundance at t. Food-(F) related mortality at t was proportional to the average

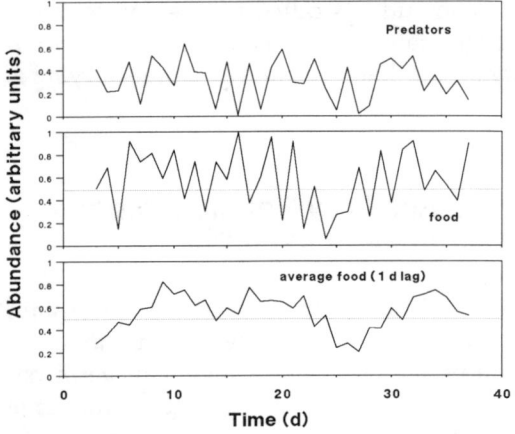
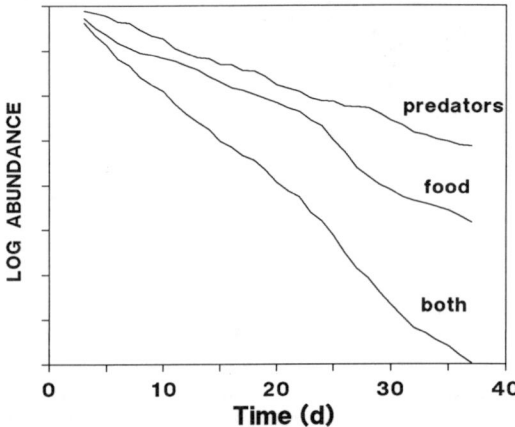

Figure 12.7. Simulation trial of daily predator and daily food abundance and 3 d averaged food abundance lagged by 1 d (average abundances are indicated by the dotted line). The predator and the lagged average food abundance estimates were used to drive the population decay (predator and food related mortality) illustrated in Figure 12.8.

Figure 12.8. Simulation trial of larval population decay resulting from daily predator related mortality only, lagged average food related mortality only (see Figure 12.7), and by both mortality factors.

food abundance over the previous three intervals $F_{ta} = [(F_{t-3} + F_{t-2} + F_{t-1})/3]$ to simulate starvation (Figure 12.7).

The effect of these two processes on the simulated population yielded time-dependent exponential decay in numbers of larvae and illustrated the greater impact of food limitation relative to predator-induced mortality (Figure 12.8). The population surviving both mortality processes was used to determine the daily mortality estimates, which were subsequently compared with the instantaneously "sampled" predator (P_t) and food (F_t) abundances. Perhaps not surprisingly, there was no significant relation between mortality and instantaneous measures of food abundance (Figure 12.9). In contrast, there was a significant relationship between mortality and predator abundance (Figure 12.10). If F_{ta}, the lagged average food abundance (not easily measured in the field), is used as the independent variable, a significant relationship with mortality is evident (Figure 12.11). Replicate trials with this model show these results to be repeatable (Figure 12.12).

In spite of the simplicity of the model and the imbedded assumptions, it serves to illustrate that the analysis of instantaneous state measures, directly related to rate processes that have distinctly different scales of operation, can lead to misleading interpretations and conclusions. Frequently, the importance of a given factor is assigned on the basis of the significance level of the chosen statistic. For example, Taggart and Leggett (1987c) were guided by this erroneous principle and assigned more significance to predator abundance than was warranted. Nevertheless, they viewed such interpretations with caution.

Predator-Prey Relationships: A Field Example

Several field studies have suggested predator-dominated control of larval fish populations based on inverse correlations between predator and prey abundances (see review in Frank and Leggett, 1985). However, such inverse correlations are often suspect due to failure to consider the influence of the physical structure and dynamics of the environment where the predator and prey populations are found (Frank and Leggett, 1985).

Several potential predators of marine fish larvae are confined to subpycnocline waters and occur in large quantities in the

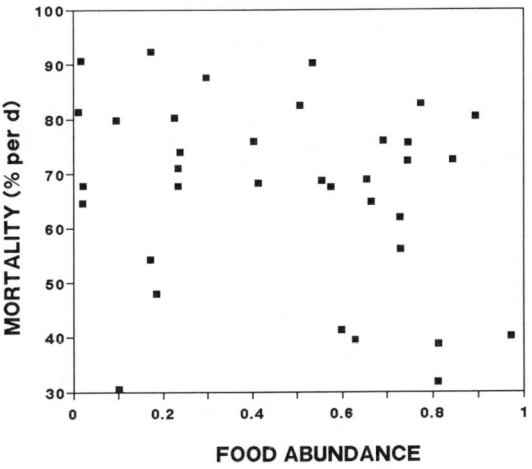
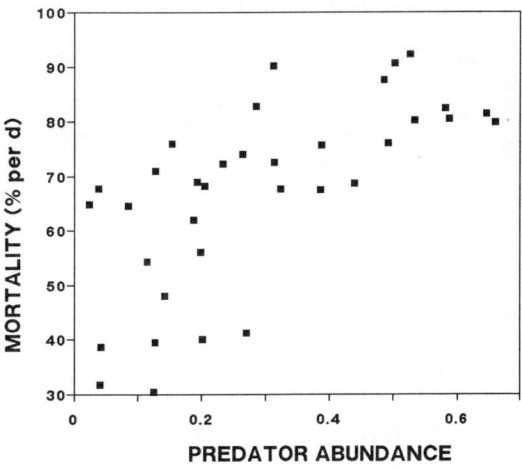

Figure 12.9. The relationship between simulated mortality calculated daily from the population ("both") illustrated in Figure 12.8 and "food" illustrated in Figure 12.7.

Figure 12.10. The relationship between simulated mortality calculated daily from the population ("both") illustrated in Figure 12.8 and "predators" illustrated in Figure 12.7.

nearshore during the upwelling of cold water masses. Upwelling favorable winds (SW) are the prevailing winds in summer in eastern Newfoundland, and large quantities of chaetognaths, jellyfish, and ctenophores (known from laboratory and field studies to consume significant quantities of fish larvae; e.g., deLafontaine and Leggett, 1988) occur in the nearshore during the course of wind-driven upwelling. Their occurrence and daily variation in abundance is a function of the wind field, with greatest abundances occurring during SW winds (Frank and Leggett, 1982a; 1985). Capelin spawning occurs on beaches in this area during June, and the mass emergence of larvae from the beach sediments is timed to periods of warm surface water mass intrusions driven by NE winds (Frank and Leggett, 1981a; Taggart and Leggett, 1987a). Winds from the NE sector are associated with low pressure systems passing across eastern Newfoundland with an approximate 5 d periodicity. Warm water intrusions in the nearshore are coupled with increased wave turbulence, and together they facilitate the emergence of capelin larvae from the beach sediments.

Warm surface waters are characterized by a prey size spectrum in the optimum size range for first feeding larvae and a depauperate predator community (Frank and Leggett, 1986). Conversely, cold water upwelling in the nearshore during SW winds inhibits larval emergence through a temperature-related reduction in larval activity and the absence of wave turbulence. The survival potential of larvae would be considerably lower in the upwelling water mass due to an unfavorably large prey size spectrum coupled with the abundant predator suite. Selective occupation of the warm surface water mass through episodic mass emergences greatly improves the growth and survival potential of capelin larvae. The characteristic fauna associated with each of these water masses produces an asynchronous temporal pattern of occurrence in the nearshore resembling a coupled predator/prey oscillation that has no biological basis whatsoever (Frank and Leggett, 1985). The actual relationship between the larvae, their predators, and their prey was identified only by resolving the biological processes at and below the scales of the physically driven hydrological processes. High-resolution approaches of this kind can reveal fairly precise larval fish, predator, and zooplankton community structures that are clearly related to

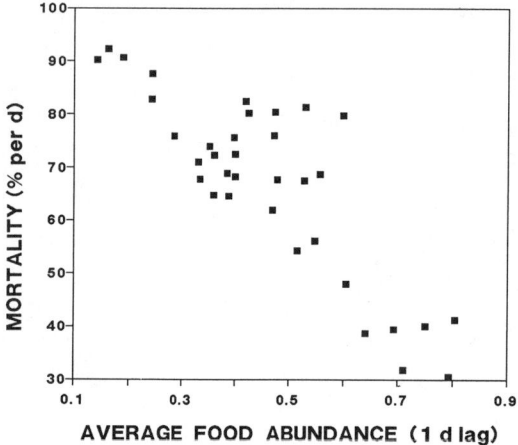

Figure 12.11. The relationship between simulated mortality calculated daily from the population ("both") illustrated in Figure 12.8 and "average food (1 d lag)" illustrated in Figure 12.7.

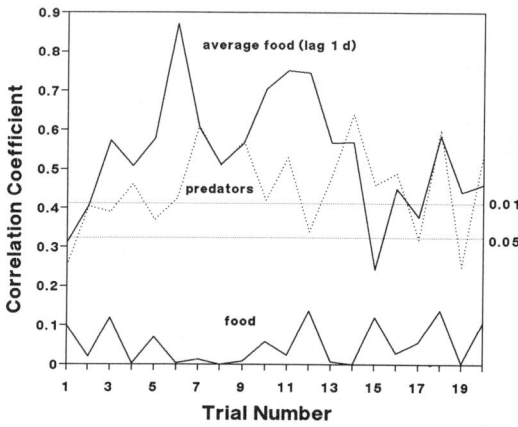

Figure 12.12. Variation in the correlation coefficients between daily mortality and food, predators, and lagged average food. Each series was calculated from 20 different trials (60 trials in all). Each trial was similar to those illustrated in Figures 12.7 through 12.11. Significance levels ($p = 0.01$ and 0.05) are noted.

the physical characteristics of dynamic water masses (Taggart et al., 1989).

Scaling Upward to Large Marine Ecosystems

Capelin spawn on beaches having a similar orientation throughout their distributional range in eastern Newfoundland (hundreds of km), and the wind field driving the nearshore water mass dynamics occurs on an equivalent or larger scale. Therefore, any one spawning beach can be used to monitor the growth and survival patterns of the early life stages of capelin. This has, in fact, been confirmed and has culminated in a robust recruitment model for capelin (Leggett et al., 1983, 1984). The physical/biological coupling observed at different trophic levels in eastern Newfoundland has also led to accurate predictions of both net-fouling conditions and catch variation in the inshore cod fishery (Taggart and Frank, 1987), as well as the distribution of humpback whales in eastern Newfoundland (Whitehead and Carscadden, 1985). A similar approach was taken to the problem of predicting catch variation in cod along the north shore of the Gulf of St. Lawrence (Rose and Leggett, 1988), where variation in onshore movements and inshore catch rates of cod are associated with wind-driven upwelling along the coast.

Nikolsky (1965) stated that "every adaptation is at the same time a limitation" to an organism, and this philosophy underlies the capelin recruitment model alluded to previously. When the interval between onshore winds exceeds the long-term average, thereby prolonging the residence of capelin larvae in the beach gravel, the energy reserves of the larvae are depleted, their physical condition is reduced, and postemergent swimming ability is seriously impaired (Frank and Leggett, 1981a, 1982b). The conclusion was that the duration of beach residence, itself a direct function of the interval between meteorological events that trigger emergence, was a major regulator of early larval survival of capelin. In addition, the capelin recruitment model required knowing when median hatching occurred each year to define the relevant period to assess the wind field. This was possible only with the knowledge of spawning times and locations, and incubation temperatures in the beach gravel. Empirical models were then devised for the latter by using the appropriate meteorological and hydrographic factors as independent variables (Frank and Leggett, 1981b). The recruitment model for the Northwest Atlantic Fisheries Organization Subarea 2J3K stock explained greater than 70% of the interannual variation in the abundance of two-year-old capelin by incorporating the effects of wind and water temperature (Leggett et al., 1983), a remarkable result considering that the study began in an isolated embayment in Conception Bay, Newfoundland. Collectively, these findings suggested that abiotic factors operating at critical periods in larval development may be more important than spawning stock biomass as regulators of year-class strength. Critical periods must be identified and quantified at time scales relevant to individual larvae if reliable forecasting of year class strength is to be achieved, and the large spatial scale of the effects suggests that the space scale of individual larvae is of less importance in such analyses. This implies that long-term averages (e.g., annual) of abiotic factors will not be sufficient to explain variations in biological processes that are event-driven.

Analysis of age-composition data has revealed parallelism in strong year classes in the beach-spawning stock of capelin and a separate, offshore stock of capelin on the Southeast Shoal of the Grand Bank, some 350 km from the nearest spawning beach and a region where capelin annually spawn during June/July on the bottom at depths averaging 50 m (Carscadden, 1983). Recent field studies have been based on the working hypothesis that the same environmental factors operating during the immediate post hatching period are responsible for strong year classes in both stocks of capelin (Frank and Carscadden, 1989). During field studies in 1986 on the Southeast Shoal, the dominant cohort of larval capelin was produced during the passage of a major meteorological low pressure system (tropical storm Charley) that coincided with sharp increases in bottom temperature and currents over the spawning grounds. This sequence of changes in the water column structure appeared to reflect an episode of destratification due to in situ mixing. The passage of the tropical storm near the study area was not an unusual event considering that the annual occurrence of tropical cyclones during the period 1969–1986 was between three and 12 (Neumann et al., 1978). Between one and five of these storms pass over the southern Grand Bank annually, primarily during August and September of the 18-year period, and August is the principal hatching period for capelin in this region (Frank and Carscadden, 1989). It was hypothesized that emergence, timed to periods of destratification, is beneficial to larval survival due to the rapid ascent of larvae from the bottom waters. During the initial stages of larval drift associated with the passage of storms the normally high predator densities that occur during stratified conditions are diluted, and feeding conditions are enhanced (Frank and Carscadden, 1989). Once again it appears that the temporal scale of variability is dominant, and storms serve an important role in this respect, acting to obliterate any predispersal spatial variability and to establish a common starting condition (both in terms of the physical and biological state of the water mass) in the early stages of larval dispersal in capelin.

Species whose life-stage transitions depend on such temporal events during the early life history should be more amenable to development of robust recruitment models. Recent examples in the literature support this view. Red drum (Sciaenops ocellatus) spawning coincides with the hurricane season in the Gulf of Mexico, with the result that the timing of hurricanes can influence the numbers and/or survival of juvenile red drum advected into nearshore nursery areas from offshore spawning sites (Matlock, 1987). The above-average, nearshore-directed larval transport to nutrient-rich estuarine nurseries that occurs during hurricanes appears to result in strong year classes of red drum (Matlock, 1987). Atlantic menhaden (Brevoortia tyrannus) also appear to have evolved a reproductive strategy to optimize the survival and shoreward transport of eggs and larvae. Schooling adults located along the western edge of the Gulf

Stream and south of Cape Hatteras time their offshore spawning to the occurrence of winter storms (Checkley et al., 1988). These winter storms result in upwelling induced spawning and buoyancy-driven shoreward transport of menhaden eggs and larvae to estuarine nursery areas. It has been hypothesized that variation in these physical conditions may explain interannual variation in menhaden recruitment and, possibly, recruitment in other species that spawn shoreward of warm boundary currents during winter months (Checkley et al., 1988). Finally, Myers and Drinkwater (1989) have shown that warm-core ring activity has the potential of reducing recruitment in 15 of 17 groundfish stocks (six species, including cod, pollock, redfish, yellowtail flounder, and silver hake) spawning on the continental shelf from the mid-Atlantic Bight to the Grand Bank.

Summary

We hope that this chapter will serve a useful purpose in warning researchers of the various pitfalls of inappropriate scaling when dealing with estimates of mortality during the early life of fishes. Sampling limitation, whether avoidable or not, underlies our collective inability to decide among several competing hypotheses concerning the growth and mortality of fishes. The implications of inappropriate scaling and the actual errors that might be incurred with regard to recruitment predictions for species in large marine ecosystems remains to be explored explicitly.

We recognize that there are several limitations to what we have presented and that our chapter can be criticized on several grounds. The literature used to support our arguments is not as broad as is perhaps possible. For example, McGurk (1989) has examined similar scale-related problems in the early life-stage dynamics of Pacific herring (*Clupea harengus*) from a different perspective, but he reached a strikingly similar conclusion (although we were unable to extract the simple measures and parameters to incorporate into our analyses.) It was also necessary to develop the relationship in Figure 12.2 by combining data for different species and different studies. The argument proposed would have been much stronger if examined on a species-by-species basis, but the data are limited. With more data, relationships such as that in Figure 12.4 (herring) may prove their utility and allow a prediction of biases in mortality estimates.

The collective experiences of many researchers (some cited herein) attempting to estimate in situ mortality and abundance has provoked some new approaches to recruitment determination. In the past, the majority of research into recruitment problems has focused almost exclusively on determining the abundance and mortality of eggs and larvae, of which the overwhelming majority never recruit to the fishery. We propose that the new insights needed to predict recruitment variations will come from the alternative approaches of studying the phenological, genetic, phenotypic, biochemical, and physiological traits of those individuals that survive and eventually recruit. This new emphasis on the characteristics of the survivors represents a unique departure from traditional thinking. The opportunity to embark in this new direction now exists as part of a Canadian National Sciences of Engineering Research Council (NSERC) multi-institutional and interdisciplinary programme called the Ocean Production Enhancement Network (OPEN).

Acknowledgments

We are grateful to W. C. Legget, S. Kerr, and an anonymous reviewer for their helpful insights, criticisms, and suggestions that led to improvements in the manuscript.

References

Andrewartha, H. G. 1957. Using conceptual models in population ecology. Cold Spring Harbor Symp. Quant. Biol. 22:219–232.

Bakun, A., and Parrish, R. 1980. Environmental inputs to fishery population models for eastern boundary currents. In Workshop on the effects of environmental variation on the survival of larval pelagic fishes. Ed. by G. D. Sharp. FAO Intergovernmental Oceanographic Commission Workshop Report No. 28.

Bakun, A., Beyer, J., Pauly, D., Pope, J. G., and Sharp, G. D. 1982. Ocean sciences in relation to living resources. Can. J. Fish. Aquat. Sci. 39:1059–1070.

Bardach, J. E., and Santerre, R. M. 1981. Climate and the fish in the sea. BioScience 31:206–215.

Bell, F. H., and Pruter, A. T. 1958. Climatic temperature changes and commercial yields of some marine fisheries. J. Fish. Res. Board Can. 15:625–683.

Bradford, M. J., and Peterman, R. M. 1989. Incorrect parameter values used in virtual population analysis (VPA) generate spurious time trends in reconstructed abundances. In International symposium on recruitment and errors in stock assessment models. Ed. by R. J. Beamish and G. A. McFarlane. Can. Spec. Publ. Fish. Aquat. Sci. 108:87–99.

Carscadden, J. E. 1983. Population dynamics and factors affecting the abundance of capelin (Mallotus villosus) in the northwest Atlantic. FAO Fish. Rep. No. 291 (3):789–811.

Checkley, D. M., Raman, S., Maillet, G. L., and Mason, K. M. 1988. Winter storm effects on the spawning and larval drift of a pelagic fish. Nature 335:346–348.

Cushing, D. H., and Dickson, R. R. 1976. The biological response in the sea to climatic changes. Adv. Mar. Biol. 14:1–122.

Dahlberg, M. D. 1979. A review of survival rates of fish eggs and larvae in relation to impact assessments. Mar. Fish. Rev. 41:1–12.

deLafontaine, Y., and Leggett, W. C. 1988. Predation by jellyfish on larval fish: An experimental evaluation employing in situ enclosures. Can. J. Fish. Aquat. Sci. 45:1173–1190.

Dragesund, O., and Nakken, O. 1971. Mortality of herring during the early larval stage in 1967. Rapp. P-v. Reun. Cons. int. Explor. Mer 160:142–146.

Fortier, L., and Leggett, W. C. 1985. A drift study of larval fish survival. Mar. Ecol. Prog. Ser. 25:245–257.

Frank, K. T., and Carscadden, J. E. 1989. Factors affecting recruitment variability of capelin (Mallotus villosus) in the Northwest Atlantic. J. Cons. int. Explor. Mer 45:146–164.

Frank, K. T., and Leggett, W. C. 1981a. Wind regulation of emergence times and early larval survival in capelin (Mallotus villosus). Can. J. Fish. Aquat. Sci. 38:215–223.

Frank, K. T., and Leggett, W. C. 1981b. Prediction of egg development and mortality rates in capelin (Mallotus villosus) from meteorological, hydrographic, and biological factors. Can. J. Fish. Aquat. Sci. 38:1327–1338.

Frank, K. T., and Leggett, W. C. 1982a. Coastal water mass replacement: Its effect on zooplankton dynamics and the predator-prey complex associated with larval capelin (Mallotus villosus). Can. J. Fish. Aquat. Sci. 39:991–1003.

Frank, K. T., and Leggett, W. C. 1982b. Environmental regulation of growth rate, efficiency, and swimming performance in larval capelin (Mallotus villosus) and its application to the match/mismatch hypothesis. Can. J. Fish. Aquat. Sci. 39:691–699.

Frank, K. T., and Leggett, W. C. 1985. Reciprocal oscillations in densities of larval fish and potential predators: A reflection of present or past predation? Can. J. Fish. Aquat. Sci. 42:1841–1849.

Frank, K. T., and Leggett, W. C. 1986. Effect of prey abundance and size on the growth and survival of larval fish: An experimental study employing large volume enclosures. Mar. Ecol. Prog. Ser. 34:11–22.

Frank, K. T., Perry, R. I., and Drinkwater, K. F. 1988. Changes in the fisheries of Atlantic Canada associated with global increase in atmospheric carbon dioxide: A preliminary report. Can. Tech. Rep. Fish. Aquat. Sci. No. 1652: v+52 p.

Graham, J. J., and Townsend, D. W. 1985. Mortality, growth, and transport of larval Atlantic herring Clupea harrengus in Maine coastal waters. Trans. Amer. Fish. Soc. 114:490–498.

Gulland, J. A. 1952. Correlations on fisheries hydrography. Letters to the editor. J. Cons. Perm. int. Explor. Mer 18:351–353.

Gulland, J. A. 1965. Survival of the youngest stages of fish and its relation to year-class strength. Spec. Publ. Int. Comm. Northw. Atlant. Fish. 6:363–371.

Harper, J. L. 1977. Population biology of plants. Academic Press, New York. 892 p.

Hempel, 1978. North Sea fisheries and fish stocks – a review of recent changes. Rapp. P.-v. Reun. Cons. int. Explor. Mer 173:145–167.

Hewitt, R. P., Theilacker, G. H., and Lo, N. C. H. 1985. Causes of mortality in young jack mackerel. Mar. Ecol. Prog. Ser. 26:1–10.

Kawasaki, T. 1985. Fisheries. In Climate impact assessment. SCOPE 27. Ed. by R. W. Kates, J. H. Ausubel, and M. Berberain. John Wiley and Sons Ltd., Toronto.

Lambert, T. C. 1984. Larval cohort succession in

herring *(Clupea harengus)* and capelin *(Mallotus villosus)*. Can. J. Fish. Aquat. Sci. 41:1552–1564.

Lasker, R. 1978. Ocean variability and its biological effects — a regional review, northeast Pacific. Rapp. P.-v. Reun. Cons. int. Explor. Mer 173:168–181.

Leggett, W. C., Frank, K. T., and Carscadden, J. E. 1983. Estimating year class strength in capelin *(Mallotus villosus)* from abiotic variables. Northwest Atlantic Fishery Organization SCR Doc. 83/IV/52 Ser. No. N710, 19 p.

Leggett, W. C., Frank, K. T., and Carscadden, J. E. 1984. Meteorological and hydrographic regulation of year-class strength in capelin *(Mallotus villosus)*. Can. J. Fish. Aquat. Sci. 41:1193–1201.

Matlock, G. C. 1987. The role of hurricanes in determining year-class strength of red drum. Contr. Mar. Sci. 30:39–47.

McGurk, M. D. 1986. Natural mortality of marine pelagic fish eggs and larvae: The role of spatial patchiness. Mar. Ecol. Prog. Ser. 34:227–242.

McGurk, M. D. 1989. Advection, diffusion, and mortality of Pacific herring larvae *Clupea harengus pallasi*, in Bamfield Inlet, British Columbia. Mar. Ecol. Prog. Ser. 51:1–18.

Munk, P., Christensen, V., and Paulsen, H. 1986. Studies of a larval herring *(Clupea harrengus)* patch in the Buchan area. II. Growth, mortality, and drift of larvae. DANA 6:11–24.

Myers, R. A., and Drinkwater, K. F. 1989. The influence of Gulf Stream warm core rings on recruitment of fish in the northeast Atlantic. J. Mar. Res. 47:635–656.

Neumann, C. J., Cry, G. W., Caso, E. L., and Jarvinen, B. R. 1978. Tropical cyclones of the north Atlantic Ocean, 1871–1977. NOAA, National Climatic Center, Asheville, NC.

Nikolsky, G. V. 1965. Theory of fish population dynamics. Nauka Press. 350 p.

Peters, R. H. 1980. Tautology in evolution and ecology. Amer. Nat. 110:1–12.

Peterson, I., and Wroblewski, J. S. 1984. Mortality rate of fishes in the pelagic ecosystem. Can J. Fish. Aquat. Sci. 41:117–1120.

Rose, G. A., and Leggett, W. C. 1988. Atmosphere-ocean coupling and Atlantic cod migrations: Effects of wind-forced variations in sea temperatures and currents on nearshore distributions and catch rate of *Gadus morhua*. Can. J. Fish. Aquat. Sci. 45:1234–1243.

Rothschild, B. J. 1986. Dynamics of marine fish populations. Harvard Univ. Press, Cambridge, MA. 277 p.

Sheperd, J. G., Pope, J. G., and Cousens, R. D. 1984. Variations in fish stocks and hypotheses concerning their links with climate. Rapp. P.-v. Reun. Cons. int. Explor. Mer 185:255–267.

Sissenwine, M. P. 1984. Why do fish populations vary? *In* Exploitation of marine communities. Ed. by R. M. May. Dahlem Konferenzen. Springer Verlag, New York.

Smith, P. E. 1978. Biological effects of ocean variability: Time and space scales of biological response. Rapp. P.-v. Reun. Cons. Int. Explor. Mer 173:117–127.

Taggart, C. T., and Frank, K. T. 1987. Coastal upwelling and Oikopleura occurrence ("slub"): A model and potential application to inshore fisheries. Can. J. Fish. Aquat. Sci. 44:1729–1736.

Taggart, C. T., and Leggett, W. C. 1987a. Wind-forced hydrodynamics and their interaction with larval fish and plankton abundance: A time-series analysis of physical-biological data. Can. J. Fish. Aqaut. Sci. 44:438–451.

Taggart C. T., and Leggett, W. C. 1987b. Short-term mortality in post-emergent capelin *Mallotus villosus*. I. Analysis of multiple in situ estimates. Mar. Ecol. Prog. Ser. 41:205–217.

Taggart C. T., and Leggett, W. C. 1987c. Short-term mortality in post-emergent capelin *Mallotus villosus*. II. Importance of food and predator density, and density dependence. Mar. Ecol. Prog. Ser. 41:219–229.

Taggart, C. T., Drinkwater, K. F., Frank, K. T., McRuer, J., and LaRouche, P. 1989. Larval fish, zooplankton community structure, and physical dynamics at a tidal front. Rapp. P.-v. Reun. Cons. int. Explor. Mer. 191:184–194.

Thompson, K. R., and Page, F. H. 1989. Detecting synchrony of recruitment using short autocorrelated time series. Can. J. Fish. Aquat. Sci. 46:1831–1838.

Walters, C. J., and Ludwig, D. 1981. Effects of measurement errors on the assessment of stock-recruitment relationships. Can. J. Fish. Aquat. Sci. 38:704–710.

Ware, D. M., and Lambert, T. C. 1985. Early life history of Atlantic mackerel *(Scomber scombrus)* in the southern Gulf of St. Lawrence. Can. J. Fish. Aquat. Sci. 42:577–592.

Whitehead, H., and Carscadden, J. E. 1985. Predicting inshore whale abundance — whales and capelin off the Newfoundland coast. Can. J. Fish. Aquat. Sci. 42:976–981.

Part Three:
Theory and Management of Large Marine Ecosystems

Introduction

Since the introduction of the large marine ecosystem (LME) approach at the 1984 AAAS annual meeting in New York, considerable development has taken place with respect to the concept. This development involves at least four main areas: (i) natural processes occurring with LMEs, (ii) variations in the productivity of the systems, (iii) identification and geographic extent of individuals LMEs, and (iv) techniques of comprehensive management. All four areas are represented in this section.

The first three chapters (13–15) address natural processes. There are discussions of processes and patterns within an LME at different scales of time and place, with the notation that "local processes are always imbedded in a matrix of larger scale processes" (Ricklefs, p. 169). Emphasis is placed on the interconnections between local systems, particularly the dynamics of predator-prey relationships, with the Antarctic krill fisheries utilized as a case study for hypothesizing and testing ecological theories. Multispecies analysis, as developed through integrated studies of large marine ecosystems, is providing new insights into fishing mortality. "Multispecies modeling enables a more complete analysis of energy flow through a species by its explicit consideration of predation, either as cannibalism or interspecific predation" (Bax and Laevastu, p. 188).

Variation in biomass yields in LMEs is also a new form of approach to fisheries management (Chapter 16). In some LMEs, population shifts and depletions appear to be the result of naturally occurring predation and excessive fishing mortality. For others, the predominant variable is environmental change in current dynamics and natural productivity. For still other LMEs, the cause of population change seems to be coastal pollution. An ecosystem research strategy is then suggested that will include both natural variabilities in terms of currents, water masses, and so on, and human perturbations in terms of not only fishing, but also marine pollution. These factors must be considered within the overall framework of the global greenhouse effect and an increase in atmospheric carbon dioxide levels.

The identification and mapping of individual LMEs on a global scale has been an ongoing process over the past several years. It has been suggested (Chapter 17) that some form of hierarchical system be devised for LMEs, including biogeographic units such as one covering the range of tuna migration, regional marine ecosystems of which there may be 25, and subsystems such as the Gulf of Maine–Georges Bank area. Whatever the arrangement, attention must be paid to jurisdictional economic and legal patterns affecting fisheries, in terms of their confluence with, or overlap of, the LME units themselves.

The techniques of LME management are addressed in Chapter 18 by Martin Belsky in terms of legal norms. "As a rule of international law, the ecosystem management approach is binding on nation-states. It therefore must be applied by those states in their domestic law and practice ... [and] in their foreign policy." Evidence is presented to support this view, with references to treaty law, convention law, and to the "ocean enclosure" movement, whereby nearly all of the commercial fisheries of the world today have been brought within the limits of national jurisdiction. The "ecosystem model" is with us in terms of ocean management; not only should it be accepted as a legal principle, but funds must be provided "to secure information and monitoring activities, and to assure appropriate application of the ecosystem approach."

—*Lewis M. Alexander and Barry D. Gold*

Chapter 13

Scaling Pattern and Process in Marine Ecosystems

Robert E. Ricklefs

Abstract

In all ecosystems, processes and the patterns they produce occur over a spectrum of spatial and temporal scales ranging from day-to-day activities of individuals to the evolution, diversification, and geographical dispersal of entire biotas over geological time. General issues revealed by this perspective include (1) the role of extrinsic physical variables in determining biological pattern, (2) the capacity of biological systems to ameliorate physical perturbations occurring on different scales, (3) the ability of biological systems to generate pattern and process at different scales, and (4) the coupling of patterns and processes with different spatial and temporal frequencies. Within marine ecosystems, special considerations arise from the varied life histories and dispersal modes of different organisms, from the patchiness and movement of the oceanic environment, and, geologically, from the changing configuration of the ocean basins. In determining causation within the marine ecosystem — whether for the local demography of a single population or the assemblage of species at a particular locality — one must consider events removed in time and space. Elucidating the rules of coupling over temporal and spatial scales provides a fundamental challenge.

Introduction

Every phenomenon, regardless of its scale in space and time, includes finer scale processes and patterns and is imbedded in a matrix of processes and patterns having larger dimensions. Many investigators consider this hierarchy of phenomena as an inconvenience and ignore the relationship of their study to phenomena on both larger and smaller scales. Increasingly, however, biologists realize that processes and patterns on different scales of time and space may be linked, and that they must therefore consider the broader context of each phenomenon. But although most studies make reference to this problem of scale, it has yet to be dealt with explicitly and generally in ecology (Allen and Starr, 1982; O'Neill et al., 1986). This chapter will outline some general considerations pertaining to the generation of temporal and spatial scale of patterns and processes. I would hope that readers might be convinced that problems relating to the hierarchy of scale may be generalizable to all systems and deserve study in their own right.

The Problem of Scale

By scale I mean the characteristic distance or time associated with variation in natural systems. Values of temperature, substrate, texture, nutrient concentration, and population density, for example, vary through both time and space. Distances or periods between peaks or troughs in values define the scale of pattern (Figure 13.1). Any given measurement may exhibit variation on many scales simultaneously. This variation may be considered as "information" concerning the state of the system. A few examples will illustrate the tremendous range of scale over time and space.

As they swim through water, water fleas (*Daphnia pulex*) leave a trailing plume of in-

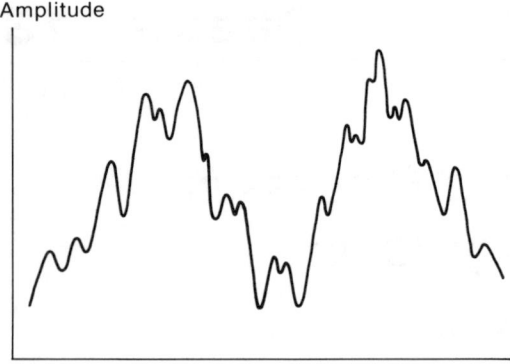

Figure 13.1. Schematic diagram of simultaneous variation in a system variable over different scales of time or space.

organic phosphorus and other small, soluble compounds (Lehman and Scavia, 1982). The physical dimensions of the plume are on the order of millimeters. Concentrations of nutrients usable by algae remain in the wake of the water flea for a matter of seconds, until they are dispersed by diffusion and turbulent mixing. The water fleas, which are grazers, also clear a path through the phytoplankton population, which closes behind after some short period. In this particular case, the pattern generated by the water flea is transient and moves with the generating source.

In her studies of settling by planktonic larvae, Butman (1987) has emphasized the generation of pattern by different factors operating on different scales. Patterns of distribution on the order of kilometers appear to be determined primarily by physical processes, although a larva may adjust its settling rate to mimic that of particle sizes in its preferred habitat. On smaller scales of centimeters, larvae can actively select microhabitats by successively sampling the substrate before committing themselves to settling.

Many patterns of biological activity in the marine realm parallel patterns in the physical environment, particularly the distribution of different masses of water. For example, on a 300-kilometer (km) transect between Portland, Maine, and Yarmouth, Nova Scotia, Huntley and Boyd (1984) found coincident peaks in the abundance of zooplankton and concentration of chlorophyll separated by more than 100 km and which corresponded to low values of water temperature at four meters depth. Many commercial fish landings and years of strong recruitment into fish stocks exhibit periodicities in the 3- to 10-year range (Bakun, 1986). At least some physical phenomena, such as the El Niño-Southern Oscillation (ENSO), have similar periods (Rasmussen, 1985) and impose tremendous variability on susceptible biological systems (Barber and Chavez, 1983; Schreiber and Schreiber, 1984). With respect to spatial scale, ENSO events have almost global repercussions, but these certainly do not occupy the large end of the scale hierarchy. Raup and Sepkoski (1986) claim to have demonstrated periodicity in the extinction rates of marine animal families and genera on the order of 26,000,000 years.

Breeding productivity of seabirds in the Galápagos Archipelago demonstrates the interaction of processes on many different scales. The Galápagos are influenced by warm water from the north, a cold surface current from the coast of South America to the southeast, and a deep current from the west forced to the surface as it strikes the archipelago (Houvenaghel, 1984). In addition, the position of these water masses changes seasonally so that most of the archipelago experiences both cold-dry and warm-wet seasons each year. The oceans and climate are further complicated by occasional ENSO events of great magnitude, during which altered current patterns may sweep normally productive waters away from particular islands (Feldman et al., 1984).

The complicated bathymetry of the Galápagos creates heterogeneous mixing of surface and deep currents throughout the archipelago. Within this oceanographic context, the breeding productivity of birds that rely on marine resources can vary tremendously both spatially and temporally, presumably reflecting changing fish stocks within foraging range of the breeding colony (Ricklefs et al., 1984). Radio tracking has shown that masked boobies (*Sula dactylatra*) breeding at a colony at Punta Cevallos, Española, in the southeastern part of the archipelago, feed primarily in a small area of undersea mounts about 60 km to the east (Anderson and Ricklefs, 1987). Within this area, birds spend most of their time dispersed widely (approximately 100 individuals per square kilometer), either flying about or sitting on the water surface. Occasionally, a single bird starts to forage, presumably owing to the presence of a local school of fish driven to the surface by predatory tuna. Within minutes, the foraging flock may include 100 individuals feeding within a hectare or less. After 20 or 30 minutes, the flock disperses, presumably having decimated the prey, dispersed them, or driven them below the surface. In this example, we see pattern in time and space imposed both by process and pattern in physical environment and also by the behavior of biological components of the system creating their own peculiar scales of pattern.

Why Is Scale Important?

Considerations of scale are critical to ecological studies for two reasons. First, every process and pattern has a temporal and spatial extent; one cannot study a phenomenon without fully appreciating its scale. In particular, local processes are always imbedded in a matrix of larger scale processes. One cannot study local population dynamics, for example, without considering the movement of individuals from one area to another. By extension, factors influencing populations in surrounding areas may influence local dynamics because dispersing individuals carry information about distant processes. As long-term data have accumulated, it has also become abundantly clear that one cannot appreciate the dynamics of a system without following it through a sample of its characteristic variability.

Second, patterns and processes that occur on different scales of time and space are linked. The general rules governing these relationships have yet to be discovered; if they exist, they will provide the only guideline we can have to determine the extent of ecosystem boundaries. Influence runs up and down the hierarchy of temporal and spatial scale according to factors that carry information between levels. In order to comprehend pattern and process on a particular scale, one must consider the transmission of influence between levels. In some cases, biological pattern may follow directly from physical pattern, and the problem of scale becomes one of measuring variation in physical environment. In most cases, however, biological systems transmit information between levels in the hierarchy of scale and may generate new scale in a system, as elaborated in the following section. Thus, an explicit consideration of scale seems necessary to any interpretation of pattern.

The Relationship Between Time and Space

Temporal and spatial scales of patterns appear to be closely related. The lore of most fields of natural history relates that events of short duration generally occupy small areas. Schopf (1972) pointed out that processes associated with individuals occur over periods of days to years and areas on the order of square meters or less (at least for benthic marine organisms). Population patterns, including predator-prey and competitive interactions, have dynamics on the order of tens to thousands of years extending over areas on the order of kilometers. Evolutionary processes, including speciation and faunal buildup, extend over millions of years and may occupy substantial portions of the Earth's surface. Steele (1978) has noted that variation associated with phytoplankton, zooplankton, and fish has correlated scales of time and space extending over perhaps three orders of magnitude. He also noted that the temporal and spatial scales of variation in these biological systems corresponded to the scales of physical processes in marine systems. Furthermore, the

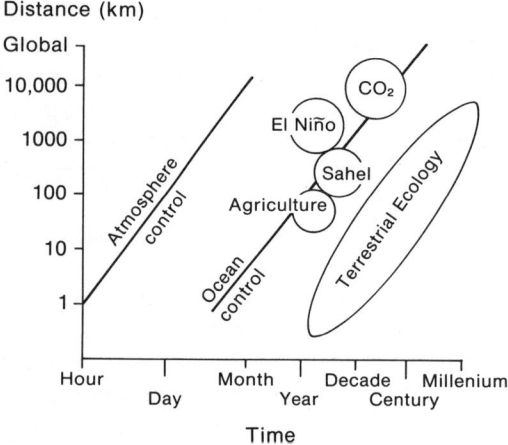

Figure 13.2. The general relationship between scales of time and space for phenomena under general control of atmospheric, marine, and terrestrial processes (after Steele, 1989). Atmospheric phenomena include weather patterns from tornadoes to extratropical cyclones; marine-influenced phenomena include shifts in oceanic currents and major climate changes; terrestrial phenomena include population variability and ecological succession.

relationship between scales of time and space appear to differ in different systems (Figure 13.2). In particular, for a given geographical scale of pattern, temporal scales are briefest in atmosphere-controlled systems (tornadoes, fronts, hurricanes, and other cyclones), longer in ocean-controlled systems (including many weather patterns), and perhaps longest in terrestrial systems. The causes for these differences will be difficult to understand unless one can understand the general relationship between spatial and temporal scales.

Two factors would seem to be important. First, the dissipation of energy in large physical systems occurs in part by progressively finer subdivision of motion or other form of energy into smaller patterns of shorter duration. Thus, a large current eventually dissipates its energy in smaller, often more transient eddies, particularly in the open ocean. Atmospheric circulation behaves in the same way. Because water is more dense and viscous than air, it seems reasonable that the scales of physical processes are of longer duration in the ocean than in the atmosphere. Second, the extent of a biological pattern over space depends on the temporal persistence of a phenomenon and its rate of spread. Thus, for a given rate of spread, scales of temporal and spatial pattern will be highly correlated. For example, a local change in population size caused by local climate or oceanographic conditions extends its influence over space as individuals actively disperse either away from or toward the local area. More persistent local changes will affect larger areas as dispersal of individuals reaches out through the population. Biological systems may also actively propagate pattern, as through the spread of disease organisms or new genotypes through populations and communities. In such cases, the energy or information that initially generates the pattern is not dissipated, but rather may be self-accelerating.

The Generation of Scale

External templates

Patterns in many systems closely resemble patterns in external forcing variables. Seasonal changes in terrestrial insect populations closely parallel seasonal changes in temperature. The productivity of phytoplankton populations follow upon the concentrations of nutrients in the water. Whenever a population responds rapidly compared to the period of variation in the environment, one can expect a system to track both temporal and spatial variation in the environment. This relationship has been worked out explicitly for some simple models. For example, when population change obeys logistic growth, such that $dN/dt = rN(1 - N/K)$ (where N is population size, r the rate of unrestricted exponential growth of the population, and K the carrying capacity of the environment for that population), and K varies as a sine function with period T, N will track K closely when $r > 2\pi/T$ (Nisbet and Gurney, 1982). When r is much less than that value, N varies in concert with K, but with such low amplitude that, for all practical purposes, population size remains constant. The principle of response and tracking also applies to spatial variation (e.g., Endler, 1977), for which rate of dispersal provides the link between temporal response of populations and spatial variation.

Because of the dominating influence of water in the marine environment, many biological systems directly track variation in temperature, light availability, nutrient concentrations, and current speeds. Thus, the pattern of such systems matches closely the template of the physical environment. Clearly, however, such biological interactions as predation, disease transmission, competition, migration, and larval settling may modify the scale of pattern and even establish new patterns.

The natural resonances of systems

New scales of pattern and process are often generated when an external event of long interval stimulates the natural resonance of a system, thereby establishing a dynamic with period determined by intrinsic qualities of the system rather than the pattern of the forcing function. In biological systems, the best known of such resonances are limit cycles in predator-prey systems (May, 1972, 1975), but the simpler model of Lotka-Volterra interaction will suffice to illustrate the point. Let H be a population of herbivores and P a population of predators. In the simplest case imaginable, their dynamics are defined by the equations

$$dH/dt = aH - bHP$$

and

$$dP/dt = c(bHP) - dP$$

where a, b, c, and d are constants, and the term bHP represents the removal of prey individuals by predators. This system has a steady-state at $H = d/cb$ and $P = a/b$, but any perturbation in H or P will cause the populations to oscillate about the equilibrium point indefinitely with a period approaching $T = 2\pi/\sqrt{ad}$. Thus, the dynamics of the system are controlled by intrinsic qualities rather than by external factors. Furthermore, the period of oscillation depends only on the birth rate of the prey and the death rate of the predator; it is, perhaps counterintuitively, independent of the rate of predation.

Many populations exhibit periodic variation (see, for example, Ricklefs, 1979). Experiments with laboratory populations have demonstrated that most systems have intrinsic qualities, primarily associated with time lags in the responses of the system to environmental conditions, that create natural resonance frequencies. Such lags occur through density-dependent interactions with resources in single-species populations (e.g., Nicholson, 1958) and in systems of interacting populations (e.g, Huffaker, 1958). Lags may occur through the period required for birth processes to match a population to the carrying capacity of the environment or through the storage of materials or energy, as illustrated in the following section.

The summation and intersection of patterns

Patterns of long temporal duration or large spatial scale may be generated by the summation or intersection of patterns of shorter period. A simple physical example of this is the slow beat produced by two or more faster vibrations. When two sine waves of different period are added, their sum can exhibit a unique period equal to the product of the periods of the underlying patterns (Figure 13.3). This phenomenon causes the rogue waves encountered in the open ocean, where many wave patterns having different periods may sum to form occasional monsters.

In biological systems, variation often results from the favorable or unfavorable intersection of variation in component processes. An example frequently referenced in the marine literature is that of the diets of fish larvae changing as the fish grow and interact briefly with a variety of variable food sources. Exceptionally good year classes require

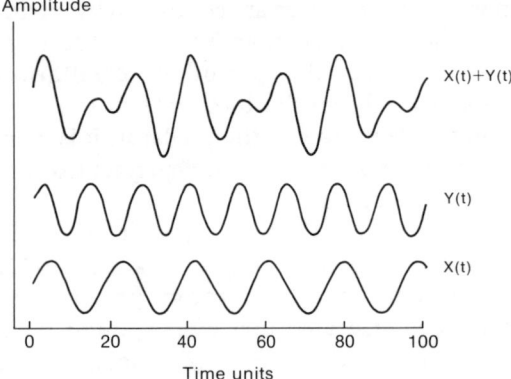

Figure 13.3. An illustration of the summation of two oscillating functions (sine waves) of dissimilar period (T = 2x and 3x, respectively) to produce an additional cycle of longer period (6x).

the intersection of the development pattern, whose period is more or less biologically fixed, with unusual abundance of the critical food sources (Lasker, 1978). Hence, the windows of sensitivity controlled by development must closely match favorable variations in a variety of resources to produce strong recruitment. When such combinations occur rarely, then a new pattern of strong age classes with long period results. In this sense, the temporal pattern of development acts as a threshold sensor that removes much high-frequency variation (both year-to-year and phenological within years) to reveal underlying variation of longer period.

Random processes and pattern

All system variables exhibit stochastic variation caused by innumerable small influences; this variation has a characteristic frequency distribution. If some process responds only to particularly high values of a variable, as does recruitment to food availability, then pulses of the response variable will occur infrequently. For example, when a forcing variable with a normal frequency distribution elicits a response only at values exceeding 1.5 standard deviations above the mean, responses will occur with a frequency of 0.067 (hence, a period of about 15 time units). The distribution of intervals between such response will also be randomly distributed according to a geometric series, but these responses nonetheless have established a pattern of different scale from the processes responsible for them.

Statisticians have long recognized that random variations can establish patterns of long period (Lotka, 1925; Feller, 1957). A particularly striking example of this point is that of a population growing by random increments: $N(t+1) = N(t) + RND$, where RND is a random normal deviate (mean = 0, standard deviation = s). Starting with $N(0) = 100$ and $s = 5$ produces runs illustrated in Figure 13.4. These simulations show striking variation which one might imagine to exhibit periodicities on the order of 25 time units. Other simulations with subdivided populations have demonstrated the generation of spatial variation of a similar nature (Endler, 1977; Rohlf and Schnell, 1971).

Roles of Movement and Storage in Scaling Pattern and Process

Intrinsic temporal and spatial periods of systems depend in part on the movement of material or individuals from one place to the next and on the storage of energy and materials (essentially movement through time). All systems have a memory of their prior states; responses to external variation and intrinsic resonances depend in part on the persistence of that memory, that is, storage. Movement and exchange add a spatial component to persistence.

Both movement and storage may either enhance variation or dampen it, depending on how they interact with external forcing functions. Movement between areas tends to even out the difference between them. When individuals respond faithfully to external forcing functions, movement reduces spatial variation, effectively filtering out spatial periodicities of environmental variation much shorter than the distance an individual dis-

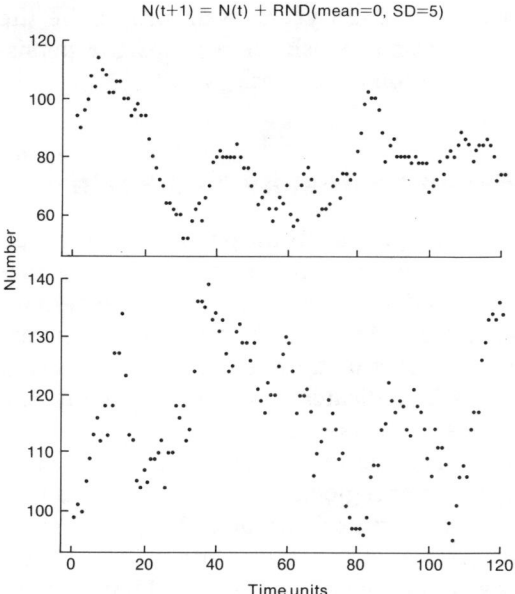

Figure 13.4. Two simulations of a process in which a number (N) is changed by a random increment (RND) during each time interval.

perses over a temporal period much shorter than its response time. But movement can also propagate biologically generated variation over large areas, particularly when a genetic novelty or a disease is involved. Both of these have great persistence because their properties reside in the dispersing system rather than being imposed by external forcing variables. The spread of genes in part causes the uniformity of species over vast areas. The efficacy of dispersal has recently been demonstrated by the rapid spread of a disease that kills the urchin *Diadema* throughout the Caribbean (Lessios et al., 1984). The dispersing agent is undoubtedly carried by water currents, but its persistence — hence, its broad geographical distribution — stems from its intrinsic, genetically determined qualities.

Storage also has a dual role of damping externally forced variation and establishing lags in system responses that result in intrinsic resonance frequencies. The relevant parameter determining the behavior of the system probably reflects the ratio between storage and response. On one hand, fat re-serves sufficient to support a long-lived individual through fasts of days or weeks reduce the impact of short-term environmental variation on population processes. On the other hand, stored reserves sufficient to sustain reproduction through periods of scarcity on the order of half a life span will likely create lags in the response of a population to environmental variation and thereby establish population oscillations. For example, the water flea *Daphnia* (Crustacea: Cladocera) stores lipid droplets during periods of food abundance, which sustain individuals during periods of high population density when the food supply is reduced. Lipids are also passed on to the offspring, extending their survival during periods of scarcity. Thus, individuals enter the population in times of declining resources, establishing a lag in the response of the population to food supply. As a result, laboratory-reared *Daphnia* populations may undergo violent oscillations with periods of 15–20 days (Goulden et al., 1982). According to theory (May, 1975; Nisbet and Gurney, 1982), the period of oscillation is approximately 4 to 5 times the lag. Hence, lipid stored sufficient for 3 to 5 days of sustenance could account for the observed oscillations. Another species of cladoceran, *Bosmina*, which does not store appreciable lipid, does not undergo comparable population cycles in the laboratory.

Scale in Aquatic and Terrestrial Systems

The issue of storage suggests an explanation for the differences between the spatial scales of pattern with similar temporal scale in marine terrestrial ecosystems. In aquatic systems, much of the pattern associated with variation resides in the water itself and, to a large extent, patterns of production are molded on this template. Populations of phytoplankton and zooplankton, in particular, respond so quickly that many attributes of biological systems follow directly from those of the physical system. In terrestrial

Table 13.1. Energy turnover in aquatic and terrestrial habitats. (From Whittaker and Likens, 1973.)

Habitat	Net primary productivity $(g \cdot m^{-1} \cdot yr^{-1})$	Biomass $(g \cdot m^{-2})$[1]	Transit time (yr)
Open ocean	125	3	0.02
Continental shelf	360	10	0.03
Estuaries	1800	1000	0.56
Temperate grassland	500	1500	3
Boreal forest	800	20000	25
Tropical forest	1800	42000	23

[1]Above ground in the case of terrestrial habitats.

habitats, storage of biomass and nutrients filters out much of shorter term variability in the system, but may also establish resonances of long period. Turnover times of energy (and, approximately, of nutrients) are approximately two orders of magnitude longer in terrestrial systems than in marine systems (Table 13.1). Storage (persistence) of pattern is responsible for cyclic succession in several environments, such as *Calluna* heaths (Watt, 1947) and balsam fir forests (Sprugel, 1976). In these cases, steady winds create waves of destruction and regeneration of vegetation leading to temporal cycles of succession at a single point in space and spatial patterns of succession at a given point in time. Consideration of storage and response times are undoubtedly responsible for the well-known population cycles of some mammals and birds, which vary over many orders of magnitude with periods of 4 to 10 years (Elton, 1942; Keith, 1963), and outbreaks of the spruce budworm with periods of 30–60 years (Ludwig *et al.*, 1978).

Cycles of comparable duration in marine systems, such as the 50–100 year cycles of abundance of many fish stocks (Steele, 1985), may also reflect intrinsic oscillations of systems independent of the periodicity of external forcing functions. If time lags caused the oscillations, they would be on the order of 10–25 years, well within the life spans of some fish, or perhaps the persistence of some diseases.

Evolution, History, and Scale

Heredity represents the persistence of genetic information, hence, stability of the phenotype and taxonomic status. Evolutionary changes and speciation events occur on very large scales of time and space. Increasing evidence indicates, however, that many ecological processes have response times that extend to evolutionary scales. Thus, the composition of regional and local biotas, as well as the characteristics of biological systems, may respond to patterns and processes having periods of millions of years. I have argued elsewhere (Ricklefs, 1987) that lack of convergence of species diversity, in local communities developing under similar physical conditions, emphasizes the role of regional and long-term processes in shaping ecological systems.

Some major ecological transitions in biological communities, particularly those associated with intensified herbivory or predation, have required tens of millions of years (Thayer, 1979; Steneck, 1983; Lidgard, 1986; Vermeij, 1987). During these transitions, vulnerable species were replaced by species that better resisted new methods of exploitation. Such transitions represent the resolution of predator-prey interactions, which one normally thinks of as occurring on short, ecological time scales. The geological record clearly indicates, however, that the spread of evolutionary novelty through the world biota may be very slow indeed. The reasons for this are not clear. Evidently, fitness changes related to major adaptations of feeding method and predator defense are small compared to evolutionary adjustments required for a trait to spread among coadapted communities of organisms. Ecological space is finely subdivided by biological associations. Novelty arising in one part of the biosphere may become established relatively rapidly, but the temporal pattern of its spread, requiring the invasion of other habitats, is tied to long-term evolutionary processes and speciation.

Evolutionary inertia may be thought of as a time lag. I am doubtful that it is responsible for the periodic bursts of extinction apparently recorded in the fossil record, although Ricklefs and Cox (1972) suggested that such a mechanism might cause cycles in geographical extent and ecological amplitude of populations on evolutionary time scales. It is nonetheless clear that biological responses to changes in the environment may occur on temporal scales characteristic of major geological processes, such as sea-floor spreading and continental drift. The interaction of geological and biological processes of this extent is evident in the geographical distribution of taxonomic groups of high rank (Brown and Gibson, 1983). To the extent that ecologically relevant change may also require millions of years, we may also expect the ecological characteristics of communities to vary over the surface of the earth. At the extreme, one could argue that some ecological traits of ecosystems cannot demonstrate convergence because evolutionary equilibration requires longer than the longest characteristic scale of environmental change.

Scale is worthy of consideration in its own right. Principles governing the generation of pattern operate at every level on the hierarchy of scale in time and space. Any phenomenon may be connected to others of greatly different dimension through mechanisms that generate variation. Practical considerations limit the scope of every study, but the hierarchical dimensions of natural systems demand that one adopt a broad perspective.

References

Allen, T. F. H., and Starr, T. B. 1982. Hierarchy. Perspectives for ecological complexity. Univ. Chicago Press, Chicago.

Anderson, D. J., and Ricklefs, R. E. 1987. Radio-tracking masked and blue-footed boobies (*Sula* spp.) in the Galápagos Islands. Nat. Geogr. Res. 3:152–163.

Bakun, A. 1986. Definitions of environmental variability affecting biological processes in large marine ecosystems. *In* Variability and management of large marine ecosystems. pp. 89–108. Ed. by K. Sherman and L. M. Alexander. AAAS Selected Symposium 99, Westview Press, Boulder, CO.

Barber, R. T., and Chavez, F. P. 1983. Biological consequences of El Niño. Science 222:1203–1210.

Brown, J. H., and Gibson, A. C. 1983. Biogeography. Mosby, St. Louis.

Butman, C. A. 1987. Larval settlement of soft-sediment invertebrates: The spatial scales of pattern explained by active habitat selection and the emerging role of hydrodynamical processes. Oceanogr. Mar. Biol. Ann. Rev. 25:113–165.

Elton, C. 1942. Voles, mice and lemmings. Problems in population dynamics. Clarendon Press, Oxford, UK.

Endler, J. A. 1977. Geographic variation, speciation, and clines. Princeton Univ. Press, Princeton, NJ.

Feldman, G., Clark, D., and Halpern, D. 1984. Satellite color observations of the phytoplankton distribution in the eastern equatorial Pacific during the 1982–83 El Niño. Science 226:1069–1071.

Feller, W. 1957. An introduction to probability theory and its applications, Vol 1 (2nd. ed.). Wiley, New York.

Goulden, C. D., Henry, L. L., and Tessier, A. J. 1982. Body size, energy reserves, and competitive ability in three species of Cladocera. Ecology 63:1780–1789.

Houvenaghel, G. T. 1984. Oceanographic setting of the Galápagos Islands. *In* Galápagos. pp. 43–54. Ed. by R. Perry. Pergamon Press, Oxford, UK, and New York.

Huffaker, C. B. 1958. Experimental studies on predation: Dispersion factors and predator-prey oscillations. Hilgardia 27:343–383.

Huntley, M., and Boyd, C. 1984. Food-limited growth of marine zooplankton. Amer. Nat. 124:455–478.

Keith, L. B. 1963. Wildlife's ten-year cycle. Univ. Wisconsin Press, Madison.

Lasker, R. 1978. The relation between oceanographic conditions and larval anchovy food in the California Current: Identification of factors contributing to recruitment failure. Rapp. P.-v. Reun. Cons. int. Explor. Mer 173:212–230.

Lehman, J. T., and Scavia, D. 1982. Microscale patchiness of nutrients in plankton communities. Science 216:729–730.

Lessios, H. A., Robertson, D. R., and Cubit, J. D. 1984. Spread of *Diadema* mass mortality through the Caribbean. Science 226:335–337.

Lidgard, S. 1986. Otogeny in animal colonies: A persistent trend in the bryozoan fossil record. Science 232:230–232.

Lotka, A. J. 1925. Elements of physical biology. Williams and Wilkins, Baltimore.

Ludwig, D., Jones, D. D., and Holling, C. S. 1978. Qualitative analysis of insect outbreak systems: The spruce budworm and forest. J. Anim. Ecol. 47:315–332.

May, R. M. 1972. Limit cycles in predator-prey communities. Science 177:900–902.

May, R. M. 1975. Stability and complexity in model ecosystems (2nd ed.). Princeton Univ. Press, Princeton, NJ.

Nicholson, A. J. 1958. The self-adjustment of populations to change. Cold Spring Harbor Symp. Quant. Biol. 22:153–173.

Nisbet, R. M., and Gurney, W. S. C. 1982. Modelling fluctuating populations. Wiley, New York.

O'Neill, R. V., DeAngelis, D. L., Waide, J. B., and Allen, T. F. H. 1986. A hierarchical concept of ecosystems. Princeton Univ. Press, Princeton, NJ.

Rasmussen, E. M. 1985. El Niño and variations in climate. Am. Sci. 73:168–177.

Raup, D. M., and Sepkoski, J. J., Jr. 1986. Periodic extinction of families and genera. Science 231:833–836.

Ricklefs, R. E. 1979. Ecology (2nd ed.). Chiron Press, New York.

Ricklefs, R. E. 1987. Community diversity: Relative roles of local and regional processes. Science 235:167–171.

Ricklefs, R. E., and Cox, G. W. 1972. Taxon cycles in the West Indian avifauna. Am. Nat. 106:195–219.

Ricklefs, R. E., Duffy, D., and Coulter, M. 1984. Weight gain of blue-footed booby chicks: An indicator of marine resources. Ornis Scand. 15:162–166.

Rohlf, F. J., and Schnell, G. D. 1971. An investigation of the isolation-by-distance model. Am. Nat. 105:295–324.

Schopf, T. J. M. 1972. Varieties of paleobiologic experience. *In* Models in paleobiology. pp. 8–25. Ed. by T. J. M. Schopf. Freeman, Cooper & Co., San Francisco.

Schreiber, R. W., and Schreiber, E. A. 1984. Central Pacific seabirds and the El Niño/Southern Oscillation: 1982 to 1984 perspectives. Science 225:713–716.

Sprugel, D. G. 1976. Dynamic structure of wave-regenerated *Abies balsama* forests in the north-eastern United States. J. Ecol. 64:889–911.

Steele, J. H. 1978. Some comments on plankton patches. *In* Spatial pattern in plankton communities. Ed. by J. H. Steele. Plenum, New York.

Steele, J. H. 1985. A comparison of terrestrial and marine ecological systems. Nature 313:355–358.

Steele, J. H. 1989. A view from the ocean. Oceanus 32:4–9.

Steneck, R. S. 1983. Escalating herbivory and resulting adaptive trends in calcareous algal crusts. Paleobiology 9:44–61.

Thayer, C. W. 1979. Biological bulldozers and the evolution of marine benthic communities. Science 203:458–461.

Vermeij, G. J. 1987. Evolution and escalation. An ecological history of life. Princeton Univ. Press, Princeton, NJ.

Watt, A. S. 1947. Pattern and process in the plant community. J. Ecol. 35:1–22.

Whittaker, R. H., and Likens, G. E. 1973. Primary production: the biosphere and man. Human Ecol. 1:357–369.

Chapter 14

Physical and Biological Scales and the Modelling of Predator-Prey Interactions in Large Marine Ecosystems

Simon A. Levin

Abstract

This chapter discusses the dynamics of predator-prey systems, beginning with classical models of the Adriatic fisheries to more recent work involving spatial pattern in diffusion reaction systems, and includes a discussion of techniques for analyzing and evaluating spatial and temporal patterns across multiple physical and biological scales. Concluding remarks involve models for the krill fisheries of the Southern Oceans.

Introduction

Large marine ecosystems (LMEs) are unmatched as testing grounds for ecological theory. The tremendous importance of such systems as fountains of renewable resources; their remarkable spatial, temporal, and biological complexity; and the intricate interplay between their relevant physical and biological factors make theory a necessity in guiding decision making and present theorists with challenging opportunities.

The fundamental theoretical problem in LMEs is to understand the factors that govern the recruitment of the major fish stocks. What are the relative influences of physical phenomena, such as the movement of fronts and sea-surface contiguous zones, and of biological factors such as predation and food availability? How would increased exploitation affect stocks?

Such analyses are complicated by the fact that for many stocks, such as the Antarctic krill (primarily *Euphausia superba*), we do not have reliable estimates even of the total stock size. Every available censusing method has its drawbacks, but theoretical approaches to stock assessment and prediction can facilitate estimation of the size of the resource.

Because of the interconnections among local systems, analysis and prediction in large marine ecosystems must be based on an approach that is regional or broader. It must be comparative, building on the natural experiments provided by different systems, because of the difficulty or impossibility of carrying out experiments on such large systems. It must integrate observations and statistical analyses of broad-scale patterns with experimental and theoretical studies of component mechanisms; modelling provides the necessary linkages between these scales of resolution. Finally, the complexity of biological interactions and the need to integrate physical and biological factors over broad geographical scales mandate the ecosystem approach that is represented in this volume, since it is virtually impossible to isolate any single factor from the context of the whole system.

Early Theoretical Approaches to Large Marine Ecosystems

Theoreticians have been trying to model the dynamics of marine fisheries ever since the Italian biologist, Umberto d'Ancona, interested his father-in-law, the distinguished mathematician Vito Volterra, in the fluctuations of the Adriatic fisheries. Volterra (1926) showed that interactions between only two variables — the densities of predator and prey — were sufficient to generate permanent oscillations in both. His equations took the form

$$\frac{dx}{dt} = f(x,y) \qquad (1)$$

$$\frac{dy}{dt} = g(x,y)$$

in which x and y represent prey and predator densities, and in which specific assumptions about the functions f and g were made to represent the nature of the interaction between predator and prey. The oscillatory character of the particular equations Volterra considered was not robust; but later extensions by Kolmogorov (1936; see also May, 1972; and Albrecht et al., 1973) demonstrated that a slight generalization of Volterra's equations could exhibit stable and characteristic oscillations. It is nonetheless worth mentioning that Volterra's assumption — that consumption rates be proportional both to predator and to prey densities — is an assumption that to this day underlies many predator-prey models, as well as estimates of consumption rates in real systems (see chapter 15, this volume). This assumption may be valid under certain conditions, but other forms may be equally appropriate (Rosenzweig, 1971). Mechanistic studies are needed as a basis for refining this assumption.

Volterra certainly recognized that his equations oversimplified the picture. Indeed, that was his point, as is the point with much theoretical work: to demonstrate that, even with virtually all detail stripped away, and under the simplest assumption concerning consumption rates, the essential oscillatory character of the system remained. Volterra himself extended these equations to certain multispecies communities; and his own specialty, integro-differential equations, provided him with a vehicle for examining the effects of age structure. Furthermore, other investigators have considered a variety of different functional relationships between predator and prey, and studied how these would affect conclusions (see, for example, Rosenzweig, 1971). Virtually all such studies, however, have viewed the biological interactions as autonomous and ignored the effects of physical processes.

The Interplay Between Physics and Biology, and the Issue of Scale

Volterra's focus on biological interactions, to the total exclusion of the details of the physical environment, reflects a dichotomy that to this day divides research and researchers on communities and ecosystems (see, for example, Kingsland, 1985). One tradition, emanating from the work of Volterra, and developed by MacArthur and others, looks for the elements of pattern that arise from biological interactions; the other regards the systems of interest as being driven by physical forces, with the biology more or less tracking changes in the environment.

In some instances, including much of the oceanographic literature, attempts have been made to study the interplay between these factors in structuring communities. In food web theory, for example, one of the key questions involves the determination of whether the lengths of trophic chains are determined by physical limits (energy availability) or by biological limits (community stability). In intertidal studies, it is a generally accepted maxim that populations are limited at one range extreme (tidal exposure) by physical tolerances, and at the other by biological interactions. In most community studies, however, while the importance of the physical environment is recognized, the tendency is to regard physical factors as hidden variables, and to express their effects as white noise, or as inputs from a distant source, or simply as local uniqueness. Surely, terrestrial ecologists will need to pay more attention to the interplay between physics and biology as they turn their attention to global environmental issues. It is an awakening long overdue, as reflected in the inadequate training that most ecologists receive in the physical sciences.

In oceanography, the need to wed these two points of view long has been evident, if not fully met. Steele (1978; see also Haury et al., 1978) has emphasized that the issue is one of appropriate scales. A rough rule of thumb (Denman and Powell, 1984) is that the temporal scales in oceanographic systems are biologically determined and the spatial scales are physically determined. But this is a generalization that provides little more than a starting point. In truth, any particular aspect of a system may be determined on some scales by biological factors and on others by physical factors. The importance of biological detail, such as how individuals search for food, or their aggregation behavior, may dominate pattern on small scales, but be dissipated on broader scales, due both to the law of large numbers and to the existence of dominant physical phenomena.

Thus, the first step in any comprehensive theory of predator-prey dynamics in the ocean must involve an examination of scales (Denman and Powell, 1984). Scale is conveniently and roughly defined as the distance (e.g., spatial or temporal) over which some variable of interest changes significantly (Denman and Powell, 1984). Spectral analyses and more sophisticated statistical analyses of temporal and spatial data provide information on the scales of variability inherent in any data set; and coherence spectra, involving several variables, provide suggestive information on the distributional correlations between variables. It is not surprising that in the analyses of data sets from large marine ecosystems, one often finds good correlations between physical factors (e.g., temperature) and primary productivity (e.g., fluorescence) on inter-

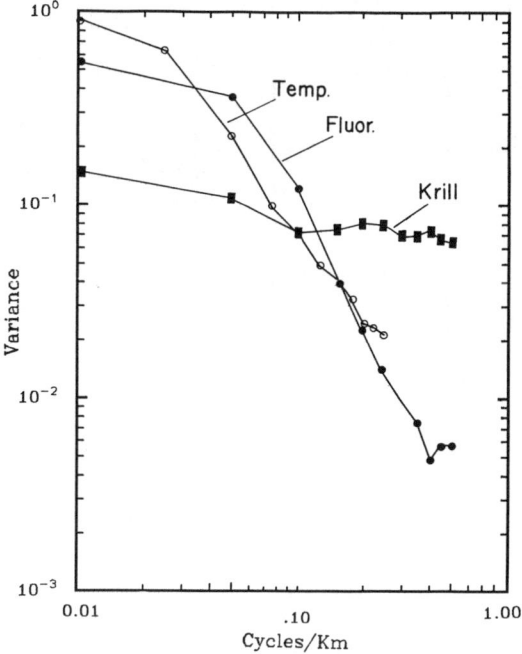

Figure 14.1. Mean spectral plots for krill, in vivo fluorescence, and temperature. [Reprinted from Weber et al., Deep Sea Research 33:10, 1327 (1986), with permission.]

mediate and possibly broader scales (see, for example, Figures 14.1 and 14.2, taken from Weber et al., 1986). In contrast, some data sets involving species higher in the food chain, such as data describing the distribution of krill in the Scotia Sea (again, see Weber et al., 1986), congruence between species and their resources or physical factors is not evident; indeed, the spectrum for krill is virtually flat, indicating a random distribution, at least over smaller spatial scales (Figure 14.1). The suggestion is that the distribution of krill is determined by its own active movements, at least on small scales (e.g., ≤ 20 km); it may well be the case that the broader aspects of its distribution more closely reflect physical factors such as the major currents; but that is impossible to determine from Figures 14.1 and 14.2.

Regarding the data present in Figures 14.1 and 14.2, which were taken in austral summer 1981 in the Southern Ocean, the correlation between fluorescence and physi-

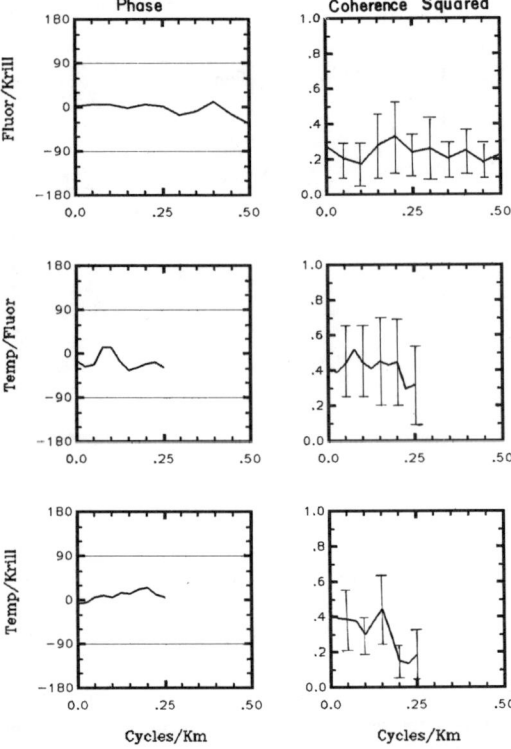

Figure 14.2. Mean crossphase and squared coherence spectra for fluorescence-krill, temperature-fluorescence, and temperature-krill. Vertical bars indicate the 95% confidence limits about the mean coherence squared estimates ($\bar{y} \pm (2.2)$ (S.D./$\sqrt{12}$)). For clarity, confidence limits for temp-fluor and temp-krill are only shown at every other computed frequency. [Reprinted from Weber et al., Deep Sea Research 33:10, 1327 (1986), with permission.]

cal factors is clear on intermediate scales (4–20 km); but Weber et al. (1986) interpret the slightly steeper slope of the fluorescence spectrum (Figure 14.1) and the coherence between fluorescence and krill (Figure 14.2) as suggesting that biological factors (grazing) must contribute to the distribution of phytoplankton on smaller scales. However, it seems evident that this can be regarded only as a suggestion, and that further studies are needed to resolve the issue; component studies at lower scales of resolution could be extremely useful here.

In conclusion, resolution of the importance of physical and biological factors is inextricably intertwined with the issue of scale. Even for a given system, it is to be expected that the importance of the various factors will shift as one changes scale. On intermediate scales, for example, we have seen that the distribution of fluorescence — and hence, presumably, primary productivity, mirror physical spectra of turbulence — since water movements overcome the influences of biological processes. On broader scales, however, divergence between the spectra of temperature and chlorophyll in some systems has been taken as evidence for the importance of biological phenomena (Denman and Platt, 1975). The simplest models for this (Kierstead and Slobodkin, 1953; Okubo, 1978, 1980) couple diffusion with population growth, and determine that sufficiently large bursts of nutrients can create patches by allowing biological growth processes to accentuate environmental inhomogeneities. For these models, indeed, the underlying trigger is physical; but biological exaggeration of the physical trigger leads to a discrepancy between the spectra. Other models of broad-scale patchiness resulting from biological (predator-prey) interactions also have been proposed (Levin, 1974; Okubo, 1974; Levin and Segel, 1976; Steele and Henderson, 1984); these rely on more complicated and nonlinear interactions, but do not depend on a critical favorable physical region. That is, the scales of interaction arise entirely from biological interactions and from the diffusive movements of species.

Perceptual scale and the measurement process influence our view of any system (Levin, 1987, 1988, 1989). The degree to which a system must be regarded as controlled by external influences, such as immigration of new recruits, depends on the scale under consideration, not just on inherent biological properties. Scale influences decisions as to which process and interactions must be regarded as important. Perhaps most importantly, notions of equilibrium and variability cannot be separated from notions of scale; such considerations underlie the analyses that have been described in this section, and that should be applied to any ecological system (Levin, 1988, 1989). Finally, the measurement process itself sets limits on the patterns that are discernible because of aliasing and related effects.

Correlations and Causation, and the Development of Component Models

A first step in determining explanations for patterns of temporal and spatial variability is statistical analysis (such as spectral analysis) of scales of variation of the critical variables, and comparing of those with natural scales of lower level processes such as growth, reproduction, and movement. But (see Armi and Flament, 1985), the more critical information is likely to be contained in coherence spectra (such as Figure 14.2).

There is a body of ecological literature that suggests that correlations and multiple regressions are sufficient for prediction (Peters, 1986). However, as Lehman (1986) has pointed out, relying on regressions gives no hint as to why they work, or when they will fail; thus, correlational studies are limited in their ability to provide the basis for predicting the responses of systems to new stresses or new conditions. Correlations do not comprise an adequate basis for determining causation. Hence, a modelling framework is strongest when it can relate phenomena on different scales, providing "mechanistic" bases underlying observed patterns. On the smallest scales, that means that models for the distribution of organisms and materials should be based upon models of the movement and aggregation of organisms, and upon how they search for, detect, and capture prey. Numerous studies, including several reported in this volume, indicate that on very small scales, individual larvae are capable of directed behavior to influence settlement, and that they respond both passively and actively to local topographic features and local water movements.

Models for passive settling behavior of larvae and its relationship to physical forces and terrain complexity are discussed elsewhere in this volume (chapter 12); similar models are being developed for seed and pollen dispersal in terrestrial environments (Niklas, 1984; Okubo and Levin, 1989). It should be possible to extend these models to include active searching behavior and habitat selection by organisms, since these features have been treated in rather sophisticated models in the existing ecological theory of optimal foraging (see, for example, Mangel and Clark, 1986). The integration of physics and biology in models for such processes is an open challenge, but one that is well within our grasp.

Aggregation responses involve a special kind of directed movement. Given the importance of aggregation for some organisms (e.g., krill), component models of aggregation responses represent another important goal. Okubo (1986; see also Okubo and Anderson, 1984) already has developed generic formulations and applied them to various organisms. His current investigations of models appropriate to krill will provide a critical component, as will Osborn's studies of Lagrangian models (chapter 8, this volume).

One of the most important, general, and intriguing lines of inquiry has involved development of encounter models, which underlie quantification of the consumption term in integrated reaction-transport models. Laurence (see, for example, chapter 11, this volume) has used encounter models as components of stochastic growth models for populations. Gerritsen and Strickler (1977), building on models used in search theory (Koopman, 1956; Kohlas, 1967), consider the relative speeds of predator and prey, and assume that capture takes place if the prey passes within a detection volume of the predator. They show that, to a first approximation, Z_P, the rate of encounter of prey, is given by

$$Z_P = \begin{cases} \dfrac{\pi R^2 N_H}{3}\left(\dfrac{\overline{u}^2 + 3v^2}{v}\right) & \text{for } v \geq \overline{u} \\ \dfrac{\pi R^2 N_H}{3}\left(\dfrac{v^2 + 3\overline{u}^2}{u}\right) & \text{for } \overline{u} \geq v, \end{cases} \quad (2)$$

where R is the encounter radius of the predator, N_H is the density of prey, $\bar{u}$ is the mean prey swimming velocity, and v is the predator swimming velocity. Gerritsen and Strickler show how these results must be modified if not all prey swim at the same speed; furthermore they develop the reciprocal term for an individual prey's predator risk. These terms can be integrated to yield consumption rates and used as the basis for corresponding appropriate terms in predator-prey models. However, such approaches basically estimate the parameters in bilinear (Lotka-Volterra) consumption terms. A critical challenge is to determine the nonlinear dependence of consumption on densities, for example, as determined by feeding studies. Rothschild and Osborn (1986, 1988) carry realism a step further by demonstrating how small-scale turbulence influences encounter rates, normally by increasing the probability that a prey will fall within the detection volume of a predator. As suggested earlier, none of these models yet incorporates behavioral features such as handling times, which can make important differences to the dynamics because they determine nonlinearities (Levin, 1974; Levin and Segel, 1976). Again, however, it is these aspects that have been treated with the greatest success in classical ecological theory, and their integration with the newer physically based models should be straightforward.

In summary, there is great progress being made in the development of component models that incorporate physical and or biological factors, and the integration of these should be achievable. Such models can help us to discriminate among various mechanisms by providing us with quantitative estimates of the influence of those component processes. For example, the work of Rothschild and Osborn can be used to determine how small-scale turbulence influences consumption rates, and for what organisms and systems such influences are likely to be important. Furthermore, it is to be hoped that once the validity of component models is established for the behavior of individual organisms, such models can serve as the basis for the statistical description of the behavior of populations and aggregations of such organisms, and for ensemble models of population dynamics and spread.

Biological Complexity

I have not touched on the issue of trophic complexity, and the need to reduce dimensionality; my treatment here will be brief. Yet this is a problem of surpassing importance, given the extremely high diversity of marine systems. Huge models that incorporate every imaginable detail may be useful to help organize our thinking (e.g., the North Sea models of Andersen and Ursin (1977)), but they are not be suitable for prediction. First of all, any choice of level of detail must be arbitrary; it is always possible to find structure within structure until one reaches the limiting absurdity in which each class consists of a single organism, whose limited past experience would be the only basis for making predictions. Well short of the theater of the absurd, one confronts the problem of an inadequate data base for parameter estimation, and the problems of errors becoming magnified through multiple pathways (Levin, 1987).

How then, do we reduce complexity? Rothschild and Osborn (1986) discuss conventional approaches — selecting data sets that seem most correlated, as for example in the coherence studies presented earlier in this chapter — and point out the problems that may be caused when the neglected variables misbehave. This is exactly the point of Lehman (1986) mentioned earlier — namely, that relying only on correlations does not tell us when the correlations will break down. Rothschild and Osborn (1986) formulate the problem as an "inverse problem" for the best estimation of parameter sets.

The usual dynamical approach is similar in spirit; that is, as discussed earlier, one considers the interactions among processes that operate on similar scales, allowing slower variables to serve as parameters and averaging faster variables. Another related approach (Simon and Ando, 1961; Paine, 1980) is to look for tight linkages, and to treat tightly interacting subgroups as modules that interact more weakly with other modules. This is related to the former approach because the assumption basically is that within-group interactions are operating over shorter scales than are between-group interactions. Examples include the buffering that might be provided within a functional group of decomposers, so that removal of one species within the group passes unnoticed at the system level because of compensatory responses within the group. There is no suggestion here that such compensatory responses are group-level adaptations; they are rather epiphenomena that result from competitive release. Furthermore, modules and functional groups should not be confused with one another. One often elects to describe ecological interactions in terms of functional groupings simply because such descriptions provide the most sensible description of the system, not because there are tight linkages within the groups. In general, in understanding ecosystem dynamics, we must not be wedded to taxonomic or other descriptions for misguided reasons. It may well be that for many purposes, a functional classification system, based, for example, on organism size or feeding habit, is much more appropriate (see, for example, Denman et al., 1989; Sheldon, 1969).

Conclusions

From a purely theoretical perspective, a wide variety of possible mechanisms underlie observed patterns of patchiness, with no way to distinguish among them without investigating mechanisms. Such explanations may be purely physical, due, for example, to factors underlying water column stability; purely biological, as in models of dissipative structures (Levin and Segel, 1976); or hybrids. For oceanographic distributions, all three classes of explanation are likely to be important.

For the complexity of large marine ecosystems, the suggested approach is to follow this sequence: (i) statistical analysis of observed distributional patterns of physical and biological variables; (ii) construction of competing models for variability and patchiness, based on that statistical analysis and on knowledge of natural scales of variability of critical processes; (iii) investigation of competing models through experimental and theoretical studies of component systems; and (iv) integration of validated component models (e.g., for aggregation and consumption) in order to provide predictive models for population dynamics and redistribution.

Thomas H. Powell, Antoine Morin, Daniel Grübaum, and I are following such an approach in trying to develop predictive models for the dynamics of the Antarctic krill fisheries. Spatial and temporal scales are the organizing concepts for oceanographic studies. Our hope is that this framework will provide a way to unravel the complicated dynamics of this complex LME.

Acknowledgments

This research was supported by U.S. Dept. of Commerce Cooperative Agreement NA88EA-H-00005 between the National Oceanic and Atmospheric Administration (NOAA) and Cornell University. The Centre for Mathematical Biology and All Souls College at the University of Oxford, and the Science and Engineering Research Council of Great Britain (Grant No. GR/D/13573) provided additional sources of funds. This is publication ERC-174 of the Ecosystems Research Center, Cornell University.

References

Albrecht, F., Gatzke, H., and Wax, N. 1973. Stable limit cycles in predator-prey populations. Science 181:1073–1074.

Andersen, K. P., and Ursin, E. 1977. A multispecies extension to the Beverton and Holt theory of fishing, with accounts of phosphorus circulation and primary production. Meddelelser fra Danmarks Fiskeri-og Havundersøgelser. NS 7:319–435.

Armi, L. A., and Flament, P. 1985. Cautionary remarks on the spectral interpretation of turbulent flows. J. Geophys. Res. 90:11779–11782.

Bax, N. J., and Laevastu, T. Chapter 15, this volume.

Denman, K. L., Freeland, H. J., and Mackas, D. L. 1989. Comparisons of time scales for biomass transfer up the marine food web and coastal transport processes. In Effects of ocean variability on recruitment, and an evaluation of parameters used in stock assessment. Ed. by R. J. Beamish and G. A. McFarlane. Can. Spec. Publ. Fish. Aquat. Sci. 108.

Denman, K. L., and Platt, T. 1975. Coherences in the horizontal distributions of phytoplankton and temperature in the upper ocean. Memoires Societe Royale des Sciences de Liege 6e serie 7:19–30.

Denman, K. L., and Powell, T. M. 1984. Effects of physical processes on planktonic ecosystems in the coastal ocean. Oceanogr. Mar. Biol. Ann. Rev. 22:125–168.

Gerritsen, J., and Strickler, J. R. 1977. Encounter probabilities and community structure in zooplankton: a mathematical model. J. Fish. Res. Board Can. 34:73–82.

Haury, L. R., McGowan, J. A., and Wiebe, P. H. 1978. Patterns and processes in the time-space scales of plankton distribution. In Spatial pattern in plankton communities. pp. 277–327. Ed. by J. H. Steele. Plenum, New York.

Kierstead, H., and Slobodkin, L. B. 1953. The size of water masses containing plankton blooms. J. Mar. Res. 12:141–147.

Kingsland, S. E. 1985. Modeling nature. Univ. Chicago, Chicago. 267 pp.

Kohlas, T. 1967. Simulation von Luftkaempfin. Ph.D. Thesis. Univ. Zurich. 116 pp.

Kolmogorov, A. 1936. Sulla teoria di Volterra della lotta per l'esisttenza. Giorn. Instituto, Ital. Attuari 7:74–80.

Koopman, B. O. 1956. The theory of search. I. Kinematic bases. Oper. Res. 4:324–346.

Laurence, G. C. Chapter 11, this volume.

Lehman, J. T. 1986. The goal of understanding in limnology. Limnol. Oceanogr. 31:1160–1166.

Levin, S. A. 1974. Dispersion and population interactions. Am. Natur. 108:207–228.

Levin, S. A. 1987. Scale and predictability in ecological modeling. In Modeling and management of resources under uncertainty. pp. 2–9. Ed. by T. L. Vincent, Y. Cohen, W. Grantham, G. P. Kirkwood, and J. M. Skowronski. Lecture notes in Biomathematics 40. Springer-Verlag, Berlin.

Levin, S. A. 1988. Pattern, scale, and variability: an ecological perspective. In Community ecology. pp. 1–12. Ed. by A. Hastings. Lecture notes in Biomathematics 77. Springer-Verlag, Heidelberg.

Levin, S. A. 1989. Challenges in the development of a theory of ecosystem structure and function. In Perspectives in ecological theory. pp. 242–255. Ed. by J. Roughgarden, R. M. May, and S. A. Levin. Princeton Univ. Press, Princeton, NJ.

Levin, S. A., and Segel, L. A. 1976. Hypothesis for origin of planktonic patchiness. Nature 259(5545):659.

Mangel, M., and Clark, C. W. 1986. Towards a unified foraging theory. Ecology 67(5):1127–1138.

May, R. M. 1972. Limit cycles in predator-prey communities. Science 177:900–902.

Niklas, K. J. 1984. The motion of windborne pollen grains around conifer ovulate cones: implications on wind pollination. Am. J. Botany 71:356–374.

Okubo, A. 1974. Diffusion-induced stability in model ecosystems. Chesapeake Bay Institute. The Johns Hopkins Univ. Tech. Report 86.

Okubo, A. 1978. Horizontal dispersion and critical scales for phytoplankton patches. In Spatial pattern in plankton communities. pp. 21–42. Ed. by J. H. Steele. Plenum, New York.

Okubo, A. 1980. Diffusion and ecological problems. Mathematical models. Springer-Verlag, New York. 254 pp.

Okubo, A. 1986. Dynamical aspects of animal grouping: swarms, schools, flocks, and herds. Adv. Biophys. 22:1–94.

Okubo, A., and Anderson, J. J. 1984. Mathematical models for zooplankton swarms: their formation and maintenance. EOS. 65:731–732.

Okubo, A. and Levin, S. A. 1989. A theoretical framework for the analysis of data on the wind dispersal of seeds and pollen. Ecology. 70:329–338.

Osborn, T. R., Yamazaki, H., and Squires, K. Chapter 8, this volume.

Paine, R. T. 1980. Food webs: linkage, interaction strength and community infrastructure. The Third Tansley Lecture. J. Animal Ecol. 49:667–685.

Peters, R. H. 1986. The role of prediction in limnology. Limnol. Oceanogr. 31:1143–1159.

Rosenzweig, M. 1971. Paradox of enrichment: destabilization of exploitation ecosystems in ecological time. Science 171:385–387.

Rothschild, B. J., and Osborn, T. R. 1986. Biodynamics of the sea: preliminary observation on high dimensionality and the effect of physics on predator-prey interrelationships. Biological Oceanography Committee C.M. 1986/L:25.

Rothschild, B. J., and Osborn, T. R. 1988. The effects of turbulence on planktonic contact rates. J. Plankton Res. 10(3):465–474.

Sheldon, R. W. 1969. A universal grade scale for particulate materials. Proc. Geol. Soc. London 1659:293–295.

Simon, H. A., and Ando, A. 1961. Aggregation of variables in dynamic systems. Econometrica 29:111–138.

Steele, J. H. 1978. Some comments on plankton patches. *In* Spatial pattern in plankton communities. pp. 11–20. Ed. by J. H. Steele. Plenum, New York.

Steele, J. H., and Henderson, E. W. 1984. Modeling long-term fluctuations in fish stocks. Science 224:985–987.

Taggart, C. T, and Frank, K. T. Chapter 12, this volume.

Volterra, V. 1926. Variazioni e fluttuazioni del numero d'individui in specie animale conviventi. Mem. R. Accad. Nazionale del Lincei (Ser. 6) 2:31–113.

Weber, L. H., El-Sayed, S. Z., and Hampton, I. 1986. The variance spectra of phytoplankton, krill, and water temperature in the Antarctic Ocean south of Africa. Deep-Sea Res. 33:1327–1343.

Yamazaki, H., and Osborn, T.R. 1989. Phytoplankton distribution in the mixed layer: Implication to krill abundance. *In* Scientific Committee for the Conservation of Antarctic Marine Living Resources, Selected Scientific Papers, Part I. pp. 331–356. SC-CAMLR-SSP/5, CCAMLR, Hobart, Tasmania, Australia.

Chapter 15

Biomass Potential of Large Marine Ecosystems
A Systems Approach

Nicholas J. Bax and Taivo Laevastu

Abstract

Increasing use of the marine environment has focused attention on the need for an integrated management approach. The ecosystem provides a convenient unit for management, but it is the additional information obtained through investigating species as an integral part of their environment that will provide a basis for improved management. Single-species analyses of exploited fish populations concentrate on fishing mortality as the variable source of mortality, but mortality from fishing is commonly only a small fraction of the total fish mortality. Multispecies analyses use trophic interactions to detail the major sources of mortality for each species. A multispecies analysis that uses data from trophic interactions to improve resource survey estimates of abundance is provided; survey estimates of abundance alone are likely to be too biased to enable sensible investigation of ecosystem interactions. Examples of static analyses (e.g., sensitivity analysis and input-output analysis) and dynamic analyses (e.g., ecosystem simulation) which describe potential species interactions within the ecosystem are provided. Prediction of the future states of the marine environment is not yet possible with ecosystem models because of the chaotic variability in early life stages. However, techniques are currently available to contrast the effects of different interventions on the marine ecosystem. This is the prerequisite for responsible management of the marine environment.

Introduction

Recent years have witnessed a rapidly increasing utilization of the world's oceans, extending their uses from fishing and transportation to include waste disposal, exploitation of oil, gas, and mineral resources, and recreation. Attendant on this increase has been an expanding public awareness of the oceans and their contribution to the global environment. Strong competing demands are now being placed on ocean resources; with the lack of a clear understanding of the ocean environment, such competing demands can only lead to extreme positions being taken by all sides in order to preserve, at all costs, their own interests. Management of the oceans, rather than reaction to the most strident or best-financed user group, demands a global perspective on the ocean environment. We must know the roles and interplay of marine mammals and man as top predators in the system, be able to quantify any dangers of oil exploration and waste disposal to the marine environment, and distinguish natural from anthropogenic changes in the marine environment if we are to achieve this perspective. This can only be realized through consideration of the world's oceans as integrated systems, or ecosystems.

A marine ecosystem represents an assemblage of organisms or communities within defined physical and physico-chemical boundaries. It is a loose construct, difficult to define in terms of the organisms themselves, which may move into, out of, or through the physical area. This is a definitional problem akin to that of delimiting the geographical distribution of a fish stock. The limits depend on the research aims of the investigator and can be defined in physical, genetic, biogeographical, or management terms. Compromises will be needed: the geographical range of the gray whales which pass through the Eastern Bering Sea is more extensive than that of the sessile benthos in the same area. Pacific salmon which migrate through the area have a geographical range intermediate to the gray whale and benthos, but add a new dimension since they hatch, may rear, and eventually return to spawn and die in freshwater. The physical area defined as an ecosystem may contain one, several, or a fluctuating number of stocks of a single marine fish species. In Balsfjord, northern Norway, there is an indigenous stock of herring and at times a small fraction of large year classes of the migratory Atlanto-Scandian stock of herring. The presence of transient stocks within the ecosystem boundaries can lead to severe perturbations that cannot be explained on the basis of the resident stocks alone. When the strong 1983 and fair 1984 year classes of the spring spawning herring entered Balsfjord, it occasioned a dramatic switch in the diet of cod, the apex predator in Balsfjord, from 5% to greater than 50% contribution by weight of herring in their stomachs (Eliassen and Grotnes, 1985). Eleven hundred migratory belukha whales present in Bristol Bay, Eastern Bering Sea, from mid-June to mid-August 1983 were estimated to have consumed about 280,000 adult salmon, some 1% to 9% of the commercial catch of the different species (Frost et al., 1983). It will be a rare ecosystem indeed that can be investigated without due consideration of transport across the defined geographical boundaries.

The marine ecosystem then is not a natural construct, but an artificial one used by man to facilitate his understanding of natural processes. Its definition will depend on the goals and rationale of the investigator. Ecosystem definitions include the epifaunal community on the carapace of a decorator crab, and the noosphere of Vernadsky, which, described by Budyko (1986), includes mankind as a completely integrated unit embracing the entire earth, expanding into cosmic space, and whose parts are harmoniously linked and act together at all levels.

Large marine ecosystems (LMEs) are the focus of this chapter, with the emphasis on their value as a management unit in fisheries exploitation. It will be convenient to define them in terms of oceanographic boundaries. The value of considering ecosystems instead of individual fish stocks accrues through the additional information obtained when investigating the fish stocks in their environment rather than in the single-species universe traditionally inhabited by fisheries scientists. Consideration of the individual species in their environment requires the simultaneous analysis of different data types. In the ecosystem simulations of Laevastu et al. (1982a), the requirement is made that the biomass estimates for the different species within the ecosystem be mutually compatible as indicated by their interactions through predation: there must be sufficient pollock in the Eastern Bering Sea, for example, to sustain a predation of 1.1 million tonnes by marine mammals, predation by other fish (2.7 million tonnes), and cannibalism (7.4 million tonnes) in addition to an annual loss to the fishery of 1.1 million tonnes. Traditionally, in single-species virtual population analyses, only the loss to the fishery is included explicitly, the remaining majority of mortalities being often lumped together and assumed constant over both age and time. Multispecies virtual population analyses (summarized by Gislason and Sparre, 1987) use interspecific predation in the ecosystem to improve estimates of biomass by providing for a time- and age-variable mortality. When all major sources of mortality in the ecosystem are known, it is possible to estimate the species production, and hence abundance, independently of resource survey abundance estimates. It is practically impossible, however, to know even the major sources of mortality without error, and a method is required to weight the quality of abundance estimates and that of predation data in order to estimate the species abundances within the ecosystem that will satisfy (or hopefully at least approximate) both data sets (Bax and Eliassen, in press). All these methods improve the historically poor estimates of species abundance in marine ecosystems. Without this important first step, ecosystem dynamics cannot be sensibly investigated.

In this chapter, we (i) present a methodology to enable the simultaneous estimation of the abundance of all major species within a defined area and their feeding interactions; (ii) detail the fates of fish biomass in several marine ecosystems; (iii) present analyses to determine the dominant interactions within these ecosystems; and (iv) enter the more speculative area of forecasting.

Gross Properties of Large Marine Ecosystems

The near-surface layers of the world oceans can be divided into natural regions based on physical boundaries. Major determinants of the characteristics of the natural regions vary in importance from one region to another and are: the presence, size, and depth of the continental shelf; current systems and their boundaries (convergences); and temperature regimes (including annual range). A categorization based on these criteria is given in Figure 15.1 from Hela and Laevastu (1962), based on the earlier work of Schott (1931; 1942). These natural regions can be considered as ecosystems, although one region does not encompass the entire ranges of all the species occurring in it because ichthyofaunal regions are considerably larger than these physical regions (Figure 15.2).

Finfish biomasses and their quantitative interactions exhibit considerable similarities in similar regions. The three major biological determinants of the finfish biomass — phytoplankton production, zooplankton composition and production, and benthic composition and production — vary between regions. In the past, much attention has been placed on primary production as the determinant of finfish biomass; however, this only applies in regions dominated by one or two pelagic species (e.g., some upwelling regions). In most ecosystems, only a small frac-

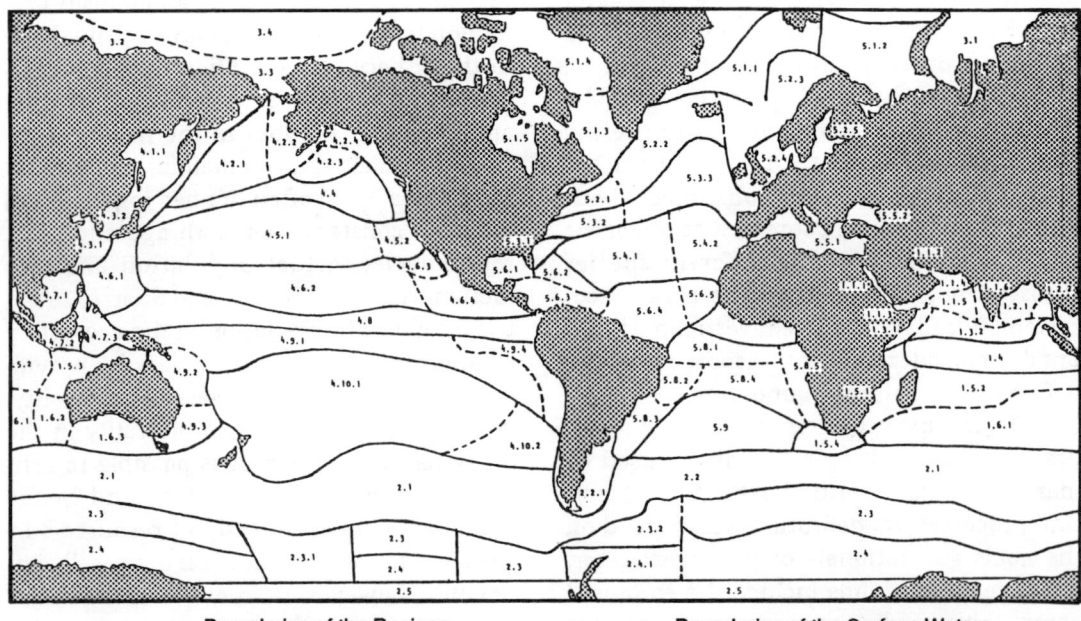

Figure 15.1. Natural regions of the oceans. (Redrawn from Hela and Laevastu, 1962.)

Key to **Figure 15.1.** Natural regions of the ocean *(facing page)*.

1.0 Indian Ocean
1.1 Arabian Sea Region
    1.1.1 Red Sea
    1.1.2 Persian Gulf
    1.1.3 Gulf of Aden
    1.1.4 Gulf of Oman
    1.1.5 C. Arabian Sea
    1.1.6 Laccadive Sea
1.2 Bay of Bengal
    1.2.1 Bay of Bengal
    1.2.2 Andaman Sea
1.3 Indian Ocean N. Eq. Cur. Region
    1.3.1 Somali W.
    1.3.2 Indian Ocean N. Eq. Cur. W.
1.4 Indian Ocean Eq. Count. Cur. Region
1.5 Indian Ocean S. Eq. Cur. Region
    1.5.1 Mozambique St.
    1.5.2 Indian Ocean Eq. Cur. W.
    1.5.3 N. Australian W.
    1.5.4 Agulhas W.
1.6 Indian Ocean Horselat. Region
    1.6.1 Indian Ocean S. Gyrals
    1.6.2 W. Australian W.
    1.6.3 Great Australian Bight

2.0 Westwind Drift Region
2.1 Indo-Pac Westwind Drift Region
2.2 Atlantic Westwind Drift Region
    2.2.1 Patagonian W.
2.3. Antarctic N. Region
    2.3.1 Area N. of Ross Sea
    2.3.2 Scotia and S. Georgia Area
2.4. Antarctic Int. Region
    2.4.1 Weddell Sea
2.5 Antarctic S. Region

3.0 Arctic Region
3.1 Kara Sea
3.2 N. Siberian W.
3.3 Chuktschee Sea
3.4 High Arctic

4.0 Pacific Ocean
4.1 Kamtschatka Region
    4.1.1 Okhotsk Sea
    4.1.2 Kamtschatka-Kurile W.
4.2 Alaska Region
    4.2.1 W. Bering Gyral
    4.2.2 Alaska Coastal W.
    4.2.3 Alaska Gyral
    4.2.4 NW American Coastal W.
4.3 N. China and Japan Seas Region
4.4 N. Pacific Drift Region
4.5 C. N. Pacific Region
    4.5.1 N. Pacific Gyral W.
    4.5.2 San Francisco W.
4.6. Pacific N. Eq. Cur. Region
    4.6.1 Philippine W.
    4.6.2 Pacific N. Eq. Cur. W.
    4.6.3 California Coastal W.
    4.6.4 W. Mexican W.
4.7 Indonesian Region
    4.7.1 S. China Sea
    4.7.2 Java and Flores Sea
    4.7.3 Sula and Celebes Sea
4.8 Pacific Eq. Count. Cur. Region
4.9 Pacific S. Eq. Cur. Region
    4.9.1 N. Polynesian W.
    4.9.2 Coral Sea
    4.9.3 N. Tasman Sea W.
    4.9.4 Peru-Galapagos W.
4.10 Pacific S. Gyrals Region
    4.10.1 Pacific S. Gyrals
    4.10.2 SW Chilean W.

5.0 Atlantic Ocean
5.1 Atlantic Subarctic Region
    5.1.1 E. Greenland W.
    5.1.2 Barents Sea
    5.1.3 Labrador W.
    5.1.4 Baffin Bay
    5.1.5 Hudson Bay
5.2 N. Atlantic Int. Boreal Region
    5.2.1 Newfoundland W.
    5.2.2 Irminger Gyral
    5.2.3 Norwegian Sea-Faroes W.
    5.2.4 N. Sea, Irish Sea, English Ch.
    5.2.5 Baltic Sea
5.3 Gulf Stream-Atlantic Drift Cur. Region
    5.3.1 Florida W.
    5.3.2 Gulf Stream W.
    5.3.3 Atlantic Drift Cur. W.
5.4 Cent. N. Atlantic Region
    5.4.1 Sargasso Sea
    5.4.2 Azoren W.
5.5 Mediterranean Region
    5.5.1 Mediterranean Sea
    5.5.2 Black Sea
5.6 Atlantic N. Eq. Cur. Region
    5.6.1 Gulf of Mexico
    5.6.2 Bahama W.
    5.6.3 Caribbean W.
    5.6.4 Atlantic N. Eq. Cur. W.
    5.6.5 Cape Verde W.
5.7 Guinea Region
5.8 Atlantic S. Eq. Cur. Region
    5.8.1 Atlantic S. Eq. Cur. W.
    5.8.2 E. Brazilian W.
    5.8.3 SE Brazilian W.
    5.8.4 Bengueal Cur. W.
    5.8.5 SW African W.
5.9 Atlantic S. Gyrals Region

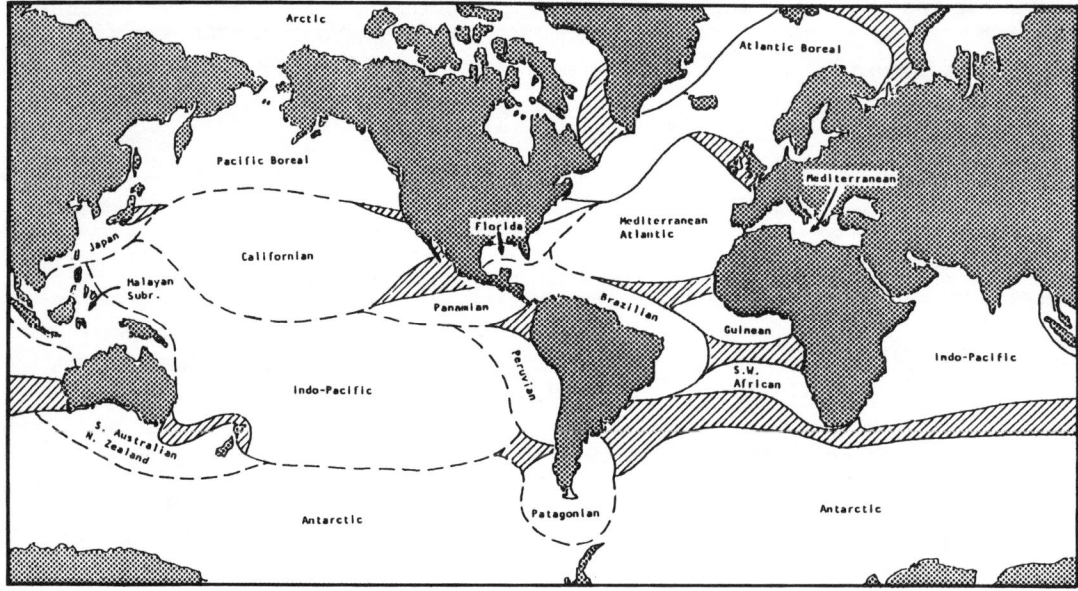

Figure 15.2. Ichthyofaunal regions of the oceans. (Redrawn from Laevastu, 1961.)

tion of the phytoplankton is used directly (much settles to the bottom where it enters the detrital cycle or the benthos) and there is no direct relation between basic organic production and finfish biomass. Finfish biomasses are higher over the continental shelf than in the open oceans due to both an increased organic production and an active benthic recycling (Table 15.1). High biomasses over the continental shelf also reflect the immigration of juveniles of many species which have grown in offshore waters before returning to coastal areas as they mature. Productivity of an ecosystem is determined to a large degree by the dynamics of the fish species it contains.

Multispecies Analysis

Multispecies analysis requires that the data derived from an individual species (or group of trophodynamically similar species) be analyzed in concert with the data available for other species. In this manner, the different data sets can be tested for compatibility. The precise design of the analysis will per- force depend on the data available for the system to be analyzed. In a multispecies analysis of the Eastern Bering Sea, Laevastu et al. (1982a,b) discounted the value of available estimates of abundance for the different species and derived independent estimates of abundance from a consideration of the removals from the system by apex predators and fishermen, the rate of production of the different species, and their interdependent feeding requirements. This strategy required an iterative approach; each change in the estimate of abundance altered the fish predation matrix, which led to new estimates of the annual loss of biomass from each species and, hence, new estimates of abundance necessary to sustain these losses. This iterative procedure continued until changes in the abundance estimates between iterations was negligible.

A similar iterative approach is used in multispecies virtual population analysis, applied in the North and Baltic Seas (reviewed by Gislason and Sparre, 1987). In this method, age-specific estimates of abundance of a few selected species are iteratively computed until the abundances are sufficient to have

Table 15.1. Estimated biomasses of fish, exploitable biomasses, and estimated annual yields in different ocean regimes. (From Laevastu and Hayes, 1981.)

Type of area characteristics	Total finfish biomass	Exploitable biomass	Sustainable annual yield (intensive fishery)
		Tonnes/km^2	
(1) Open continental shelves with upwelling-type circulation			
Tropics	25 to 45	8 to 15	3 to 7
Medium latitudes	40 to 60	12 to 20	4.5 to 8
Higher latitudes[1]	30 to 40	11 to 17	3.5 to 5.5
(2) Open continental shelves, no upwelling-type circulation			
Tropics	15 to 30	4 to 10	1.8 to 4
Medium latitudes	25 to 45	8.5 to 12	4 to 6
Higher latitudes	20 to 35	8 to 14	2 to 4
(3) Wide marginal seas (e.g., North Sea)	25 to 45	9 to 18	6 to 7.5
(4) Semi-closed seas, Mediterranean-type circulation	12 to 25	4 to 8	1.2 to 2.0
(5) Semi-closed seas, Baltic-type circulation	18 to 28	5.5 to 9.5	2.2 to 3.5
(6) Open ocean			
Low latitudes	3 to 6	0.5 to 1.2	$(< 0.3)^2$
High latitudes	5 to 12	1.5 to 3	$(< 0.6)^2$

In items 1 to 5, the biomass and yield estimated refer to areas shallower than 500 m.
[1] Assuming no great quantities of marine mammals present. [2] These yields cannot be obtained due to dispersed nature of the resources.

provided all the recorded loss to fisheries, predation by the other age classes and species in the model, together with a small residual mortality. This method has the advantage of using (and fitting the model to) a series of catch statistics over a period of years. It provides a powerful tool where fishing mortality is a significant proportion of the mortality and where age-specific data are available for all included species. It is not clear at the moment whether the predation patterns, derived from stomach content analyses, will have to be empirically determined on an annual basis or whether there is a consistent relationship between predation pressure and numbers at different abundance levels, which could be used in the model to determine year-to-year variations in the predation on each species.

Polovina (1984) adapted the Laevastu et al. (1982a) model to express it as a set of simultaneous equations that could be solved using standard computer software. Each equation represents the losses and gains in biomass for each species. The equation set is solved for the biomass of each species, which equates growth in biomass with losses from fishing, nonpredation mortality, and interspecific predation. Bax and Eliassen (in press) used this approach, but introduced a prey-switching algorithm to enable the representation of the seasonal variability in diet typical of nontropical ecosystems. It is this method that is presented below. The method as first presented requires absolute confidence in the diet composition data, a requirement that is not always reasonable. An extension to the method which balances the information obtained from diet composition data with that obtained from estimates of abundance is presented second. Representation of spatial patterns in multispecies analysis is discussed at the end of the section.

Simple analytic model

The primary assumption for the methods of Laevastu et al. (1982a), Polovina (1984), and Bax and Eliassen (in press) is that the biomasses of the different species were the same at the end as at the beginning of the period over which the data were collected. Specifically, in the application of Bax and Eliassen (in press), it is required that the biomasses of the different species did not change permanently over the course of a year.[1] Annual biomass growth must now equal annual biomass loss

$$G = F + M1 + M2 \quad (1)$$

where G is the biomass growth, F the loss to fishing, M1 is natural losses not attributable to predation (disease, senescent mortality, emigration), and M2 is the mortality resulting from predation by the species included in the model. In multispecies analysis M2 can be further specified

$$M2_i = \sum_j STOC_{ji} * FOOD_j * BIOM_j \quad (2)$$

where $STOC_{ji}$ is the proportion of the diet of species j derived from group i, $FOOD_j$ is the food requirement of j per unit time and biomass, and $BIOM_j$ the biomass of j. Indices i and j go from 1 to n, the number of species in the model. Substituting for M2 in equation 1 gives

$$G_i = F_i + M1_i + \sum_j STOC_{ji} * FOOD_j * BIOM_j \quad (3)$$

a set of n equations which can be solved to give a unique solution when $STOC_{ji}$ is known for all j,i and when four of the remaining variables are also known. Commonly $BIOM_j$ is the variable to be solved for; empirical estimates are used for the remaining variables. A nontrivial solution is required when the F_i's are given as absolute catch and at least one is nonzero.

Equation 3 applies when the stomach compositions remain constant over the time period under study. However, in temperate ecosystems, there are substantial seasonal variations in abundance, causing seasonal variations in diet.[2] To represent seasonal diet variability, $STOC_{ji}$ was first transformed to a suitability index (Sparre, 1980), which can be derived from the Holling (1959) disk equation as shown by Pepin (1987). Thus

$$SUIT_{ji} = \frac{STOC_{ji}/BIOM_i}{\sum_i STOC_{ji}/BIOM_i}. \quad (4)$$

If stomach contents were directly proportional to the biomasses of the different species, all suitability indices would equal the reciprocal of the number of species. The suitability indices, together with current biomasses, are then used to compute the stomach composition for each time period, k

$$STOC_{kji} = \frac{SUIT_{ji} * BIOM_{ki}}{\sum_i SUIT_{ji} * BIOM_{ki}}. \quad (5)$$

Incorporating equation 5 into 3 gives

$$G_{ki} = F_{ki} + M1_i$$
$$+ \sum_j \frac{SUIT_{ji} * BIOM_{ki}}{\sum_i SUIT_{ji} * BIOM_{ki}} * FOOD_{ki} * BIOM_{ki}, \quad (6)$$

a set of equations which can again be solved for a unique, and nontrivial, solution as was described for Equation 3. However, a further constraint is made to express the biomass of species i in time period k+1 as a direct function of its biomass in time period k after growth and mortalities have occurred

---

1 If it is known, or suspected, that the biomass of a species did change over the period of data collection, then this change can be included as a growth or loss term in the equation, and the equations solved in the standard manner.
2 Bax and Eliassen (in press) used a time step of one month.

$$BIOM_{k+1\,i} = BIOM_{ki} + G_{ki} - F_{ki} - M1_i$$
$$- \sum_j STOC_{kji} * FOOD_{kj} * BIOM_{kj}. \quad (7)$$

An iterative technique is now necessary to solve the equation set 6. Initial estimates of beginning year biomasses ($BIOM_{0i}$) are obtained from survey data, and the equation set 7 projected through a one-year period. Resulting year-end biomasses ($BIOM_{12i}$) are compared with initial estimates, and if these differ from initial estimates, then new seed values for biomass ($BIOM'_{01}$) are calculated as

$$BIOM'_{12i} = BIOM_{0i} - \frac{BIOM_{12i} - BIOM_{0i}}{Constant}.$$

Iterations continue until year-end biomasses equal beginning of the year biomasses. Typically, 100 iterations are required to find the biomass solution; however, 500 iterations are performed to ensure that the solution is indeed stable. The value of the constant increases slowly throughout the iterative process, making finer adjustments with later iterations.

Suitability indices ($SUIT_{ji}$) depend on both empirical stomach content data ($STOC_{ji}$) and biomass ($BIOM_i$) as shown in equation set 4. When biomass estimates change through the iterative process, suitability indices require a corresponding recalculation at the end of each iteration using the updated estimate of mean annual biomass and the empirical stomach content data. Mean annual food composition of the species in the model will thus continue to equal those estimated empirically.

In equation sets 6 and 7, values for food requirements, growth, and fishing mortality have a seasonal dimension, k. Growth and food requirements are made functions of temperature; fishing mortality is specified to correspond to the observed distribution of commercial catches throughout the year.

Extensions to simple model

Incorporating information from resource surveys. In the set of equations defining the simple model (3 or 6), there are n equations used to solve for n biomasses. The solution is nontrivial when the catch of at least one species is nonzero. In obtaining this solution, it is assumed that the remaining variables (catch, nonpredation mortality, interspecies predation mortality, and growth) are known, and that the estimates are without error. At the same time, no use is made of empirical estimates that might be available for the biomass of some or all of the species. In the application of the simple model to Balsfjord, Bax and Eliassen (in press) found that the biomass estimates differed considerably from those obtained by an acoustic and trawl survey of the same area. They concluded that the stomach composition data used to determine the interspecies predation were in error, perhaps because the sampling had failed to cover sufficient temporal and spatial dimensions of the fjord. It was not possible to use the survey estimates of biomass in the model and solve for the expected stomach composition because there are n^2 data points for this variable and only n equations. Consequently, they developed a method to solve the set of equations applying different weightings to the survey estimates of biomass and the stomach composition data.

The concept of limiting the proportion of biomass of each species available to be consumed each month was introduced to make final diet composition a function of biomass. Proportion available was made a function of the growth in biomass of each species

$$\text{Proportion Available} =$$
$$CMX * [EXP(G_i - M1_i) - 1]$$

where G_i and $M1_i$ are the mean growth rates and nonpredation natural mortality rates of species i, and CMX is a constant. Availability was then added as a constraint to equation 5, such that

$$STOC_{kji} * FOOD_{kj} * BIOM_{kj} \leq$$
$$CMX * [EXP(G_i - M1_i) - 1].$$

When the total food required from a species exceeded the specified availability, then its representation in the diet of all predators was reduced proportionately so that availability was no longer exceeded. This resulted in insufficient food being obtained by the predators, whose growth rate was correspondingly reduced.[3]

A unique solution to the set of equations (6) is still obtained; however, the solution depends on the proportion of biomass growth made available for predation. When this availability is high, the solution is the same as that for the unconstrained solution; that is, the diet composition equals that determined empirically and the solution biomasses are very different from the survey estimates (Figure 15.3; sum of squared differences close to zero for diet, close to maximum for fish biomass estimates). This situation is reversed when availability is low and solution biomasses approximate those estimated from the survey, whereas the solution diet differs from that determined empirically (Figure 15.3). At the same time, the proportion of the full diet that is obtained is reduced.

The final specification of the proportion of growth available for predation is based on the degree of confidence the investigator has in the two data sets. In the example shown for Balsfjord (Figure 15.3), an obvious choice is an availability of between 0.75 and 0.80, the position joint minimization of sum of squared differences for both biomasses and diet composition.

Accounting for spatial variability. Events in marine ecosystems occur over many different time and space scales, from the movements of an individual organism pursuing (or fleeing) a potential prey (predator) which may last less than a second, to the large-scale animal migrations or oceanographic events that may take years. It is not possible to account for all spatial and time scales within a single model; a model describing the dynamics of plankton is unlikely to be of much direct use to fisheries managers, and a model that describes annual

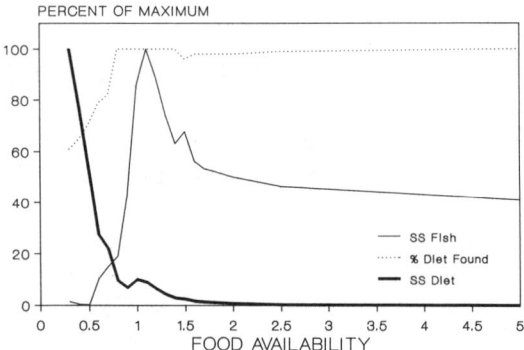

Figure 15.3. Percentage of maximum sum of squared differences between empirical and solution values for diet composition and abundance estimates. Percentage of maximum food requirements that are obtained. Data from multispecies analysis of Balsfjord by Bax and Eliassen (in press).

changes in fish or mammal populations is unlikely to shed much light on the population dynamics of plankton patches. The feeding habits of a single species will not be constant over the area defined as ecosystem. Thus, in Balsfjord, a relatively small area of 240 km^2, Klemetsen (1982) found that cod taken at three different stations had different diets. These differences in the diets remained consistent over the year of study. Laevastu and co-workers have investigated northeast Pacific ecosystems using both box and gridded models to account for the spatial variability. In the box models (Figure 15.4a), the ecosystem is divided into several areas that are studied individually. Migrations between boxes can be accounted for in only an approximate manner, and species dynamics must be averaged over the area of each box. More complicated are the gridded models (Figure 15.4b), which enable the species interactions to be calculated at a finer scale including at least part of the time and space-dependent predator/prey overlap. Thus, the modelled diets will reflect seasonal variations in prey availability caused by their migrations within the ecosystem. Gridded models require more effort to design and run, but there may be few alternatives if one

---

3 Although growth was reduced, no corresponding reduction was made to availability. Availability remained a constant proportion of potential growth.

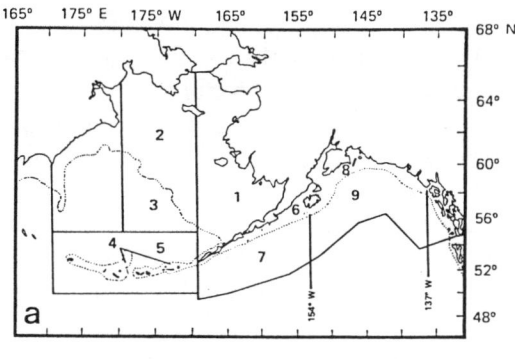

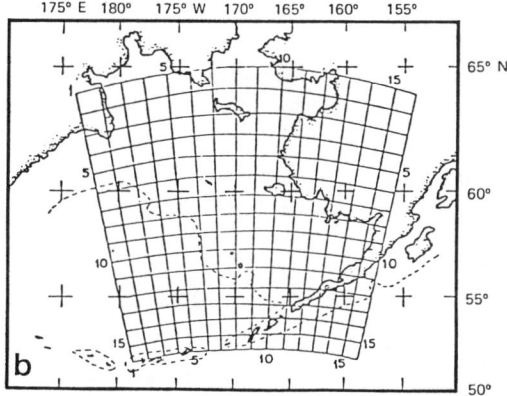

Figure 15.4. Representation of spatial variability in ecosystem simulation of the Eastern Bering Sea. (**a**) box model; (**b**) gridded model. (From Laevastu and Larkins, 1981.)

desires to investigate events that have important spatial characteristics on a scale smaller than the defined ecosystem. Schneider et al. (1987) note that both seabirds and fisheries require a resource that is highly aggregated, and suggest that significant negative interactions between the two might occur before any major reduction in prey stock was observed. They conclude that management models omitting spatial structure are unlikely to estimate the full impact of aggregated consumers on one another.

Fates of Finfish Biomass in the Ecosystem

A feature common to the multispecies analyses of different ecosystems is the "discovery" that there are more fish out there, par-

ticularly juvenile and smaller fish, than was previously thought. In single-species analyses of abundance (e.g., VPA), it was often assumed, despite all evidence to the contrary, that natural mortality was constant over age. Multispecies analyses, where predation is computed instead of assumed, quickly pinpointed the problem of assuming constant natural mortality. New, and higher, estimates of juvenile abundance were required to balance this increased predation. At the same time, multispecies analyses have highlighted the importance of mortality other than the fisheries to the dynamics of fish stocks. Much of the earlier biological advice on mesh selectivity or yield per recruit based on a constant natural mortality for all ages has been wrong (Gislason and Helgason, 1985).

There is nothing inherently wrong in analyzing the ecosystem one species at a time, provided that the biology of the species, including potential interactions, is considered. Much of the erroneous information generated by single-species analyses has resulted from undue emphasis on effects of fishing, while ignoring, or assuming constant, all other sources of biomass loss. This simply makes no sense in most ecosystems, where fishing mortalities form only a small fraction of the total mortalities. Extensions to single-species analyses to account for some of the likely fates of the fish provide a more complete model of the species dynamics. For example, consideration of the biology of walleye pollock in the Eastern Bering Sea highlights the importance of cannibal-

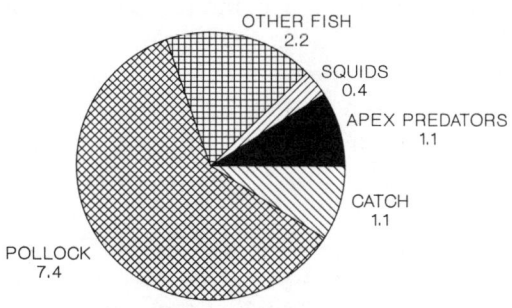

Figure 15.5. Sources of mortality of pollock in the Eastern Bering Sea ecosystem.

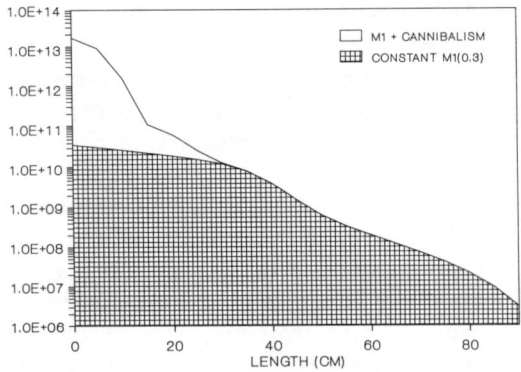

Figure 15.6. Estimated numbers at length for Eastern Bering Sea pollock 1981–1983 with and without cannibalism. (From Honkalehto, in prep.)

ism in this species (Figure 15.5). Some 60% of the annual mortality in this species results from its own cannibalism, whereas the annual catch constitutes less than 10% of total mortalities. Taking this into account, Honkalehto (in prep.) adapted the expanded single-species VPA of Lleonart et al. (1985), which includes cannibalism, to the Bering Sea. This approach yielded considerably higher estimates of juvenile pollock abundance than would be found from traditional VPA, where natural mortality is designated constant over age (Figure 15.6). Projections of effects of fishing on this stock will be noticeably different once the effects of cannibalism are taken into account, because fishing removes the principal predator of the younger fish — adult pollock.

The roles of birds, marine mammals, and man as apex predators in the ecosystem are frequently mentioned, but it is predation by fish that causes most of the mortalities. Relative losses of biomass, or energy, to predation are given for five ecosystems in Figure 15.7. In each ecosystem, fish predation accounts for the majority of mortalities. Only in the North Sea do mortalities caused by fishing approach those due to fish predation (at least for the fish species represented in the North Sea MSVPA). Often, one species dominates the predation interactions (i.e., there is a keystone predator). In the Eastern Bering Sea, it is the pollock, on Georges Bank, the silver hake, and in Balsfjord, it is the cod. The fates of the other species cannot be analyzed in isolation from this dominating predation pressure, and changes in the physical environment may be more important for their indirect effects through predation pressure than for any direct impact on the subordinate species itself (Skud, 1982).

Size is an important determinant of the fate of fish. Thus, small pollock in the Eastern Bering Sea are the prey of larger pollock. These smaller fish, together with birds consuming smaller fish, may benefit from increased commercial fisheries, which remove larger, predatory fish (Furness, 1982). However, where birds and commercial fisheries exploit fish of the same size, bird abundance can parallel the commercial harvest, raising the possibility of competitive interactions. The abundance of Cape cormorants off Nambia closely matches trends in the catch of anchovies (Crawford et al., 1987), and Brown (1981; cited by the same authors) notes that the collapse of the anchoveta resource off Peru was followed by a considerable reduction in the numbers of cormorants. Competitive interactions can only be said to occur in these instances if either the birds or the commercial fisheries contribute to significant changes in abundance of the

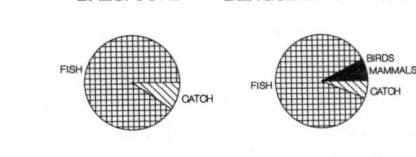

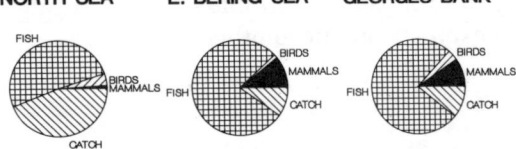

Figure 15.7. Sources of major mortalities of fish in five different ecosystems. Data sources: Balsfjord (from Bax and Eliassen, in prep.); Benguela Current (from R. Crawford, Dept. Environ. Affairs, Cape Town, SA); North Sea (from Sparholt, 1987); Eastern Bering Sea (from Laevastu and Larkins, 1981; and Laevastu et al., 1982a); Georges Bank (from Sissenwine, 1986).

fish. This is often not certain, and population fluctuations are often attributed to abiotic events such as an El Niño. Seals and commercial fisheries that exploit fish in the same area may have negative interactions, regardless of the size of their prey or total resource abundance, because of physical harm to the seals or damage to fishing gear.

As conceptual models include more facets of the ecosystem, it becomes more difficult to predict the end result of any intervention. It is at this point that we are required to analyze ecosystem interactions in a more formalized manner.

Species Interactions in the Ecosystem

Ecosystem models have a well-deserved reputation for being large, complicated, and difficult to understand. It is rarely possible to examine just the output from such a model and reach an understanding of its mechanisms. Single-species analyses will remain popular because of the transparency of their assumptions and their portability. A major role of multispecies analyses will be to detail important biological features of a species, its place in the ecosystem, and the most influential factors affecting it in order to provide single-species analyses with an adequate framework in which to operate. Species interactions can be analyzed either through the analysis of biomass or energy flow in the ecosystem, or through simulating potential impacts on the ecosystem with an ecosystem model. The first method is discussed in this section, the latter method in the next section.

Static analyses

Sensitivity, or error, analyses are an important part of developing a model of an ecosystem. These analyses identify the major interactions within the model. It then requires the investigator's judgement to determine

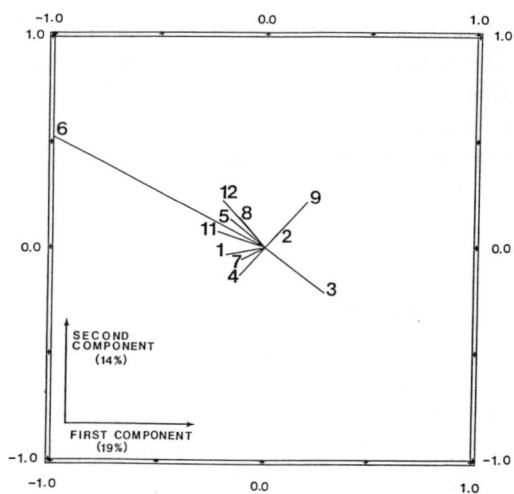

Figure 15.8. Degree of first two principal components of ecosystem variability for the Georges Bank accounted for by food requirements of the species groups. Results are from Monte Carlo error analysis (from Bax, 1985). Group 6 is silver hake.

whether these interactions are a property of the model design, or whether they are a true reflection of ecosystem interactions. Monte Carlo error analysis, where all input parameters are simultaneously and randomly perturbed within a prescribed error distribution, was used by Bax (1985) to investigate an ecosystem model of Georges Bank for the late 1960s. Predation was shown to be a more important controlling mechanism than food limitation in this area, and silver hake was identified as a keystone predator for this time period.[4] Thus, food requirements of silver hake were identified as having the most impact on variability in the Georges Bank ecosystem, as described by the first two principal components of system variability (Figure 15.8). Silver hake occupied a role in the Georges Bank ecosystem comparable to that of walleye pollock in the Eastern Bering Sea and cod in the North Sea, or Balsfjord; they are all the keystone predators in their system, displaying varying levels of cannibalism. Sissenwine (1986) describes the fish community of Georges Bank as being cannibalistic at the community level, and suggests that this moderated the effects of over-

---

[4] These conclusions are not unique to this analysis and have been dealt with in detail by, e.g., Sissenwine (1986).

fishing in the late 1960s and early 1970s. The potential for feedback mechanisms of this nature is an example of ecosystem properties that can be used to enhance single-species models by defining the forces driving fluctuations in abundance of the species.

Much of marine ecosystem research can be couched in terms of flow of energy through the system. Single-species modelling is directly concerned with the flow of energy through a species and what disruptions might occur to this flow following the removal of energy (in biomass) by the fishery. Multispecies modelling enables a more complete analysis of energy flow through a species by its explicit consideration of predation, either as cannibalism or interspecific predation. Predation is usually represented as the direct feeding interactions of one species on another. Indirect interactions — for example, several species with a common prey or predator — may be as important as these direct interactions (e.g., May et al., 1979), but are more difficult to define. These indirect interactions will be present in detailed ecosystem models, but may be obscured by the level of detail in these representations of the ecosystem. Input-output analysis (IOA), which has been applied to terrestrial ecosystems (reviewed by Szyrmer and Ulanowicz, 1987), provides a concise descriptive method of the direct and indirect flows of energy through ecological communities. Results from IOA are expressed in terms of total flow; that is, the flow from one compartment to another, which includes any and all flow passing through other compartments on the way. As an example, the sum of the direct flows of energy to any species (through feeding) will equal the total flow of energy to that species from the primary producers; it is only the primary producers that introduce new energy (as biomass) into the system.

Bax and Eliassen (in prep.) applied IOA to Balsfjord to compare the direct and total flows of energy (expressed as biomass) from

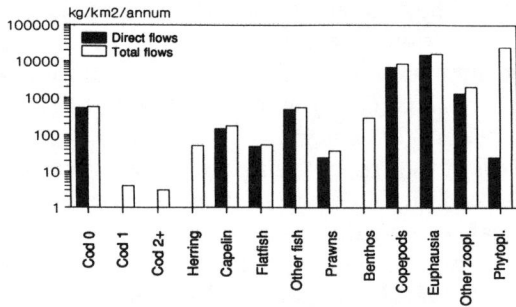

Figure 15.9. Comparison of direct and total flows of biomass to herring from each of the other trophic groups in Balsfjord. (From Bax and Eliassen, in prep.)

each compartment to herring (Figure 15.9). Direct flows identify the herring as dependent on the pelagic environment, with no energy flows to herring from the benthic environment.[5] However, when total flows are computed, connections to further species are identified. This includes a total flow from adult cod, which represents the recycling of waste matter and mortalities from adult cod through benthos and subsequently other species before being directly available to the herring. IOA describes the input to each compartment that will balance the designated output. In the above example, output was defined as the total biomass loss from each compartment, and included metabolic losses in addition to biomass taken from Balsfjord as catch. When output is restricted to commercial catch, the inputs will only represent those necessary to produce this catch. Outputs from each species necessary to produce a 100 kg · km^{-2} catch of prawn in the prawn trawler fishery are compared with those required to produce a 100 kg · km^{-2} catch of cod in the directed cod fishery (Figure 15.10). Much higher outputs are required to produce the prawn catch than the catch of cod. This is due mainly to the large bycatch associated with the prawn trawlers. The ratio of total primary production required to the resulting commercial catch is 1:4.4 for the prawn fishery compared with

---

[5] The small direct flow from flatfish represents a potential predation by herring on the pelagic flatfish eggs and larvae.

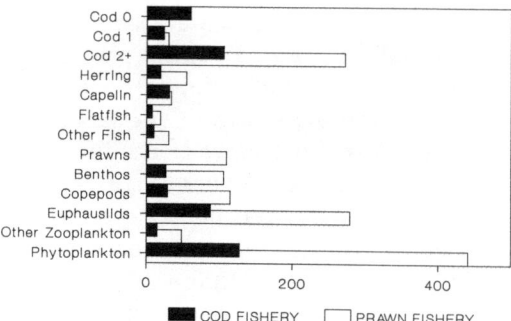

Figure 15.10. Output from each group in the Balsfjord ecosystem required to produce 100 kg · km^{-2} of cod in the directed cod fishery, or 100 kg · km^{-2} prawn in the prawn trawl fishery. (From Bax and Eliassen, in prep.)

only 1:1.3 for the directed cod fishery.[6] This indicates that in terms of energy removal it is far more expensive to produce the 100 kg · km^{-2} of prawn than the 100 kg · km^{-2} of cod, a point worthy of consideration in a subarctic fjord where food may be limiting.

The examples of ecosystem analysis provided in this section have dealt with a static view of the system and have attempted to describe the species interactions at one point in time. These interactions can be used to enhance single-species models by providing a biological framework. Ecosystems, however, are dynamic, characterized more by fluctuations than by stability. To be able to forecast the results of perturbations to the ecosystem — be it the effects of oil development or the effects of an unusually large year class of fish — ecosystem simulations will be required. This is the topic of the next section.

Dynamic simulation

If we are to evaluate the potential effects of different actions on the marine ecosystem, we need to be able to predict future changes in abundance of the populations. Unfortunately, this is not currently possible except in the most restricted circumstances. Populations fluctuate in abundance without any intervention by mankind, and in practice, it is difficult, nigh impossible, to separate these natural fluctuations from those of anthropogenic origin. However, from a management perspective, decisions have to be made despite our lack of predictive capabilities. We can enjoy some expectation of success in predicting what will happen in the ecosystem following a particular event, *if all else remains constant.* When we attempt to predict the consequences of an event using a single-species model, we have to assume that all remaining species in the ecosystem which interact with that species (at all life-history stages) remain constant. With a multispecies model, this assumption is relaxed somewhat because interactions with other species are represented, but even here, it is unlikely that the model will be able to capture the chaotic variability in early life stages. Regardless, there are still changes induced by the physico-chemical environment to consider. The best we can presently hope for is that by building and testing predictive models, we can increase our understanding of ecosystem processes and become aware of the gaps in our knowledge, while at the same time providing qualitative comparisons of the effects of proposed interventions in the ecosystem.

Earlier in this chapter we stressed the dominant role of fish predation in the marine ecosystem. As the abundances of the major piscivorous fish species change, then so will the impact of their predation on the remainder of the ecosystem. Laevastu et al. (1982a,b) detail expected changes in consumption by Pacific cod in the Eastern Bering Sea that would result from fluctuations in abundance of Pacific cod (Figure 15.11). Empirical observations on cod consumption in the Barents Sea, summarized by Torsvik (1987), show increased consumption of cod following their increased abundance.

These data further illustrate the concurrent changes to other species in the ecosystem as revealed by the contribution of these other species to the diet of cod (Figure

---

6   This primary production is only that required to produce the output to the fishery. It does not include the energy flow necessary to maintain either stock with associated metabolic costs.

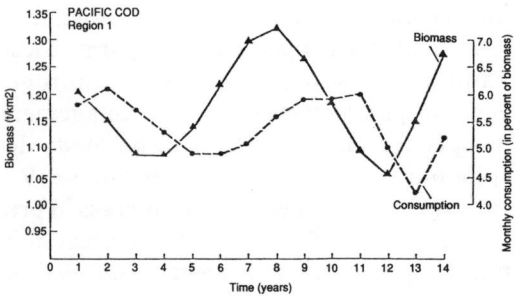

Figure 15.11. Fluctuations of the Pacific cod biomass in the Eastern Bering Sea and the consumption by Pacific cod in percent of initial biomass. (From Laevastu et al., 1982b.)

15.12). Thus, in 1984, when consumption of cod was relatively low, shrimp dominated the diet (30% or 680,000 tonnes consumed), but by 1986, the second year of higher consumption by cod, shrimp contributed only 7% (270,000 tonnes) to the cod's diet. Over the same period, the other food category (noncommercial fish, benthos, and squid) became dominant in the diet. Once species interactions of this magnitude have been recorded for an ecosystem, we can begin to consider the potential effects of the selective removal or enhancement of predators.

Potential long-term effects of changing the fishing patterns in the North Sea would be an increase in the value of the total catch by 50% or more (Gulland, 1981). This potential increase would, it was suggested, result primarily from the reduction in abundance of predators of the higher valued fish. The potential consequences of increased fishing pressure in the Eastern Bering Sea were examined by Laevastu and Marasco (1984). A substantially increased fishing pressure on pollock, cod, and yellowfin sole led to decreases in the biomasses of the target species, but caused increases in the biomasses of nontarget species (Figure 15.13). Nontarget species increased because of the reduction in predation pressure once pollock and cod were reduced. In the halibut/turbot complex (not shown), there was no change in biomass resulting from the increased fishing pressure because the decrease in predation was balanced by increased mortalities as bycatch in the trawl fisheries. Neither of these two approaches used an explicit stock and recruitment relationship, but instead concentrated on potential interactions at the juvenile and adult stages where quantitative data are available.

Consequences of increased fishing are not the only impact to be examined with multispecies methods. Losses of biomass to birds and mammals are as high as losses to man in many ecosystems (Figures 15.5 and 15.6). Furness (1982) described the competitive interactions between fish and birds in marine ecosystems. These interactions led to significant indirect impacts of man's activities on seabird populations. Furness ascribed the general increase in seabird populations in Britain and Ireland over the last 80 years to an increased abundance of smaller fish in these areas resulting from removal of the larger piscivorous fish by the commercial fisheries.

Direct competitive interactions exist between mammalian predators of the marine

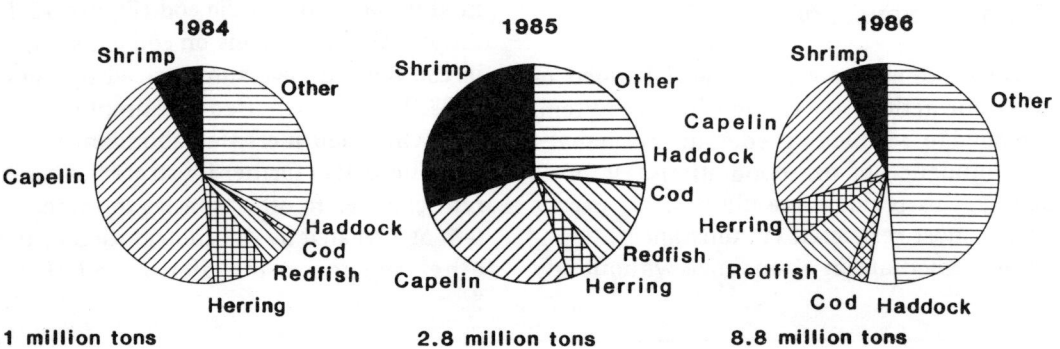

Figure 15.12. Consumption by Atlantic cod in the Barents Sea from 1984 to 1986. (Reported by Torsvik, 1987.)

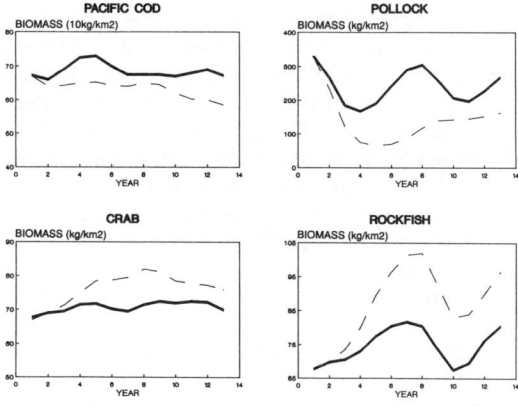

Figure 15.13. Simulated changes in the abundances of four groups in the Eastern Bering Sea following increased fishing pressure on pollock, Pacific cod and yellowfin sole. (From Laevastu and Marasco, 1984.)

environment — seals, whales, and man. Seals die in fishermen's nets and, in turn, these nets are damaged by seals. Whereas seabirds and man often exploit a different size of fish, marine mammals and fishermen can compete directly for the same fish. Gulland (1987) cites estimates by the Royal Commission on Seals and Sealing in Canada that the impact of seals on fisheries may range from negligible amounts to over C$1000 per seal per year from fish consumption alone. The impact of fisheries on seals was not given, but is certain to vary greatly for populations in different areas.

Conclusions

Throughout this chapter we have stressed the importance of predation interactions in the ecosystem. The interactions are those that can be evaluated from stomach content analyses. Little attention has been paid to early life stages, where death through processes other than predation may be as significant. This is not because theory is not available to describe these early life stages, but because the focus of this chapter is the potential interactions of man with the marine ecosystem. With this limitation, we have restricted ourselves to features that can be measured and to intervention strategies that have the potential for being realized. Models and techniques presented herein cannot hope to represent the variability apparent in the marine ecosystem, because the models do not represent important processes occurring in the first six months of a fish's life. No acceptable models are available for this early life stage where year-class strength is often determined. Available models commonly assume a predictable relationship between the numbers of parents and the number of their progeny (i.e., they follow equilibrium theory). Wiens (1984) suggests that for many communities, resource levels are essentially nonlimiting for most of the time, and under these conditions, the communities may be essentially free of biotic and even abiotic control. At such times, community fluctuations would not be predictable from equilibrium-based theory. It is only at the comparatively rare times when the environment becomes constraining that competition will become a driving force in the dynamics of the community. Marine fish populations are dominated by the rare event: the good year class which may provide a resource base for other fish as well as birds, marine mammals, and man for several years. We are not able to predict its occurrence, but we can evaluate its ramifications for the ecosystem.

It is evident that we now have sufficient information on which to base management policy at the level of the ecosystem. This does not mean that this will happen. As we consider the effects on the ecosystem of our management policies, we also have to accept responsibility for our actions. At the moment, using single-species models in fisheries management, we are usually able to restrict our attention to the value of the catch in monetary units with which we are comfortable. We have diminished our responsibility accordingly. The advent of ecosystem management will provide estimates of the direct impact on all components of the ecosystem of each proposed intervention. This will force us to consider the many and diverse effects of our interventions, and to evaluate effects for which we have, as yet, no common currency.

References

Bax, N. J. 1985. Application of multi- and univariate techniques of sensitivity analysis to SKEBUB, a biomass-based fisheries ecosystem model, parameterized to Georges Bank. Ecol. Model. 29:353–382.

Bax, N., and Eliassen, J.-E. In press. Multispecies analysis of a North Norwegian fjord. Solution and sensitivity analysis of a simple ecosystem model. J. Cons. int. Explor. Mer.

Bax, N., and Eliassen, J.-E. In preparation. An investigation of the total biomass flows through a North Norwegian fjord ecosystem using input-output analysis. 34 pp.

Brown, R. G. B. 1981. Cooperative investigation on the anchovy and its ecosystem between Peru and Canada. Boll. Inst. Mar Peru, 55–61.

Budyko, M. I. 1986. The evolution of the biosphere. D. Reidel Publishing Company, AA Dordrecht, The Netherlands. 423 pp.

Crawford, R. J. M., Shannon, L. V., and Pollock, D. E. 1987. The Benguela ecosystem. Part IV. The major fish and invertebrate resources. Oceanogr. Mar. Biol. Ann. Rev. 25:353–505.

Eliassen, J.-E., and Grotnes, P. 1985. Feeding habits of cod (Gadus morhua) in Balsfjorden, northern Norway in relation to the distribution and availability of potential food species. ICES C.M. 1985/G:58 [Mimeo]. 15 pp.

Frost, K. J., Lowry, L. F , and Nelson, R. R. 1983. Investigations of Belukha whales in coastal waters of western and northern Alaska, 1982–1983: Marking and tracking of whales in Bristol Bay. Final Rep., Alaska Dept. Fish. and Game, Fairbanks, AK. 104 pp.

Furness, R. W. 1982. Competition between fisheries and seabird communities. Adv. Mar. Biol. 30:225–307.

Gislason, H., and Helgason, Th. 1985. Species interaction in assessment of fish stocks with special application to the North Sea. Dana 5:1–44.

Gislason, H., and Sparre, P. 1987. Some theoretical aspects of the implementation of multispecies virtual population analysis in I.C.E.S. ICES C.M. 1987/G:51 [Mimeo]. 39 pp.

Gulland, J. A. 1981. Long-term potential effects from the management of the fish resources of the North Atlantic. J. Cons. int. Explor. Mer 40:8–16.

Gulland, J. A. 1987. The impact of seals on fisheries. Mar. Policy 3:196–204.

Hela, I., and Laevastu, T. 1962. Fisheries hydrography. Fishing News Books, London. 137 pp.

Holling, C. S. 1959. The components of predation as revealed by a study of small mammal predation on sawfly. Can. Entomol. 91:293–32.

Honkalehto, T. In preparation. A simple length-cohort analysis for walleye pollock with cannibalism and predation by mammals. Compass Systems Inc., 4640 Jewell St., San Diego, CA 92109.

Klemetsen, A. 1982. Food and feeding habits of cod from the Balsfjord, northern Norway during a one-year period. J. Cons. int. Explor. Mer 40:101–111.

Laevastu, T. 1961. Natural bases of fisheries in the Atlantic Ocean: Their past and present characteristics and possibilities for future expansion. In Atlantic Ocean Fisheries. pp. 18–39. Ed. by G. Borgstrom and A. J. Heighway. Fishing News Books Ltd, Surrey, UK.

Laevastu, T., and Hayes, M. L. 1981. Fisheries oceanography and ecology. Fishing News Books Ltd., Surrey, UK. 161 pp.

Laevastu, T., and Larkins, H. A. 1981. Marine fisheries ecosystem: Its quantitative evaluation and management. Fishing News Books Ltd., Surrey, UK. 161 pp.

Laevastu, T., Favorite, F., and Larkins, H. A. 1982a. Resource assessment and evaluation of the dynamics of the fisheries resources in the Northeastern Pacific with numerical ecosystem models. In Multispecies approaches to fisheries management advice. pp. 70–81. Ed. by M. C. Mercer. Can. Spec. Publ. Fish. Aquat. Sci. 59.

Laevastu, T., and Marasco, R. 1984. Some analyses of consequences of fisheries expansion in the Gulf of Alaska and Eastern Bering Sea. NOAA/NMFS/NWAFC Proc. Rep. 84–14. 30 pp.

Laevastu, T., Marasco, R., and Hayes, M. L. 1982b. Evaluation, harvesting and management of fluctuating stocks. In OCEANS 82. pp. 761–765. IEEE Coun. Ocean. Engin., Mar. Tech. Soc.

Lleonart, J., Salat, J., and MacPherson, E. 1985. CVPA, an expanded VPA with cannibalism. Application to a hake population. Fish. Res. 3:61–79.

May, R. M., Beddington, J. R., Clark, C. W., Holt, S. J., and Laws, R. W. 1979. Management of multispecies fisheries. Science 205:267–277.

Pepin, P. 1987. Influence of alternative prey abundance on pelagic fish predation of larval fish: A model. Can J. Fish. Aquat. Sci. 44:222–227.

Polovina, J. 1984. Model of a coral reef ecosystem. Coral Reefs 3:1–11.

Schneider, D. C., Hunt, G. L., Jr., and Powers, K.

D. 1987. Energy flux to pelagic birds: A comparison of Bristol Bay (Bering Sea) and Georges Bank (Northwest Atlantic). *In* Seabirds: Feeding, ecology and role in marine ecosystems. Chapter 11. Ed. by J. P. Croxall. Cambridge Univ. Press, Cambridge, UK. 408 pp.

Schott, G. 1931. Geographie des Indischen und Stillen Ozeans. Boysen, Hamburg.

Schott, G. 1942. Geographie des Atlantischen Ozeans. Boysen, Hamburg.

Sissenwine, M. P. 1986. Perturbation of a predator-controlled continental shelf ecosystem. *In* Variability and management of large marine ecosystems. Chapter 5. Ed. by K. Sherman and L. M. Alexander. AAAS Selected Symposium 99, Westview Press, Boulder, CO. 319 pp.

Skud, B. E. 1982. Dominance in fishes: The relation between environment and abundance. Science 216:144–149.

Sparholt, H. 1987. A suggestion for residual natural mortalities, M1, to be used in the North Sea MSVPA model. ICES C.M. 1987/G:53 [Mimeo]. 17 pp.

Sparre, P. 1980. A goal function of fisheries (legion analysis). ICES C.M. 1980/G:40 [Mimeo]. 81 pp.

Szyrmer, J., and Ulanowicz, R. 1987. Total flows in ecosystems. Ecol. Model. 35:123–136.

Torsvik, N. 1987. Sjopattedyra Og Torsken Konkurrent om Foda i Barentshavet. Fiskets Gang 22(45):628–630.

Wiens, J. A. 1984. On understanding a non-equilibrium world: Myth and reality in community patterns and processes. *In* Ecological communities. Chapter 25. Ed. by D. R. Strong, D. Simberloff, L. G. Abele, and A. B. Thistle. Princeton Univ. Press, Princeton, NJ.

Chapter 16

Productivity, Perturbations, and Options for Biomass Yields in Large Marine Ecosystems

Kenneth Sherman

Abstract

Biomass changes measured in multimillion-metric-ton shifts in dominance among fish stocks have been reported from large marine ecosystems (LMEs) around the globe. Increasing attention has been focused on synthesizing the biological and environmental information to examine probable causes for these population shifts and depletions that are occurring at a time of growing awareness of global change in atmospheric CO_2, methane, and ozone levels. Of the 18 LMEs for which recent syntheses have been completed, the controlling variable in four — the Yellow Sea (Huanghai Sea), Gulf of Thailand, Great Barrier Reef, and the U.S. Northeast Continental Shelf — appears to be both naturally occurring predation and excessive fishing mortality (human predation). For six LMEs, the predominant variable is environmental change in current dynamics and natural productivity — Oyashio Current, Kuroshio Current, California Current, Humboldt Current, Iberian Coastal Current, and the Benguela Current — whereas in the Baltic ecosystem, changes in coastal productivity measured over the past 70 years appear driven principally by coastal pollution. For the other seven LMEs, the available information is inconclusive precluding initial determinations on the driving forces. Conservation and management of fisheries biomass responding to strong environmental signals will be enhanced with improved forecasts of pertinent physical oceanographic conditions. LMEs controlled by predation offer more options for management.

The Large Marine Ecosystem (LME) Concept

A new era in ocean use was initiated, when in 1982, the United Nations Law of the Sea Convention established Exclusive Economic Zones up to 200 nautical miles from the baselines of territorial seas, granting coastal states sovereign rights to explore, manage, and conserve the natural resources of the zones (LOS, 1983). Within the boundaries of the new economic zones are LMEs that are being subjected to increased stress from growing exploitation of renewable resources, dumping of urban wastes, and fallout from aerosol contaminants. LMEs are defined as relatively large regions of the world ocean, generally on the order of $\geq 200,000$ km^2, characterized by unique bathymetry, hydrography, and productivity within which marine populations have adapted reproductive, growth, and feeding strategies (Sherman and Alexander, 1986, 1989).

Although the designation of LMEs is, at present, an evolving scientific and geopolitical process (Alexander, 1986; Morgan, 1988), sufficient progress has been made to allow for useful comparisons to be made of the different processes influencing large-scale changes in the biomass yields of living marine resources in LMEs. To facilitate the comparisons, a series of symposia have been convened at the annual meetings of the American Association for the Advancement of Science (AAAS). Reports presented at the first AAAS symposium on LMEs argued that LMEs were tractable global units for the conservation and management of living marine resources (Sherman and Alexander, 1986). The second symposium provided additional information on the nature of variability in biomass yields within LMEs and extended the designation of LMEs globally with the application of legal, political, and geographic criteria (Alexander, 1989; Prescott, 1989; Belsky, 1989). The third AAAS symposium (this volume) on LMEs addressed the utility of the comparative ecosystem approach for determining the principal sources of variability in biomass yields.

The movement toward total ecosystem management has been growing slowly within the international community for several decades. The trend began with the deliberations of the International Council for the Exploration of the Sea (ICES) in its first meetings conducted at the turn of the century. The meetings were prompted by the realization that the capacity of the oceans to produce an inexhaustible yield of fish biomass was finite, and that overfishing could result in the serious depletion of the stocks. The first attempts to deal with the management of regional ecosystems took place in Kristiana, Norway, in 1901 where representatives from Denmark, Germany, Norway, Russia, and the United Kingdom set the course for the establishment of ICES with a series of resolutions, directed to the establishment of joint international biological and hydrographic studies. ICES, over the past 75 years, has become a vital force in the development and implementation of joint international studies of marine ecosystems. In the process, scientists participating in ICES programs have helped focus on the advantages of coordinated multidisciplinary studies of LMEs (e.g., North Sea, Baltic, Norwegian Sea, U.S. Northeast Continental Shelf ecosystem, Iberian coastal ecosystem). As the trend for management of living resources moves from single-species to multispecies assemblages, it becomes increasingly important to encompass entire ecosystems as management units. This approach will ensure that management measures designed to optimize the natural productivity of target species assemblages will also include consideration for related competitor/predator populations and their environments.

By matching sampling effort to the time and space scales of the processes that are of most direct influence to growth and survival of living marine resource populations, forecasts of biomass yield trends among the species can be improved for LMEs. Studies of changes in abundance and population renewal of resource species in general and fish stocks in particular on a large marine ecosystem scale is in agreement with the recent proposition by Ricklefs (1987) that ecologists should begin to address critical community processes on a regional basis. Ricklefs (1987) argues that "the regional historical viewpoint provides a fundamental challenge to ecologists. Broadened concepts of the regulation of local community structure, incorporation of historical, systematic, and biogeographic information into the phenomenology of community ecology, and expanded investigations that address global variation in local species richness will help unite local and regional perspectives."

Global Biomass Yields and Marine Productivity

Controversy surrounds predictions of annual global levels of fisheries biomass yields. The annual yield based on FAO statistics in 1985 was approximately 77 million metric tons (mmt). Preliminary estimates for 1987 show an increase to 82.6 mmt, which is attributed to the upward trends of sardines and anchovies (FAO, 1989). Fisheries projections given in the *Global 2000 Report* (U.S. Council on Environmental Quality, 1980) indicated that the world harvest of fish was expected to rise little by the year 2000 from 60 mmt reached in the 1970s. In contrast, in *The Resourceful Earth* Wise, (1984) argued for an annual yield of 100–120 mmt per year of conventional species by the year 2000. Estimates of annual sustained marine fish yield based on food chain energetics for the world ocean were estimated at 70 mmt · yr^{-1} in the late 1960s by Ryther (1969). The pelagic estimates of Ryther were challenged as too low by nearly 50% (Alverson et al., 1970). The last comprehensive effort to estimate global yields was published in 1971 by FAO (Gulland, 1971). Among the suggestions offered then for further research to increase the precision of global estimates was the need for studies to improve the knowledge of "...some of the basic scientific problems concerning fish production...." During the nearly two decades since the publication of the FAO global fish yield estimates, some advances have been made in improving estimates of primary production. However, little significant progress has been made in overcoming the lack of detailed information on secondary production and energy transfer efficiencies up the food chain to fish (Lasker, 1988).

Future progress in improving estimates of global fishery yields is unlikely to be forthcoming without benefit of an improved understanding of the food chain linkages in the sea at the appropriate spatial scales. The spatial and temporal scales of biological and physical observations important to pelagic fish production, depicted by Steele (1980), stress the need for a wide range of observations from kilometers to thousands of kilometers and from days to years (Figure 16.1). The importance of studying food chain dynamics as part of a systematic approach to improving forecasts of fishery yields is underscored by the increasing evidence that the success of annual population renewals or recruitment in the sea is contingent on events occurring during the first six months of life for many of the world's fish populations (ICES, 1987). For most stocks of fish, large numbers of eggs and larvae are produced, but only a small fraction survive to become reproducing adults. It is not clear what factors are principal causes of this large-scale mortality, nor is there presently a satisfactory explanation of the relationship between the size of the spawning biomass and the size of subsequent year classes of new recruits to the populations.

Ecosystem Research Strategy

Improvements in global yield levels of the fisheries can be expected as a number of regional studies now underway are completed. Several studies have been conducted where effort has been directed to examining the variability in fish biomass yield from an ecosystem perspective (Beddington, 1986; Daan, 1986; Kullenberg, 1986; Incze and Schumacher, 1986; Sherman and Alexander, 1986; Sissenwine, 1986). LMEs have been proposed as regional units for investigating the principal spatial and temporal events controlling the recruitment of fish stocks on a global basis (Sherman, 1988). The designation of LMEs is based not only on biological and physical criteria, but also on the basis of geopolitical, legal, and economic considerations (Alexander, 1986; Sherman and Alexander 1986; Christy, 1986; Morgan, 1988, 1989; Belsky, 1989; Prescott, 1989). Fisheries resources that were previously shared among nations are now under national regulation and licensing. In accordance with the tenets of the Law of the Sea, extension of national jurisdictions have been claimed over the 200-mile Ex-

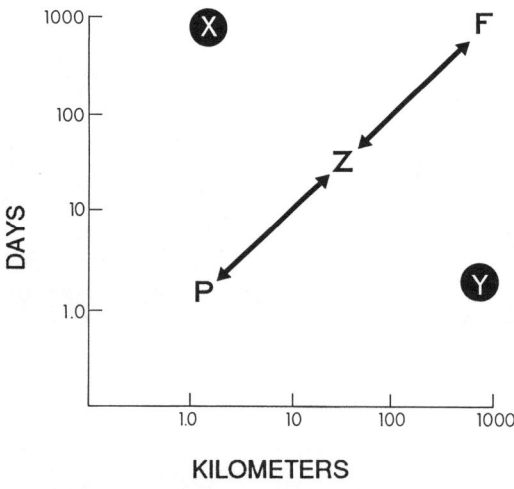

Figure 16.1. A heuristic presentation of scale relations for the food web: P (phytoplankton), Z (herbivorous zooplankton), and F (pelagic fish). Two physical processes are indicated by X, predictable fronts with small cross-front dimensions, and Y, unpredictable weather-induced effects occurring on relatively large scales. (From Steele, 1980.)

clusive Economic Zones (EEZs) of maritime nations. Restrictions on the use of fisheries resources within the EEZs are often imposed with little regard for the natural boundaries of the marine ecosystem. Most coastal populations of fish are highly mobile, migrating hundreds to thousands of kilometers within the relatively large ecosystems where they grow, reproduce, and die. Critical spawning strategies and feeding migrations have evolved since the last ice age that are difficult to understand unless observed throughout the population ranges of the stocks under investigation.

Within the EEZ of the United States, two assessment strategies are generally applied by National Marine Fisheries Service of NOAA, the federal agency responsible for monitoring the status of the U.S. fisheries resources. First, fisheries independent stock assessment surveys of fish eggs and larvae, and bottom and pelagic fish and their physical, chemical, and biological environments are conducted on mesoscale grids of 20–100 km at frequencies of two to 12 times per year to obtain estimates of the size of fish stocks. Second, bioenvironmental studies are carried out on the processes controlling the recruitment levels of new year classes.

The research strategy employed can be considered as part of a hierarchal system now evolving as a means for linking global biogeographic events occurring from millennia to decadal time scales to regional units of LMEs on time-scale changes ranging from decadal to seasonal. At the subsystem level of the hierarchy are the measurements of biological and physical variables ranging from seasonal to weekly or daily, contingent on local conditions. Within the LMEs, research is conducted on spawning, feeding, growth, recruitment mortality, and trophodynamic relationships. Two aspects of the environment are examined: (i) natural variability in relation to hydrography, currents, water masses, and weather, and (ii) human perturbations, including fishing and the effects of waste disposal, exploitation and extraction of petrogenic hydrocarbons, and the effects of aerosols and eutrophication on the capacity of the ecosystem to produce sustained yields of fish populations. To link the scientific research and development activities to the management of the living resources will require additional effort in the development and testing of bioenvironmental and socioeconomic models for improving predictions of population levels and optimizing fisheries yields (Table 16.1).

Although the new approach to management of living resources is in the developmental stages within the United States (Marchesseault, 1986; MacCall, 1986b), it is being practiced in at least one major international management agency, the Commission for the Conservation of Antarctic Marine Living Resources (CCAMLR). The CCAMLR has adopted a conservation approach that seeks to (i) prevent any harvested population from falling below the level that ensures the greatest net annual increment to stable recruitment; (ii) maintain the ecological relationships between harvested, dependent, and related populations of Antarctic living marine resources; (iii) restore depleted populations; and (iv) prevent or minimize the risk of changes in the marine ecosystem which are not potentially reversible over two or three decades.

The importance of the CCAMLR ecosystem approach to conservation and management of living marine resources is underscored by its membership. Among the 20 countries that are signatories are the world's principal fishing nations, including Chile, Japan, Peoples Republic of China, the United States, and the Soviet Union. The U.S. catch in 1985 was equal to Chile's and represented 6% of world landings. Japan was the leading fishing nation with 13% of the catch, followed by the Soviet Union (12%) and the Peoples Republic of China (8%). Whether these countries will adopt a more holistic ecosystem approach to fisheries management within their EEZs following the CCAMLR model remains an open question.

Principal Yield Controls in LMEs

Over the past five years, increasing attention

has been focused on synthesizing the available biological and environmental information as a means for improving the yields of living marine resources within LMEs (Sherman and Alexander, 1986, 1989; Belsky, 1989; Evans, 1989). The collective syntheses argue for greater emphasis in linkage studies between local and regional community processes as proposed by Ricklefs (1987) to overcome scale problems in both terrestrial and aquatic studies of ecology that are too narrowly focused.

Of the 18 LMEs for which syntheses have been completed, the principal sources of changes in the yield levels of fish species have been attributed to environmental change in six ecosystems — the Oyashio Current ecosystem, the Kuroshio Current ecosystem (Minoda, 1989; Terazaki, 1989), the California Current ecosystem (MacCall, 1986a), the Humboldt Current ecosystem (Canon, 1986), the Iberian Coastal ecosystem (Wyatt and Perez-Gandaras, 1989), and the Benguela Current ecosystem (Crawford et al., 1989). Predation by the crown of thorns starfish has been identified as the principal process causing change in the Great Barrier Reef ecosystem (Bradbury and Mundy, 1989), and recruitment overfishing considered as human predation has been reported the cause of population collapses in the Gulf of Thailand ecosystem (Piyakarnchana, 1989), the Yellow Sea ecosystem (Tang, 1989), and the U.S. Northeast Continental Shelf ecosystem (Sissenwine, 1986). Variability in the productivity of at least one LME, the Baltic, has been attributed to pollution (Kullenberg, 1986). The information on the nature of changes in the remaining seven ecosystems was insufficient to make initial determinations of probable cause (Figure 16.2). Although these initial determinations need to be verified, it is possible to project the potential consequences of the findings in relation to the management and conservation of fisheries resources. The research emphasis in ecosystems with an overriding environmental signal should focus on advancing the state of the art in examining the relationships between the physical factors controlling food chain dynamics. This should lead to improvements in lead time for providing forecasts of abundance trends for use in making economic decisions. For example, it appears that in the area of confluence between the Oyashio and Kuroshio

Table 16.1. Ecosystem R&D strategy for U.S. fisheries.

Spatial	Temporal	Unit
Global (world ocean)	Millennia-decadal	Pelagic biogeographic
Regional (Exclusive Economic Zones)	Decadal-seasonal	Large marine ecosystems
Local	Seasonal-daily	Subsystems

Biological elements
    Spawning strategies
    Feeding strategies
    Productivity, trophodynamics
    Stock fluctuations/recruitment/mortality

Environmental elements
    Natural variability (hydrography, currents, water masses, weather)
    Human perturbations (fishing, waste disposal, petrogenic hydrocarbon impacts, aerosol contaminants, eutrophication effects)

Options and advice: international, national, local
    Bioenvironmental and socioeconomic models
    Management to optimize fisheries yields

Feedback loop
    Evaluation of ecosystem status
    Evaluation of fisheries status
    Evaluation of management practices

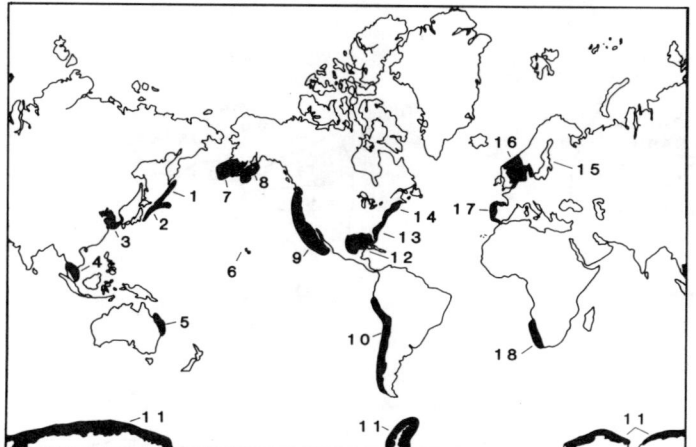

Figure 16.2. Predominant variables influencing changes in fish species biomass in large marine ecosystems. Predominant variable: Predation (X); Environment (O); Pollution (P); Inconclusive Information (+). [Reprinted from Sherman, 1988, Large marine ecosystems as global units for recruitment experiments, pp. 459–476, in Toward a theory on biological-physical interactions in the world ocean, B. J. Rothschild, Ed., with permission. Copyright 1988, Kluwer, Dordrecht, The Netherlands.]

Current ecosystems, the sardine, *Sardinops melanostica,* populations have been undergoing large-scale annual increases from less than 400,000 mt in 1975 to approximately 5 mmt in 1984 (Figure 16.3). In the period preceding the population explosion of sardine, the zooplankton biomass of the region was estimated to have increased three times between 1951 and 1975 (Table 16.2). The percentage increase in the yield of sardine to the total yield from the region increased from 22% of the catch in 1979 to 61% of the catch in 1984 (Figure 16.4). This increase in sardines followed the increase in zooplankton biomass on the spawning and nursery grounds situated within the confluence of the cold Oyashio Current ecosystem and warm Kuroshio Current ecosystem (Minoda, 1989). The upward flip in the sar-

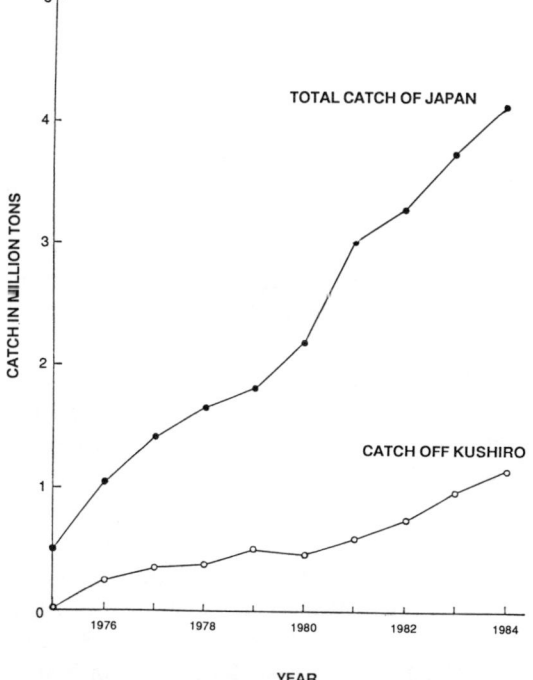

Figure 16.3. Catch in Japanese sardines of Kushiro, Hokkaido, and total catch of Japan during 1975 and 1984. (After Minoda, 1989.)

Table 16.2. Mean zooplankton biomass (g wet weight·m^{-3}) between May and July in the Oyashio, the Kuroshio, and the mixing waters south of Hokkaido for 25 years from 1951 to 1975. The area is restricted between north of 35°N and west of 146°E. (From Minoda, 1989.)

Region	Area (km^2)	Year		
		1951-57	1958-66	1967-75
Oyashio	48600	14.7	33.1	45.3
Mixed	142100	7.2	11.0	15.6
Kuroshio		4.7	5.4	12.3

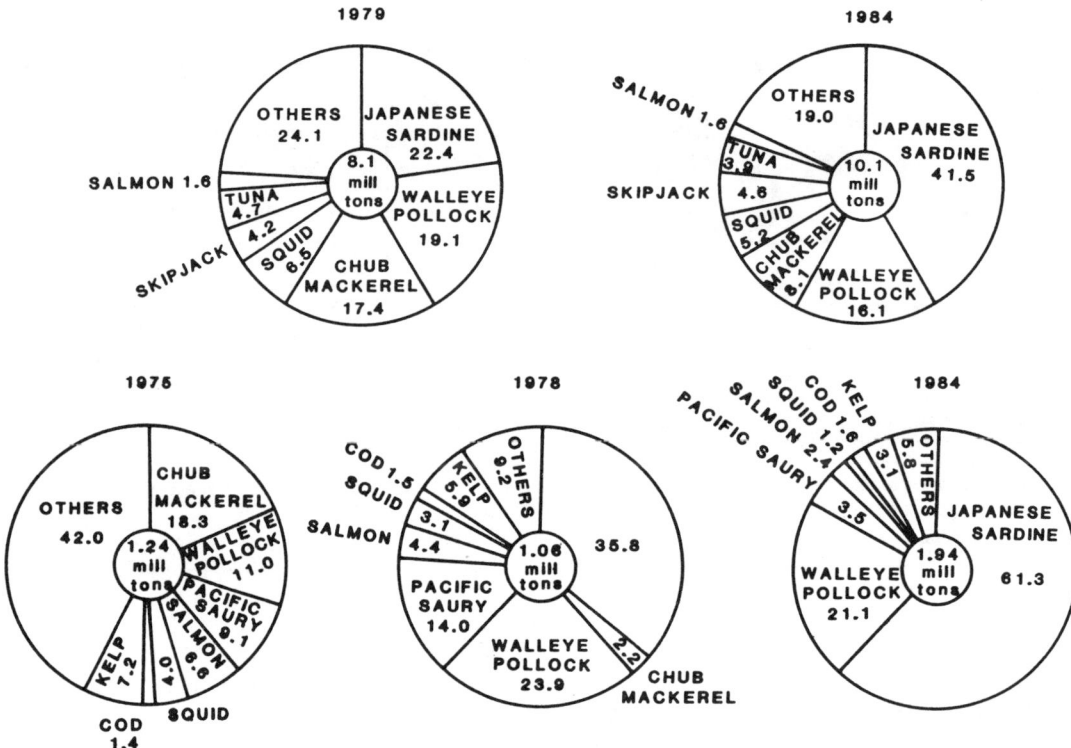

Figure 16.4. (**A**) Proportion of major commercial specimens in total catch in Japan, 1979 and 1984. (**B**) Proportion of major commercial specimens in total catch in the southeast of Hokkaido in 1975, 1978, and 1984. (From Minoda, 1989.)

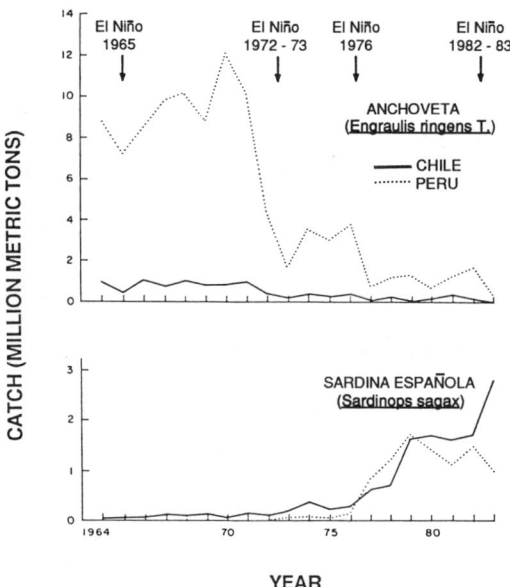

Figure 16.5. Fluctuations in the catch of anchovies and sardines in the Humboldt Current area off Chile and Peru, 1964 to 1983. (From Canon, 1986.)

dine populations off the Japanese coast have been attributed to increased egg production by the adult spawning biomass, favorable conditions for the food of larval stages, and improved environmental conditions enhancing zooplankton abundance levels within the feeding areas of the zooplanktivorous juvenile stages (Watanabe *et al.*, 1980; Terazaki, 1989).

Coincident to the population explosion of sardines in the confluence between the Oyashio and Kuroshio Current systems, a massive increase in sardine abundance has been reported for the Humboldt Current ecosystem off the Chilean coast (Canon, 1986), where yields to the fisheries of sardines *(Sardinops sagax musica)* increased from a few thousand mt in 1970 to 3 mmt in 1983 (Figure 16.5). In contrast, biomass dominance flips from sardines *(Sardinops sagax caerulea)* to anchovies *(Engraulis mordax)*

LME Biomass Productivity and Perturbations / 213

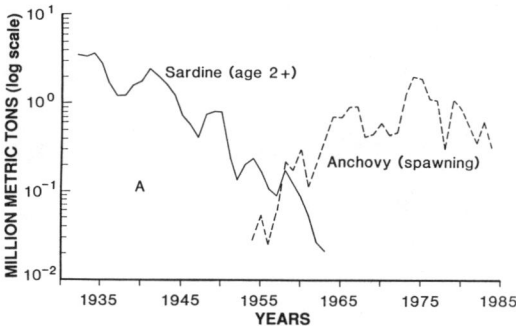

Figure 16.6. Time series of sardine (age 2+) and anchovy spawning biomass (log scale) off California and northern Baja California. "A" denotes approximate anchovy spawning biomass in 1940–1941. (From Mac-Call, 1986a.)

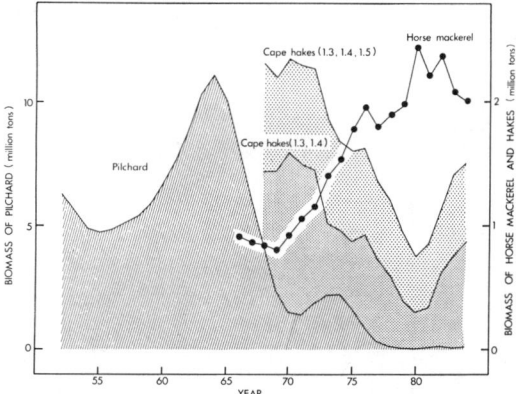

Figure 16.8. VPA estimates of the biomass of pilchard, *Sardinops ocellatus,* Cape horse-mackerel, *Trachurus capensis,* and Cape hakes, *Merluccius capensis,* and *M. paradoxus,* off Namibia, showing expansion of the horse-mackerel resource following collapse of the pilchard resource and the opposite trends in biomasses of horse-mackerel and Cape hakes. (From Crawford *et al.,* 1989.)

have been reported for the California Current (MacCall, 1986a; Figure 16.6). Biomass flips in the Iberian coastal system have been reported between sardines *(Sarda pilchardus)* and horse-mackerel *(Trachurus* spp.) (Wyatt and Perez-Gandaras, 1989; Figure 16.7). A biomass flip occurs when the species rapidly drops to a very low level and is replaced by a second species. Although more species have been implicated in dominance shifts in the Benguela Current ecosystem, the changes in abundance levels are largely attributed to environmental signals related to temperature changes in the system (Crawford *et al.,* 1989; Figure 16.8). In contrast,

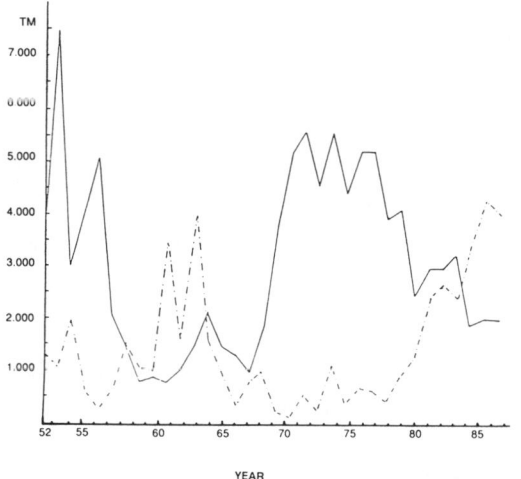

Figure 16.7. Annual catches of sardine, *Sardina pilchardus,* and jack mackerel, *Trachurus trachuras,* 1953 through 1985. (From Wyatt and Perez-Gandaras, 1989.)

large-scale, multimillion-metric-ton changes in the species biomass of fish of the U.S. Northeast Continental Shelf ecosystem are attributed to overfishing (Sissenwine, 1986). The pelagic component of the Northeast Shelf fish stocks appears to be controlled by predation. In less than a decade, during the period 1968 through 1975, high fishing mortality reduced the stocks of Atlantic mackerel *(Scomber scombrus)* and Atlantic herring *(Clupea harengus).* The release of predation pressure of these two species on the sand eel *(Ammodytes* spp.), a prey common to both Atlantic herring and Atlantic mackerel, led to a multimillion metric ton population explosion of sand eel (Figure 16.9); a congruent population shift in sand eel was reported for the North Sea ecosystem (Sherman *et al.,* 1981).

The rapid decline in the fish stock biomass of the Gulf of Thailand from a level of 150 kg · h^{-1} fisheries yield in 1963 and less than 50 kg · h^{-1} in 1984 is attributed to fisheries-imposed loss of reproductive potential, rather than to any environmental perturbation of the system (Piyakarnchana, 1989; Figure 16.10). Along the western margin of the Pacific rim, a similar decline in biomass attributed to recruitment overfish-

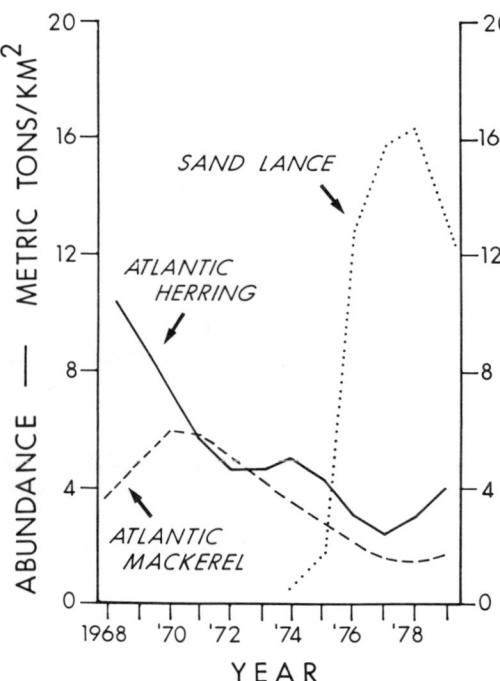

Figure 16.9. Decline of Atlantic herring and Atlantic mackerel, and apparent replacement by the small, fast-growing sand eel in the Northeast Continental Shelf ecosystem (measured in metric tons per sq km, 1968–1979). (From Sherman et al., 1983b.)

Figure 16.10. Average catch-effort (kg · h^{-1}) demersal fish in the Inner Gulf of Thailand from 1963 to 1983. (From Piyakarnchana, 1989.)

ing has been reported for the Yellow Sea ecosystem, where reductions of bottom fish species including croaker (*Pseudosciaena crocea*) declined by nearly 40% between the 1960s and 1980s (Tang, 1989; Figure 16.11). The management options are greater for LMEs in which the overriding influence in changes in fish stock dominance are controlled by fishing predation than in environmentally controlled systems. For example, given the reduction in bottom fish within the Yellow-Sea ecosystem, a management strategy was adopted that encouraged the growout of young fleshy prawn. Initial results were encouraging, with catches amounting to 1000 mt in 1984 and 2000 mt in 1985 (Tang, 1989).

The estimated 70% predation of large fish on smaller sizes of fish within the U.S. Northeast Continental Shelf ecosystem (Sissenwine, 1986) suggests that the removals through selective fishing of large sizes of highly predatory Atlantic cod (*Gadus morhua*) and silver hake (*Merluccius bilinearis*) could increase fisheries yields from the present level of 10% of the total natural fish production of the ecosystem (Fogarty et al., 1987). A multispecies, predator-prey model of the North Sea fisheries suggests that by selectively eliminating large Atlantic cod through targeted fishing, the total fishery yields could increase threefold (Andersen and Ursin, 1977). Available evidence corroborates the relative coherence in the zooplankton component of the Northeast U.S. Shelf ecosystem in having undergone no significant change in the characteristic large

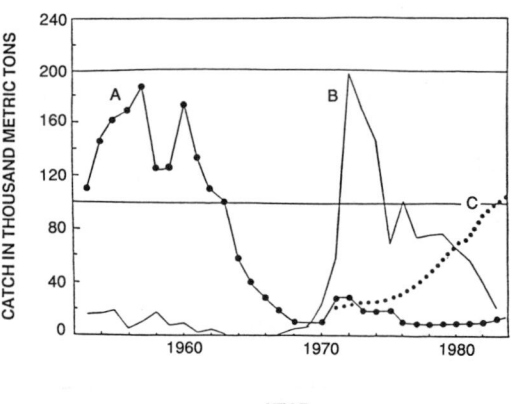

Figure 16.11. Annual catch of dominant species (**A**), small yellow croaker and hairtail, (**B**) Pacific herring and Japanese mackerel, (**C**) *Setipinna taty*, anchovy, and scaled sardine. (From Tang, 1989.)

Table 16.3. Selected hypotheses concerning variability in biomass yields of large marine ecosystems presently under investigation.

Ecosystem	Predominant variables	Hypothesis
Oyashio Current[1] Kuroshio Current[2] California Current[3] Humboldt Current[4] Benguela Current[5] Iberian Coastal[6]	Density independent natural environmental perturbations	*Clupeoid population increases:* Predominant variables influencing changes in biomass of clupeoids are major increases in water column productivity resulting from shifts in the direction and flow velocities of the currents and changes in upwelling within the ecosystem.
Yellow Sea[7] U.S. Northeast Continental Shelf[8] Gulf of Thailand[9]	Density dependent predation	*Declines in fish stocks:* Precipitous decline in biomass of fish stocks is the result of excessive fishing mortality, reducing the probability of reproductive success. Losses in biomass are attributed to excesses of human predation expressed as overfishing.
Great Barrier Reef[10]	Density dependent predation	*Change in ecosystem structure:* The extreme predation presure of crown-of-thorns starfish has disrupted normal food chain linkage between primary production and the fish component of the reef ecosystem.
East Greenland Sea[11] Barents Sea[12] Norwegian Sea[13]	Density independent natural environmental perturbations	*Shifts in abundance of fish stock biomass:* Major shifts in the abundance levels of fish stock biomass within the ecosystems are attributed to large-scale environmental changes in water movements and temperature structure.
Baltic Sea[14]	Density independent pollution	*Changes in ecosystem productivity levels:* The apparent increases in productivity levels are attributed to the effects of nitrate enrichment resulting from elevated levels of agricultural contaminant inputs from the bordering land masses.
Antarctic marine[15]	Density dependent	*Status of krill stocks:* Annual natural production cycle of krill is in balance with food requirements of dependent predator populations. Surplus production available to support economically significant yields is (is not?) within acceptable fisheries effort levels.
	Density independent natural environmental perturbations	*Shifts in abundance of krill biomass:* Major shifts in abundance levels of krill biomass within the ecosystem are attributed to large-scale changes in water movements and other physical and chemical perturbations, including ozone depletion and global warming.

[1]Minoda, 1989. [2]Terazaki, 1989. [3]MacCall, 1986. [4]Canon, 1986. [5]Crawford et al., 1989. [6]Wyatt and Perez-Gandaras, 1989. [7]Tang, 1989. [8]Sissenwine, 1986. [9]Piyakarnchana, 1989. [10]Bradbury and Mundy, 1989. [11]Buch and Hovgård, Chapter 3, this volume. [12]Skjoldal and Rey, 1989. [13]Ellertsen et al., Chapter 2, this volume. [14]Kullenberg, 1986. [15]SC-CAMLR-VII, 1988.

seasonal signal in biomass changes, species dominance patterns, or long-term changes in abundance levels since the early decades of the century (Sherman et al., 1983a). Also, laboratory experiments on prey needs of larval fish when compared to simulated in situ predator-prey encounter rates suggest that no large-scale catastrophic loss of any year class of at least two species of fish — Atlantic cod *(Gadus morhua)* and haddock *(Merluccius bilinearis)* — can be attributed to mismatch in potential contact with invertebrate planktonic prey of the appropriate size and quality to ensure average growth and survival conditions for larvae. Therefore, it is likely that recruitment to the haddock and Atlantic cod stocks of the Northeast Shelf ecosystem are controlled by predation during the juvenile stage of development (Laurence and Lough, 1985; Cohen et al., 1988).

The evidence for predator regulation of the Northeast Shelf ecosystem is growing (chapter 11, this volume). Initial efforts are underway to develop and test a theory of larval survival based on predator-prey interactions that have been modeled after queuing and encounter rate theory (Laurence, 1977;

Beyer and Laurence, 1981; Cohen et al., 1988). Consideration is also being given to augmenting the encounter rate approach by scaling the study to contact rates from larval fish and their invertebrate planktonic prey to other organisms up the food chain (Rothschild and Osborn, 1986). Success in the approach will be contingent on the use of new technology for measuring change in the systems, including hydroacoustics, image analysis, and remotely sensed satellite images of synoptic large- and small-scale ocean features. Initial results following the application of the new technologies from an ecosystem perspective have already been tested and proven successful in forecasting probable recruitment levels of Atlantic cod within the Norwegian Sea ecosystem (ICES, 1987; chapters 2 and 10, this volume). A summary of hypotheses presently under investigation that are focused on the principal driving forces influencing changes in the biomass yields of several LMEs is given in Table 16.3.

Global Changes

Now that there is a growing awareness of the potential global impacts of ozone depletion (Cicerone, 1987) and the greenhouse effect of changes in atmospheric carbon dioxide levels (National Academy of Sciences, 1979), it is particularly imperative that measurements of natural and human change on living marine resources be conducted in a systematic manner. The strategy outlined in Table 16.1 is an initial iteration of one approach. It is likely that other strategies will be forthcoming. However, it is important to consider the multidimensional scaling factors that need to be considered if cause-and-effect relationships are to be appropriately quantified and acted upon in a responsible approach to the conservation and management of living marine resources on a local, regional, and global basis.

References

Alexander, L. M. 1986. Introduction to Part Three: Large marine ecosystems as regional phenomena. In Variability and management of large marine ecosystems. pp. 239–240. Ed. by K. Sherman and L. M. Alexander. AAAS Selected Symposium 99. Westview Press, Boulder, CO. 319 pp.

Alexander, L. M. 1989. Introduction to Part Two: Geographic perspectives of large marine ecosystems. In Biomass yields and geography of large marine ecosystems. pp. 337–338. Ed. by K. Sherman and L. M. Alexander. AAAS Selected Symposium 111. Westview Press, Boulder, CO. 493 pp.

Alverson, D. L., Longhurst, A. R., and Gulland, J. A. 1970. How much food from the sea? Science 168:503–505.

Andersen, K. P., and Ursin, E. 1977. A multispecies extension to the Beverton and Holt theory of fishing, with accounts of phosphorus circulation and primary production. Medd. Danm. Fisk.-og. Havunders. N.S. 7:319–435.

Beddington, J. R. 1986. Shifts in resource populations in large marine ecosystems. In Variability and management of large marine ecosystems. pp. 9–18. Ed by. K. Sherman and L. M. Alexander. AAAS Selected Symposium 99, Westview Press, Boulder, CO. 319 pp.

Belsky, M. 1989. Developing a management regime for large marine ecosystems. In Biomass yields and geography of large marine ecosystems. pp. 443–459. Ed. by K. Sherman and L. M. Alexander. AAAS Selected Symposium 111. Westview Press, Boulder, CO. 493 pp.

Berman, M. S. Chapter 10, this volume.

Beyer, J. E., and Laurence, G. C. 1981. Aspects of stochasticity in modelling growth and survival of clupeoid fish larvae. Rapp. P.-v. Reun. Cons. int. Explor. Mer 178:17–23.

Bradbury, R., and Mundy, C. N. 1989. Large scale shifts in biomass of the Great Barrier Reef ecosystem. In Biomass yields and geography of large marine ecosystems. pp. 143–167. Ed. by K. Sherman and L. M. Alexander. AAAS Selected Symposium 111, Westview Press, Boulder, CO. 493 pp.

Canon, J. R. 1986. Variabilidad ambiental en relacion con la pesqueria neritica pelagica de la zona Norte de Chile. In La pesca en Chile. pp. 195–205. Ed. by P. Arana. Escuela de Ciencias del Mar, Facultad de Recursos Naturales, Universidad Catolica de Valparaiso. 359 pp.

Christy, F. T., Jr. 1986. Can large marine ecosystems be managed for optimum yield? In Variability and management of large marine ecosystems. pp. 263–267. Ed. by K. Sherman and

L. M. Alexander. AAAS Selected Symposium 99, Westview Press, Boulder, CO. 319 pp.

Cicerone, R. J. 1987. Changes in stratospheric ozone. Science 237:35–42.

Cohen, E. B., Sissenwine, M. P., and Laurence, G. C. 1988. The "recruitment problem" for marine fish populations with emphasis on Georges Bank. In Toward a theory on biological-physical interactions in the world ocean. pp. 373–392. Ed. by B. J. Rothschild. NATO ASI Series. Series C: Mathematical and Physical Sciences, Vol. 239. Kluwer, Dordrecht, The Netherlands. 650 pp.

Crawford, R. J. M., Shannon, L. V., and Shelton, P. A. 1989. Characteristics and management of the Benguela as a large marine ecosystem. In Biomass yields and geography of large marine ecosystems. Ed. by K. Sherman and L. M. Alexander. AAAS Selected Symposium 111, Westview Press, Boulder, CO. 493 pp.

Daan, N. 1986. Results of recent time–series observations for monitoring trends in large marine ecosystems with a focus on the North Sea. In Variability and management of large marine ecosystems. pp. 145–174. Ed. by K. Sherman and L. M. Alexander. AAAS Selected Symposium 99, Westview Press, Boulder, CO. 319 pp.

Ellertsen, B., Fossum, P., Solemdal, P., Sundby, S., and Tilseth, S. Chapter 2, this volume.

Evans, W. E. 1989. Management of large marine ecosystems. In Biomass yields and geography of large marine ecosystems. pp. 461–468. Ed. by K. Sherman and L. M. Alexander. AAAS Selected Symposium 111, Westview Press, Boulder, CO. 493 pp.

Food and Agriculture Organization of the United Nations [FAO]. 1985. Selected world fisheries landings, 1938–84. Yearbook of Fishery Statistics, Vol. 58 and preceding volumes. Excludes marine plants after 1970 and marine mammals.

FAO. 1989. FAO yearbook annuaire. Fishery statistics commodities, Vol. 65, 1987. FAO, United Nations, Rome.

Fogarty, M. J., Sissenwine, M. P., and Grosslein, M. D. 1987. Fish population dynamics. In Georges Bank. pp. 494–509. Ed. by R. H. Backus. MIT Press, Cambridge, MA. 593 pp.

Gulland, J. A. (Editor). 1971. The fish resources of the ocean. FAO, Fishing News (Books) Ltd., Surrey, UK.

International Council for the Exploration of the Sea [ICES]. 1987. Report of the Working Group on Larval Fish Ecology to the Biological Oceanography Committee of ICES, Hirtshals, Denmark, 17–19 June 1987. ICES C.M.1987/L:28.

Incze, L., and Schumacher, J. D. 1986. Variability of the environment and selected fisheries resources of the Eastern Bering Sea ecosystem. In Variability and management of large marine ecosystems. pp. 109–143. Ed. by K. Sherman and L. M. Alexander. AAAS Selected Symposium 99, Westview Press, Boulder, CO. 319 pp.

Kullenberg, G. 1986. Long-term changes in the Baltic ecosystem. In Variability and management of large marine ecosystems. pp. 19–31. Ed. by K. Sherman and L. M. Alexander. AAAS Selected Symposium 99, Westview Press, Boulder, CO. 319 pp.

Lasker, R. 1988. Food chains and fisheries; an assessment after 20 years. In Toward a theory on biological- physical interactions in the world ocean. pp. 173–182. Ed by B. J. Rothschild. NATO ASI Series. Series C: Mathematical and Physical Sciences, Vol. 239. Kluwer, Dordrecht, The Netherlands. 650 pp.

Laurence, G. C. 1977. A bioenergetic model. The analysis of feeding and survival potential of winter flounder larvae *Pseudopleuronectes americanus* during the period from hatching to metamorphosis. Fish. Bull., U.S. 75:529–546.

Laurence, G. C., and Lough, R. G. 1985. Growth and survival of larval fishes in relation to the trophodynamics of Georges Bank cod and haddock. NOAA Tech. Mem. NMFS-F/NEC-36. 150 pp.

Laurence, G. C. Chapter 11, this volume.

Law of the Sea [LOS], 1983. The Law of the Sea: official text of the United Nations Convention on the Law of the Sea with annexes and index. United Nations, New York, sales no. E.83.v.5.

Levin, S. A. Chapter 14, this volume.

MacCall, A. D. 1986a. Changes in the biomass of the California Current ecosystem. In Variability and management of large marine ecosystems. pp. 33–54. Ed. by K. Sherman and L. M. Alexander. AAAS Selected Symposium 99, Westview Press, Boulder, CO. 319 pp.

MacCall, A. D. 1986b. Rethinking research for fishery and ecosystem management. In Rethinking fisheries management. Proceedings from the tenth annual conference held June 1–4, 1986. pp. 179–193. Ed. by J. G. Sutinen and L. C. Hanson. Center for Ocean Management Studies, Univ. Rhode Island. 236 pp.

Marchessault, G. D. 1986. A New England Fishery Management Council perspective. In Rethinking fisheries management. Proceedings from the tenth annual conference held 1–4 June 1986. pp. 167–177. Ed. by J. G. Sutinen and L. C. Hanson. Center for Ocean Management Studies, Univ. Rhode Island. 236 pp.

Minoda, T. 1989. Oceanographic and biomass changes in the Oyashio Current ecosystem. *In* Biomass yields and geography of large marine ecosystems. pp. 67–93. Ed. by K. Sherman and L. M. Alexander. AAAS Selected Symposium 111, Westview Press, Boulder, CO. 493 pp.

Morgan, J. R. 1988. Large marine ecosystems an emerging concept of regional management. Environment 29(10):4–9 and 29–34.

Morgan, J. 1989. Large marine ecosystems in the Pacific. *In* Biomass yields and geography of large marine ecosystems. pp. 377–394. Ed. by K. Sherman and L. M. Alexander. AAAS Selected Symposium 111, Westview Press, Boulder, CO. 493 pp.

National Academy of Sciences. 1979. Carbon dioxide and climate: A scientific assessment. Climate Research Board, National Academy of Sciences, Washington, DC.

Piyakarnchana, T. 1989. Yield dynamics as an index of biomass shifts in the Gulf of Thailand Ecosystems. *In* Biomass yields and geography of large marine ecosystems. pp. 95–142. Ed. by K. Sherman and L. M. Alexander. AAAS Selected Symposium 111. Westview Press, Boulder, CO. 493 pp.

Prescott, V. 1989. The political division of large marine ecosystems in the Atlantic Ocean and some associated seas. *In* Biomass yields and geography of large marine ecosystems. pp. 395–442. Ed. by K. Sherman and L. M. Alexander. AAAS Selected Symposium 111. Westview Press, Boulder, CO. 493 pp.

Ricklefs, R. E. 1987. Community diversity: Relative roles of local and regional processes. Science 235:167–171.

Rothschild, B. J., and Osborn, T. R. 1986. Biodynamics of the sea: Preliminary observations on high dimensionality and the effect of physics on predator-prey interrelationships. ICES C.M. 1986/L:25.

Ryther, J. H. 1969. Photosynthesis and fish production in the sea. Science 166:72–76.

Scientific Committee for the Conservation of the Antarctic Marine Living Resources [SC-CAMLR-VII]. 1988. Report of the Seventh Meeting of the Scientific Committee, Hobart, Australia, 24–31 October 1988. 211 pp.

Sherman, K. 1988. Large marine ecosystems as global units for recruitment experiments. *In* Toward a theory on biological-physical interactions in the world ocean. pp. 459–476. Ed. by B. J. Rothschild. NATO ASI Series. Series C: Mathematical and Physical Sciences, Vol. 239. Kluwer, Dordrecht, The Netherlands. 650 pp.

Sherman, K., and Alexander, L. M. (Editors). 1986. Variability and management of large marine ecosystems. AAAS Selected Symposium 99, Westview Press, Boulder, CO. 319 pp.

Sherman, K., and Alexander, L. M. (Editors). 1989. Biomass yields and geography of large marine ecosystems. AAAS Selected Symposium 111. Westview Press, Boulder, CO. 493 pp.

Sherman, K., Jones, C., Sullivan, L., Smith, W., Berrien, P., and Ejsymont, L. 1981. Congruent shifts in sand eel abundance in western and eastern North Atlantic ecosystems. Nature 291:486:489.

Sherman, K., Green, J. R., Goulet, J. R., and Ejsymont, L. 1983a. Coherence in zooplankton of a large Northwest Atlantic ecosystem. Fish. Bull., U.S. 81:855–862.

Sherman, K., Lasker, R., Richards, W., and Kendall, A. W., Jr. 1983b. Ichthyoplankton and fish recruitment studies in large marine ecosystems. Mar. Fish. Rev. 45(10–1112):1–25.

Sissenwine, M. 1986. Perturbation of a predator-controlled continental shelf ecosystem. *In* Variability and management of large marine ecosystems. pp. 55–85. Ed. by K. Sherman and L. M. Alexander. AAAS Selected Symposium 99, Westview Press, Boulder, CO. 319 pp.

Skjoldal, H. R., and F. Rey. 1989. Pelagic production and variability of the Barents Sea Ecosystem. *In* Biomass yields and geography of large marine ecosystems. pp. 241–286, Ed. by K. Sherman and L. Alexander. AAAS Selected Symposium 111, Westview Press, Boulder, CO. 493 pp.

Steele, J. H. (Chairman). 1980. Fisheries ecology: Some constraints that impede advances in our understanding. An Ad Hoc Group of the Ocean Sciences Board, National Academy of Sciences, Washington, DC.

Tang, Q. 1989. Changes in the biomass of the Huanghai Sea ecosystem. *In* Biomass yields and geography of large marine ecosystems. pp. 7–35. Ed. by K. Sherman and L. M. Alexander. AAAS Selected Symposium 111, Westview Press, Boulder, CO. 493 pp.

Terazaki, M. 1989. Recent large-scale changes in the biomass of the Kuroshio Current ecosystem. *In* Biomass yields and geography of large marine ecosystems. pp. 37–65. Ed. by K. Sherman and L. M. Alexander. AAAS Selected Symposium 111, Westview Press, Boulder, CO. 493 pp.

U.S. Council on Environmental Quality and the Department of State, Gerald O. Barney (Director). 1980. The global 2000 report to the president: Entering the twenty-first century. Vols. I-III. U.S. Government Printing Office, Washington, DC.

Watanabe, T., Honjo, K., and Okutani, T. 1980. Fluctuation of population size of the Japanese sardine, *Sardinops melanostica* (Temminck and Schlegel) and the Kuroshio. *In* The Kuroshio IV. Proceedings of the fourth CSK symposium, 1979. Ed. by A. Y. Takenouchi. Saikon Publishing Co., Tokyo.

Wise, J. P. 1984. The future of food from the sea. *In* The resourceful earth. pp. 113–127. Ed. by J. L. Simon and H. Kahn. Basil Blackwell Inc., New York.

Wyatt, T., and Perez-Gandaras, G. 1989. Biomass changes in the Iberian ecosystem. *In* Biomass yields and geography of large marine ecosystems. pp. 221–239. Ed. by K. Sherman and L. M. Alexander. AAAS Selected Symposium 111. Westview Press, Boulder, CO. 493 pp.

Chapter 17

Geographic Perspectives in the Management of Large Marine Ecosystems

Lewis M. Alexander

Abstract

Recent research concerning large marine ecosystems (LMEs) has highlighted several issues of geographic importance. First, it has been necessary to tentatively identify LMEs at the global and regional scales and to approximate the areal extent of each system. In order to understand the nature and scope of specific LMEs, it is important to study unique hydrographic regimes and, where appropriate, submarine topography, as well as the associated trophically linked populations. Second, there should be awareness of the global positioning of LMEs, including areas where there is the potential for overlap. Third, attention must be paid to problems of overexploitation and pollution, as well as to changes in the direction of ocean currents and to other natural phenomena. Finally, assessments are required of large-scale biomass shifts within LMEs as a result of environmental changes. This chapter considers the interplay of these various issues and the potential impacts of these factors on management practices.

The Locational Component of LMEs

LMEs are multidimensional phenomena, the study of which involves a wide variety of disciplines, representing both the natural and social sciences. This chapter considers LMEs from the perspectives of geography, loosely defined as the study of natural and human phenomena of the Earth's surface. Three factors are paramount to a geographic analysis: location, interaction, and change.

An initial step in the concept of location is the identification of the element to be positioned somewhere on the surface of the Earth. It is necessary first to define LMEs, which has been done generically over the past several years; to pinpoint the location of individual units; and then to define the geographic borders of these units. At the 1984 AAAS session, a number of specific LMEs were identified — for example, the Antarctic, the Baltic, and the Eastern Bering Sea (Sherman and Alexander, 1986). Three years later, papers were presented at a second AAAS session tentatively outlining LME systems for the Atlantic and Pacific oceans (Sherman and Alexander, 1989). All listings followed the definition of LMEs as "areas characterized by unique hydrographic regimes, submarine topography, and trophically linked populations" (Sherman and Alexander, 1986).

It is the responsibility of geographers to take these systems and describe their dimensions on global and regional maps. Ultimately, it might be hoped to produce an "Atlas of LMEs" showing the distribution of these features in relation to other features of the Earth's surface, such as land masses, bottom topography, and major ocean currents.

But it takes courage to define something on a map and to outline its boundaries. One problem is orders of magnitude. Given the physical definition of LMEs, how large should the units be? Are LMEs of one area of the world of comparable size and complexity to those of other areas? For example, attention in recent years has been paid to the Northeast Continental Shelf off the United States and western Nova Scotia, the Scotian Shelf south of Nova Scotia, the Grand Banks, and the eastern Canadian Arctic. How many separate LMEs does this stretch of water between Cape Hatteras and eastern Greenland actually represent? Are the boundaries between systems clearly defined? One might remember that during the U.S./Canada *Gulf of Maine* case before a chamber of the World Court, the United States argued that the Northeast Channel between Georges and Browns Banks was an important ecological dividing line, while Canada held that Great South Channel, closer to Cape Cod, formed the biological limit. And are these really subdivisions of a larger Northeast Shelf LME?

In chapter 16, Kenneth Sherman argues for a hierarchical system for classifying fisheries. At the global level are biogeographic units, such as those covering the range of tuna migrating in the central Pacific. At the regional level are LMEs, of which he identifies 25. At the local level are subsystems. I have no quarrel with that hierarchy, if applied on a global basis. But I remain concerned about the practical application of this hierarchy to the world pattern of fishery resources. Are we really comfortable that there is agreement as to what the major LMEs of the world are? And within these LMEs, are there both subsystems and sub-subsystems? Off southeastern New England, is the Narragansett Bay system of the same magnitude as the Georges Bank system?

Morgan, a geographer who has studied LMEs, particularly in the Pacific Basin, recently (1987) wrote,

> Since the LME must consist not only of an important commercial species but of its living and non-living environment, the boundaries of the LME cannot be determined with an acceptable degree of accuracy without considerable knowledge of food supplies, predator-prey relationships, and limiting conditions of the oceanic environment, such as temperatures, salinities, and oxygen concentrations. The population dynamics of the various fish species are important for both managing and defining the LME.

Once the LME is delineated, "suitable managerial organizations, whether they be international, regional or local, can then be considered, with the objective of regulating fisheries for a maximum yield consistent with sustainability and equity" (Morgan, 1987). Again, the processes are well defined, but it seems to me that considerable more study is needed before an acceptable system of identifying and delimiting large marine ecosystems can be achieved. My concern here is not that natural and social scientists are neglecting to research individual LMEs, but rather that there might be more systematic, comparative analyses so that, first, we can get some consensus as to what the large marine ecosystems of the world ocean are and, second, that we can develop a hierarchy of LMEs and begin to identify the place of subsystems or, alternatively, of multiple systems in a world classification.

One obvious problem in this identification process is that of overlap. Except for possibly the Antarctic ecosystem, there would seem to be few cases, worldwide, where the trophically dependent populations of one LME or its subsystem are completely isolated from those of neighboring ones. And as one moves laterally across an LME, advancing seaward from the coast, is there a sharp break at the outer edge of the shelf or a gradual decrease in the densities of the populations? Conditions undoubtedly vary from one LME to another.

Interactions Within LMEs

A second concern of geographers, as well as others, is with interactions within an LME. These can be divided into two major categories — those interactions involving the natural sciences and those representing both the natural and social sciences. Most of what we have heard here during the past two and a half days concerning LMEs has dealt with the natural sciences. Within the category of social sciences, I would include law, economics, and political geography — that is, who has what kinds of jurisdiction over the fisheries resources? Let me dwell for a moment on the social science interactions.

I start with the jurisdictional patterns, because other social science elements flow from this. All but a very small percentage of commercially harvested species come from the exclusive economic or exclusive fishery zones of coastal states. This means, in some cases, that the rules and regulations of one particular state apply to the LMEs off its coast. But there are exceptions. The North Sea LME is divided jurisdictionally into two parts. In the northeast and east central parts of the Sea is the Exclusive Economic Zone (EEZ) of Norway, where Norwegian law applies. But the rest of the North Sea beyond territorial limits falls within the area of the European Economic Community's Common Fisheries Policy, which means that a number of countries share in the exploitation of the fisheries resources. One of the more vexing issues of the common fisheries arrangement is reaching agreement among the various parties on effective conservation measures for the North Sea fisheries. But this arrangement itself is unique; for most LMEs straddling national maritime boundaries, cooperative management strategies are difficult to achieve.

Many maritime limits between adjacent or opposite economic or fisheries zones are still undetermined, which raises the possibility of disputes between fishing fleets of various countries. Consider, for example, the possible impacts of the unresolved maritime boundary between Thailand and Malaysia in the western South China Sea, or, further east, the controversies resulting from disputed sovereignties over the Spratly and Paracel Islands. In the latter case, such claimant countries as China, Taiwan, and Vietnam have no formal contact whatever with one another. A somewhat analogous situation exists between the two Koreas in the western part of the Sea of Japan. Under such conditions, how can any management arrangements be effective?

Even if maritime boundary delimitations are settled, there may be no joint management. A Chamber of the World Court in 1984 drew a U.S./Canadian boundary across the northeastern portion of Georges Bank, separating what both parties had earlier agreed was a single marine biological system. In the years since the court decision, what have the two nations done toward joint conservation and management of the shelf's living marine resources?

For a country such as the United States, there is the additional problem of federal/state jurisdictions. Although most of the LME's resources are located seaward of the individual states' limits, some nursery grounds may be within state waters, thereby complicating management schemes. State and federal fisheries regulations are not always compatible.

Beyond jurisdictional interactions are those of law and economics. How are conservation and use within an LME to be managed? Will the systems adopted by one or more states be suitable for the particular characteristics of the resource? Two points are involved here. First, there must be adequate knowledge of the resource itself, so that sustainable yields can be maintained over time without serious depletion. In the course of three AAAS symposia over the past four years, we have heard of major population declines in areas such as the Gulf of Thailand and the North Sea. Do these reflect inadequate biological knowledge, or an unwillingness on the part of the states concerned to enact and implement effective conservation measures? The second point — enacting and implementing conservation measures based on sound data on the whole ecosystem — is one of the real keys to viable interactions between the natural and social sciences.

But can one, in hindsight, fault the exploiters, without really considering what the context of the scientific data was within which they operated? Sometimes it is very difficult to fix blame. Morgan, for example, notes (1987) that in the El Niño system, the annual Peruvian anchoveta catch fell from a 1970 peak of 13 million metric tons to a

1972 catch of less than 5 million tons:

> Just how much of the 'damage' to the fishery can be attributed to the natural phenomenon of El Niño and how much should be blamed on overfishing cannot be determined with our current state of knowledge of the ecosystem involved.

The Factor of Change

A third factor with which geographers must be concerned is change. And this is sometimes a difficult topic, because how does one measure and graphically display change? The change may be in the physical character of the waters, as has occurred with the perturbations of El Niño. If I plot on a chart of the Pacific the variations in the track of El Niño for the decade 1978–1988, what does this tell me about managing an associated LME? I can measure and map the pollution impacts of a large oil spill on Georges Bank, or the slow, ongoing effects of the dumping of sludge in the New York Bight. Would either of these have serious, long-term effects on an LME? Are the changes in the biomass irreversible? If the sand lance are now replacing the herring on Georges Bank, is it likely, with adequate conservation efforts, that the herring will return at a later date? What is perhaps more important, can we learn anything from the species changes off New England that could have relevance to other LME situations where heavy exploitation is occurring?

One of the techniques of geography is to classify features and processes of the Earth's surface, not so much in isolation as in types of associations, and then to test these as they are repeated in various forms and contexts in other geographic areas. To do this, we obviously have to have some knowledge of the features and processes with which we are dealing, and I am pleased that a relatively new concept of marine geography, as exemplified by the work of my colleagues, Morgan and Prescott, has gradually been taking shape (Morgan, 1987, 1989; Prescott, 1989). As noted in earlier studies, we look upon LMEs as excellent case studies in applied regionalism, and I hope and expect that geographers will continue to play a part in the further identification and assessment of LMEs as features of the world ocean.

I close with a plea for more emphasis on comparative studies of LMEs. Research to date has indicated that there are a number of similarities between individual marine ecosystems, and from these similarities, various generalizations or "principles" can, over time, be derived. Even when important differences are found to exist between LMEs, these may provide clues as to the units' functionings. We are dealing here with a new form of global system, much as meteorologists, 50 years ago, developed the concept of air mass analysis. The dilemma is, on the one hand, that natural and social scientists must do in-depth analyses of the unique features of LMEs, or even of their subsystems. But it is also necessary to compare LMEs with one another in terms of their forms and processes, in order to understand more fully how these newly identified phenomena actually function.

References

Morgan, J. R. 1987. Large marine ecosystems an emerging concept of regional management. Environment 29(10):4–9 and 29–34.

Morgan, J. R. 1989. Large marine ecosystems in the Pacific Ocean. *In* Biomass yields and geography of large marine ecosystems. pp. 377–394. Ed. by K. Sherman and L. M. Alexander. AAAS Selected Symposium, 111, Westview Press, Boulder, CO. 493 pp.

Prescott, V. 1989. The political division of large marine ecosystems in the Atlantic Ocean and some associated seas. *In* Biomass yields and geography of large marine ecosystems. pp. 395–442. Ed. by K. Sherman and L. M. Alexander. AAAS Selected Symposium 111, Westview Press, Boulder, CO. 493 pp.

Sherman, K. Chapter 16, this volume.

Sherman, K., and Alexander, L. M. (Editors). 1986. Variability and management of large marine ecosystems. AAAS Selected Symposium 99, Westview Press, Boulder, CO. 319 pp.

Sherman, K., and Alexander, L. M. (Editors). 1989. Biomass yields and geography of large marine ecosystems. AAAS Selected Symposium 111, Westview Press, Boulder. 493 pp.

Chapter 18

Interrelationships of Law in the Management of Large Marine Ecosystems

Martin H. Belsky

Abstract

A new rule of customary international law has evolved that requires a comprehensive approach to management of the ocean space. This comprehensive management approach is based on the total ecosystem model. Total ecosystem management requires a regulatory process that considers the whole system encompassing the resources of an area and the habitats for these resources. Effective total ecosystem management requires adequate information about the scope and nature of ecosystems; the impact of activities and pollutants on those ecosystems; and the best regulatory mechanisms to assure protection of the ecosystem and the resources that are part of that system.

As a rule of international law, the ecosystem management approach is binding on nation-states. It therefore must be applied by those states in their domestic law and practice. It must also be applied by states in their foreign policy. Thus, the new customary international law rule mandating total ecosystem management requires nation-states, both in their domestic laws and policies and in their international treaties and other arrangements, to establish adequate impact assessment and monitoring programs.

Introduction

The commonly accepted scientific definition of an ecosystem is that it "is a functional unit of physical and biological organization with characteristic trophic structure and material cycles, some degree of internal homogeneity, and recognizable boundaries." It consists of both living and nonliving elements (Lie, 1985). A less scientific definition is that an ecosystem is "the pattern of relationships between all biotic (living) and abiotic (nonliving) entities within a defined boundary of space and time" (Hoban and Brooks, 1987). Under either definition, it was not until very recently that international law and practice responded adequately to the clear scientific evidence that actions affecting any part of an ecosystem necessarily affected the whole ecosystem (Belsky, 1986). When marine management decisions were made, nation-states rejected an ecosystem approach in both their domestic statutes and in their international agreements (Friedheim, 1975).

This nonacceptance was founded on traditional doctrines governing international relations. National sovereignty was the major premise of international law. The right to free use (and even abuse) of the seas, and the freedom to fish (and even overfish) were sacrosanct principles (D'Amato and Hargrove, 1975).

Each nation-state had total control over the ocean areas adjacent to its coasts and over its nationals and vessels for activities on the high seas. There were few international rules governing the oceans and their resources (Knight, 1975; MacRae, 1983). As a result, each country sought to exploit resources to the maximum (Comment, 1977). Such a policy made commercial and political sense. The more of the ocean's resources that were being used by "our" citizens, the better it was for "our" economy. Growth was the key concern; conservation and pollution were not. Cooperation with other countries was just not relevant. "Foreigners" did not have a political constituency; domestic resource developers did (VanderZwaag 1983; Hoban and Brooks, 1987).

In the late 1960s, the problems of overexploitation and pollution became evident, and some nation-states began to enact legislation to govern activities in or affecting their adjacent marine areas. However, domestic legislation was on an "as needed" — or, more factually, on an "as perceived as needed" — basis. Reforms were species by species, pollutant or pollutant source by pollutant or pollutant source, and had to be documented by clear evidence (Belsky, 1986). Only with concrete evidence of a problem, and a visible popular desire for environmental and species protection controls to handle the problem, were government officials willing to put restrictions on individuals and businesses (Walsh, 1982).

This ad hoc approach also prevailed at the international level. As early as the 1950s, the international community began to recognize that some cooperative action was needed to handle and control transnational marine pollution and transboundary resources (Churchill and Lowe, 1983; Carroz, 1984). However, nation-states were concerned about the precedent of ceding authority or potential resources to any other country and about domestic political opposition to controls or restrictions set or enforced, even in part, by "outsiders" (Chapman, 1967; Comment, 1977). Thus, early agreements were minimal in scope (Friedheim, 1975). When nation-states did act collectively, they did so on an ad hoc basis and in response to a specific perceived problem (Belsky, 1986). These incremental steps at ocean management at both the national and international level did provide, however, the basis for broader policies and actions.

Moving to an Ecosystem Approach

By the 1970s, nations were considering and enacting increasingly stringent laws (i) to control pollution of coastal and ocean waters, (ii) to require policies and plans to protect their coasts and adjacent waters, and (iii) to mandate reconciliation of conflicting uses of their ocean space (Lutz, 1976). A similar process was occurring at the transnational level. A newly awakened environmental constituency made nation-states more willing to negotiate broader pollution agreements (Boczek, 1986; Speranskaya, 1986).

Nation-states desired to secure international rights to resources beyond their traditional limited territorial seas and to control activities in extended areas beyond their coastal waters. This led to unilateral claims to the right to exploit hydrocarbon and living marine resources in zones as far out as 200 miles from their coasts (MacRae, 1983; Pinto, 1985). Concerns about foreign fishing and over-exploitation led nation-states to design plans and sometimes restrictions for both endangered and commercial species (Copes, 1981; Farnell, 1981; Gordon, 1981). The major purpose of most of these fisheries laws was to provide exclusive access for a particular nation's citizens to the resources off that country's coasts. Nevertheless, the resulting legislation often also included detailed provisions for the protection of the species and its habitat and provisions to restrict fishing and preclude overexploitation (FCMA, 1976; Belsky, 1989).

Through this process, more of the ocean space came under individual nation-state control. By the 1980s, it was estimated that 38% of the oceans, over 90% of the potential commercially exploitable fish stocks, and 87% of offshore hydrocarbons exist within the collective 200-mile exclusive economic zones of all nations (Churchill and Lowe, 1983). Fewer resources and activities were in international waters and thus unregulated.

Extended jurisdiction made it more likely that an ecosystem, or at least large parts of an ecosystem, were within one nation's ocean space. Government regulators were forced to recognize the impact of exploitation of one resource on others, and the cumulative impact of individual policies on the whole ecological mosaic (Gordon, 1981).

Broader economic zones also increased the number of overlapping jurisdictional claims and the potential for conflict between adjacent coastal states. This indicated the utility of cooperative action to manage and conserve any shared resources and minimize adverse impacts on their coasts and adjacent ocean space from activities of nearby states (VanderZwaag, 1983; Comment, 1984; Belsky, 1986).

The need for new management regimes and an increased environmental sensitivity fed upon each other. A more concerned public accepted principles espoused by the scientific community as a basis for protective action. These principles were founded on the interaction of pollution and resource management and the need for a more comprehensive approach (Borgese, 1986; Hoban and Brooks, 1987).

As a result, at both the national and multinational level, nation-states began to take a comprehensive look at their actions and include this broader approach in their individual and joint arrangements for pollution control and resource management (Lutz, 1976; Carroll and Mack, 1982; Miller, 1983; Boczek, 1986).

The international rhetoric changed from one of unlimited nation-state authority and power to one of "responsible" stewardship. For example, in 1975, the United Nations established a Charter of Economic Rights and Duties of States which told nation-states to cooperate in the use of resources so as to avoid harm (Bilder, 1980). Even more significantly, in 1972, a United Nations Conference on the Environment (Stockholm Declaration, 1972) stressed the fact that everything is part of an interdependent system, and that pollution and resource management are inextricably intertwined (Smith, 1984).

On October 30, 1980, the United Nations General Assembly adopted, without vote, a resolution calling for a Draft World Charter for Nature, (United Nations, 1980), which required actions by the community of nation-states and their citizens to be conducted in such a way as not to threaten the "integrity of the ecosystems and organisms with which they coexist" (Smith, 1984). A new international law rule of "state responsibility" was emerging that required governments and citizens to prevent harm to one's own property and resources and also harm to that of adjacent resources and property (Handl, 1978, 1986).

This new international law practice was applied to the oceans. The community of nations adopted the arguments of scientists and scholars that the oceans were unique international resources and thus the responsibility of the world community (Speranskaya, 1986). Each nation-state had the international responsibility to protect marine areas and to control the activities of its nationals in and on the oceans (Boczek, 1986). Each country had a concurrent and collective obligation to safeguard the "oceans commons" (Belsky, 1986).

The "Ecosystem Model" as Binding International Law

International law differs from the law of individual nation-states in the manner in which binding rules are proposed and then established. The domestic law of each country is established by legislatures or the political executives. International rules are ordinarily not promulgated by any worldwide legislature or agency. Rather, they are established by (i) the practice of nations, (ii) acceptance by the world community as a general principle of law, (iii) judicial decisions and scholarly consensus, and (iv) international conventions. A review of the present status of the "ecosystem model" under these tests indicates its present binding character.

State practice has been moving towards a comprehensive ecosystem approach to ocean management. Nation-states increasingly accept the interconnected nature of ocean policies and the need to coordinate such policies on an ecosystem basis to assure both maximum protection of their oceans and coasts and future continued exploitation of their resources (Lutz, 1976; Boczek, 1986; Roe, 1987).

As indicated by the Stockholm Declaration and the Draft World Charter for Nature, this comprehensive approach is accepted by the world community as a general principle that should be applied by all countries. The most recent expression of this consensus can be found in the Convention for the Conservation of Antarctic Marine Living Resources (CCAMLR) and recent drafts of a proposed convention for Antarctic nonliving resources, which explicitly mandate an ecosystem management approach (CCAMLR, 1980; Laughlin, 1987).

Legal scholars have long argued that the oceans are a "commons" and that a comprehensive ecosystem approach to oceans management is both essential and the evolving rule. They point to the 1982 United Nations Convention on the Law of the Sea (UNCLOS) as strong evidence that the evolution has now been completed (Sohn, 1984; Oxman, 1986; Speranskaya, 1986).

Finally, the 1982 Convention itself indicates that the ecosystem approach is now binding international law. By definition, a treaty is binding on those states that have agreed to it. This is the doctrine of *pacta sunt servanda* (Kelsen, 1966; Vienna Convention on the Law of Treaties, Art. 26, 1969; Restatement, 1985). Thus, for those nations that have ratified UNCLOS, a comprehensive approach is mandated.

Under the 1982 Convention, nation-states are responsible for their ocean space, and, with other nations, responsible for the world's seas (Pardo, 1983; Speranskaya, 1986). Each nation is to control activities in its ports (art. 218), its coastal areas (art. 220), and its exclusive economic zone (art. 56). It also must control the activities of its nationals and vessels in all ocean areas (arts. 94, 211, 217, 117–18). These responsibilities include obligations to minimize and control pollution (arts. 194, 207, 210). They also include an obligation to manage fisheries on an ecosystem model.

Under UNCLOS, stocks in each nation's EEZ (art. 61); stocks shared by nation-states (art. 63); stocks in the high seas (art. 117–20); marine mammals (art. 65); and stocks that are highly migratory (art. 64), anadromous (art. 66), and catadromous (art. 67) must all be managed so as to avoid overexploitation. Second, UNCLOS management provisions require a "maximum sustainable yield" standard which has to be qualified by other "relevant environmental and economic factors" and takes into account the "interdependence of stocks." Finally, these specific provisions on management of living resources must be read in the light of art. 194 (5), which requires nation-states to include in pollution measures all those necessary "to protect and preserve rare or fragile ecosystems as well as the habitat of depleted, threatened, or endangered species and other forms of marine life." These three requirements, of course, constitute the ecosystem approach (Boczek, 1983).

UNCLOS mandates that these obligations are to be implemented in domestic laws, and in bilateral and multilateral treaties and other cooperative arrangements (arts. 117–18; 194, 197; 207, 213; 210, 216; 211, 217–20). Thus, the Convention obligates its signatories, individually and collectively, to be guardians of the oceans and its resources. To be appropriate stewards, they must recognize the synergistic ecosystem nature of the oceans.

Even for those nations that have not adopted the Convention, the concepts of responsibility described in the 1982 Law of the Sea Convention are binding as custom. It is now accepted that except for certain provisions not relevant here, the Convention states current law (MacRae, 1983; Malone,

1984; Sohn, 1984). The international community has confirmed its state-by-state practice, its official pronouncements, the writings of its scholars, and the consensus of the nations of the world as stated in the Convention as the new and binding rule of international law (Sohn, 1984; Speranskaya, 1986).

Implementation of the "Ecosystem Model"

Rules of international law are binding on government officials, and these political leaders have an international obligation to seek to conform their countries' actions to international law (Henkin et al., 1980; Restatement, 1985). Thus, they have the responsibility to incorporate the ecosystem model, as either a treaty or customary rule, into their own legal system; in new bilateral and multilateral agreements; and in any informal marine regulatory or management programs (Restatement, 1985; Schneider, 1985).

Additionally, in many nations, including the United States, international law is part of the domestic law unless specifically overridden by domestic law. Thus, assuming that there is no directly inconsistent domestic law, government officials must obey international law and interpret and apply other statutes and rules in a manner most consistent with an ecosystem approach. In addition, a country's representatives have the constitutional obligation to further the laws and policies of their government in their foreign relations. This means insistence on an ecosystem approach in negotiating bilateral or multilateral marine environmental protection or resource management agreements (Belsky, 1989).

Mandate for Science

At the same time as the international legal preference for ecosystem management was being established, another rule of international law was evolving — a rule that would mandate basic science, assessment, and monitoring as a basis for regulation and control of activities that affect the environment (Lutz, 1976; United Nations, 1980; Creech, 1986).

It is a fundamental scientific truism that environmental protection and careful resource management must be based on basic and applied research, which provide adequate information. Moreover, that information must focus on the comprehensive and interrelated nature of the environment (Lie, 1985; Yuru, 1985; Creech, 1986). In response, countries began to incorporate into their domestic law information requirements as a condition precedent for new regulations and for authorization of new or modified development activities (Lutz, 1976).

A new "state practice" developed that required regulators to consider the cumulative impacts of their proposed actions (NEPA, 1969; Lutz, 1976; Mayda, 1985). To obtain the necessary information, government agencies and developers had to conduct basic research and baseline studies of the environment, and then monitor both human and natural changes to that environment (Murphy and Belsky, 1980). Nation-states also provided for environmental impact reviews of proposed multistate action and assessment and monitoring in their treaty arrangements and agreements (NEPA, 1969; Carroll and Mack, 1982).

The world community also recognized the need for a collective requirement of environmental impact assessment and monitoring. As early as 1972, the General Assembly of the United Nations accepted the recommendations of the Stockholm Conference on the Human Environment (G.A. Res. 2994, Stockholm Declaration, 1972), and instructed nations to adopt rational planning mechanisms in dealing with their resources and environment. Such planning, of necessity, included the use of science to identify risks to the environment (Principles 14, 17, & 19, Stockholm Declaration, 1972). The 1972 Stockholm Conference also adopted an "Action Plan for the Human Environment," which required a broad and in-clusive "global environmental assessment program." Under that Program, nation-states, individually and collectively, are to provide for adequate environmental assessment and monitoring of all existing and proposed activities.

More recently, the United Nations Environment Program (UNEP) has stated as international "principles" that nation-states must include environmental impact assessment as part of their domestic and bilateral and multilateral actions (UNEP, 1987).

Specifically, these Principles expect nation-states (i) to assure that before decisions are made environmental effects "are to be taken fully into account" (Goal 1); (ii) to undertake environmental assessments for all "significant" activities so as to make sure that impacts are known and considered (Principle 1); and (iii) to include in such assessments, "at a minimum," a description of the whole affected environment and the impact of activities "direct, indirect, cumulative, short-term, and long-term" on the environment (Principle 4).

The premise of these requirements for assessment and monitoring is that effective management and protection of the environment requires the "addressing [of] the full range of environmental issues, from pollution abatement to the rational use and protection of all resources" (Mayda, 1985). This, of course, is the ecosystem management approach to assessment and monitoring (Mayda, 1985).

This ecosystem approach to assessment and monitoring is particularly relevant to marine management. In dealing with the oceans, improved knowledge is dependent on assessment of the ecological balance in various systems — or an ecosystem approach to scientific research and monitoring (Lie, 1985).

Adequate regulatory controls on activities and limitations on pollutants must be preceded by and based on assessment of the present status of the marine environment. Changes to that environment must then be monitored. A continual evaluation of current information must be tied into periodic revision of restrictions (Steele, 1982; Wilkinson and Connor, 1983; Yuru, 1985).

The ecological balance is, of course, a function of the organisms that live in the oceans. Ecological laws indicate that biological and nonbiological factors interact and condition each other within the same ecosystem. If fishing is not in conformity with these laws, resources may be wasted, or underexploited, or more likely depleted, or overexploited (Yuru, 1985).

Environmental studies must therefore not only analyze the impact of activities on such resources and their habitats, but also the productivity levels of organisms in particular food chains and the effects of various allowable levels of exploitation on this productivity (Sherman, 1986).

As living resource management must be based on an ecosystem model, basic research and then follow-up assessment and monitoring of the oceans must also focus on the whole ecological mosaic in a region — pollution and its impact on a particular species and other species in its chain, resource catch levels, and the impact of exploitation of one species on other species. In addition, these scientific studies must also analyze the impact of particular controls on pollution and catch levels and the impact of fishery limitations on the reproductivity of a particular species and other species affected by the productivity of that species (Federal Inter-agency Committee, 1981; NOAA, 1986).

Conclusion

As I have indicated in earlier papers, it is now customary law that nation-states, both in their domestic law and multinational arrangements, apply an ecosystem approach to management of large marine ecosystems (Belsky, 1986, 1989). This chapter argues that the ecosystem management approach is complemented by a new international norm mandating ecosystem science, assessment, and monitoring.

International law, when not made applicable through specific treaty, is established by state practice, scholarly consensus, internationally accepted general principles, and custom. Based on these standards, ecosystem-based science has evolved into a "norm" or customary rule of international law (Lutz, 1976; Boczek, 1986).

First, nation-states have recognized the need for such scientific studies and are increasingly including requirements for them in their domestic laws. They have also incorporated assessment and monitoring requirements in their bilateral and multilateral pollution and resource management agreements. These requirements either explicitly mandate ecosystem-based research or are being increasingly interpreted as requiring ecosystem-based research (Convention for the Protection of the Natural Resources and Environment of the South Pacific Region, 1987; Roe, 1987).

Second, legal scholars cite this state practice as supporting their thesis that the international obligation to protect the marine environment includes ecosystem-based studies and monitoring (Mayda, 1985; Schneider, 1985).

Third, United Nations Resolutions and Reports detail the international obligation for such ecosystem-based science (Stockholm Declaration, 1972; UNEP, 1987).

Finally, and perhaps of greatest significance to the development of this new international rule, is the recent codification of marine law in the United Nations Convention on the Law of the Sea. As noted previously, except for some provisions not relevant to this discussion, the world community accepts the Convention as binding customary international law, or at least as best evidence of such law. UNCLOS establishes a "framework for the effective conservation and management of the marine environment... [which] has the consent and support of the organized world community" (Johnston, 1984).

In addition to providing for an ecosystem approach to marine management, that framework also requires science, assessment, and monitoring to implement that ecosystem approach. Nation-states in their exclusive economic zone are to acquire "the best scientific evidence" to assure "proper conservation and management measures" for living marine resources. This evidence must include analysis of the effects of harvesting on related species (Art. 57, UNCLOS, 1982).

Nation-states are to control their nationals and coordinate their activities when resources, and their associated species, occur within more than one zone, or within a zone and in the high seas, or totally within the high seas. Such coordination includes assuring that the "best scientific evidence" is obtained for conservation measures and that the interdependence of stocks is considered (Art. 119).

These resource management mandates, applicable to domestic activities, bilateral and multilateral arrangements, and activities by nationals on the high seas (Arts. 63–67, 117–119), are to be in accord with the general obligation of states to "protect and preserve the marine environment" (Art. 192). Thus, nation-states are to consider relevant environmental factors in their resource management assessments and decisions (Arts. 61(3), 119(1)(a)). They are to take such measures as are necessary to preserve ecosystems and the habitat of marine life (Art. 194(5)). Such measures shall include environmental assessment of risks and monitoring of risks and effects (Art. 204). Such assessment and monitoring is to be done directly by each nation-state and indirectly and cooperatively through international organizations (Arts. 200–201, 204)

International law is a unique, almost bizarre system. Aggressive support for a particular legal theory can become the basis for existence of that legal theory—whether through incorporation in domestic law, or pressure by legal scholars, or state practice or codification in multilateral agreements (Belsky, 1986).

An ecosystem management approach, based on adequate information assessment and monitoring, is clearly the preferred international policy. It is in accord with the scientific consensus. It builds on the specific provisions of the United Nations Convention on the Law of the Sea and on the policy of the world community expressed in that Convention that the ocean and its resources are to be protected to the maximum extent possible.

Moreover, it is being applied not only in domestic law but in bilateral and multilateral agreements. As I have indicated in earlier papers (Belsky, 1986, 1989), one implication of 200-mile exclusive economic zones is the opportunity for nation-states to apply enlightened environmental policies to this wider jurisdiction. Under UNCLOS, this enlightened policy should be an ecosystem approach. This policy necessarily includes a mandate for adequate research funding — to secure information and monitor activities, and to assure appropriate application of the ecosystem approach.

References

Belsky, M. H. 1986. Legal constraints and options for total ecosystem management of large marine ecosystems. In Variability and management of large marine ecosystems. pp. 241–261. Ed. by K. Sherman and L. Alexander. AAAS Selected Symposium 99, Westview Press, Boulder, CO. 319 pp.

Belsky, M. H. 1989. Developing an ecosystem management regime for large marine ecosystems. In Biomass yields and geography of large marine ecosystems. pp. 443–459. Ed. by K. Sherman and L. M. Alexander. AAAS Selected Symposium 111, Westview Press, Boulder, CO. 493 pp.

Bilder, R. B. 1980. International law and natural resources policies. Nat. Res. J. 20:451, 453.

Boczek, B. 1983. The protection of the Antarctic ecosystem: A study in international environmental law. Ocean Dev. Int. L. 13:347.

Boczek, B. 1986. The concept of regime and the protection of the marine environment. In Ocean yearbook 6. pp. 271–297. Ed. by E. Borgese, N. Ginsberg, J. Baylson, N. Dunning, and D. Dzurek. Univ. Chicago Press, Chicago.

Borgese, E. 1986. The future of the oceans: A report to the Club of Rome.

Carroll, J. E., and Mack, N. B. 1982. On living together in North America: Canada, the United States, and international environmental relations. Denver J. Int. L. and Policy 12:1.

Carroz, J. E. 1984. Institutional aspects of fishery management under the new regime of the oceans. San Diego L. Rev. 21:513.

Chapman, W. 1967. Fishery resources in offshore waters. In The law of the sea: Offshore boundaries and zones. pp. 87–105. Ed. by L. Alexander. Ohio State Univ. Press, Columbus. 321 pp.

Churchill, R. R., and Lowe, A. V. 1983. The Law of the Sea. Manchester Univ. Press. Dover, NH. 321 pp.

Comment. 1977. An environmental assessment of emerging international fisheries doctrine. Columbia J. Environ. L. 4:143.

Comment. 1984. Compensating private parties for transnational pollution injury. St. John's L. Rev. 58:528.

Convention for the Conservation of Antarctic Marine Living Resources [CCAMLR]. 1980. Done 7 May 1980. T.I.A.S. No. 8826. Reprinted in Int. Legal Materials 19:841.

Convention for the Protection of the Natural Resources and Environment of the South Pacific Region. 1987. November 25, 1986. Reprinted in Int. Legal Materials 26:38.

Copes, P. 1981. Marine fisheries management in Canada: Policy objectives and development constraints. In Center for Ocean Management Studies, Comparative Marine Policy. pp. 135–136. Praeger, Brooklyn, NY. 260 pp.

Creech, H. 1986. In search of an ocean information policy. In Ocean yearbook 6. pp. 15–28. Ed. by E. Borgese, N. Ginsberg, J. Baylson, N. Dunning, and D. Dzurek. Univ. Chicago Press, Chicago.

D'Amato, A., and Hargrove, J. 1975. An overview of the problem. In Who protects the ocean? pp. 1–35. Ed. by J. Hargrove. West Publishing Co., St. Paul, MN. 250 pp.

Farnell, J. 1981. EEC fisheries management policy. In Center for Ocean Management Studies, Comparative marine policy. pp. 137–144. Praeger, Brooklyn, NY. 260 pp.

Federal Interagency Committee on Ocean Pollution Research, Development & Monitoring. 1981. National marine pollution program plan.

Fishery Conservation and Management Act [FCMA] of 1976 (later retitled Magnuson Fishery and Conservation Management Act). 1976. Pub. Law No. 94–265. 90 Stat, 331 (1976)(codified in 16 U.S.C. Sections 1801–1882 (1976, 1982 and Supp. III 1985).

Freidheim, R. L. 1975. Ocean ecology and the world political system. In Who protects the ocean? pp. 151–190. Ed. by J. Hargrove. West Publishing Co., St. Paul, MN. 250 pp.

Gordon, W. 1981. Management of living marine

resources: Challenge of the future. *In* Center for Ocean Management Studies, Comparative marine policy. pp. 145–167. Praeger, Brooklyn, NY. 260 pp.

Handl, G. 1978. The principle of "equitable use" as applied to internationally shared natural resources: Its role in resolving potential international disputes over transboundary pollution. Rev. Belge De Droit Int. 14:40.

Handl, G. 1986. National uses of transboundary air resources: The international entitlement issue reconsidered. Nat. Res. J. 26:405.

Henkin, L., Pugh, R., Schachter, O., and Smit, H. 1980. International law — Cases and materials. West Publishing Co., St. Paul, MN. 1152 pp.

Hoban, T., and Brooks, R. 1987. Green justice: The environment and the courts. Westview Press, Boulder, CO.

Johnston, D. 1984. Conservation and management of the marine environment: Responsibilities and required initiatives in accordance with the 1982 U.N. Convention on the Law of the Sea. *In* Law of the Sea Institute, The developing order of the oceans. pp. 133–179. Ed. by R. Krueger and S. Riesenfeld. Law of the Sea Institute, Honolulu. 749 pp.

Kelsen, H. 1966. Principles of international law. pp. 454–456. Ed. by R. Tucker, 2nd edition. Rinehart, New York, NY. 461 pp.

Knight, H. 1975. International fisheries management: A background paper. *In* The future of international fisheries management. pp. 1–49. Ed. by H. Knight. West Publishing Co., St. Paul, MN. 253 pp.

Laughlin, T. 1987. The Antarctic treaty system as a conservation system. Paper presented at the Center for Oceans Law and Policy Seminar, "The Polar Regions."

Lie, U. 1985. Marine ecosystems: Research and management. *In* Managing the oceans: Resources, research, law. pp. 311–328. Ed. by J. Richardson. Lomond Publications, Inc., Mt. Airy, MD. 407 pp.

Lutz, R. 1976. The laws of environmental management: A comparative study. Am. J. Compar. L. 24:447.

MacRae, L. M. 1983. Customary international law and the United Nations Law of the Sea Treaty. California Western Int. L. J. 13:181.

Malone, J. 1984. Who needs the sea treaty? Foreign Policy 54:44.

Mayda, J. 1985. Environmental legislation in developing countries: Some parameters and constraints. Ecol. L. Q. 12:997.

Miller, K. 1983. The earth's living terrestrial resources: Managing their conservation. *In* Environmental protection: The international dimension. pp. 240–266. Ed. by D. Kay and H. Jacobson. Allanjeld, Osmun & Co., Totowa, NJ. 340 pp.

Murphy, J., and Belsky, M. H. 1980. OCS development: A new law and a new beginning. Coastal Zone Management J. 7:297.

National Environmental Policy Act [NEPA] of 1969. 1986. Public Law No. 91–90. 83 Stat. 852 (1969)(codified at 42 U.S.C. Sections 4321–4370.

National Oceanic and Atmospheric Administration [NOAA]. 1986. NOAA fishery management study.

Oxman, B. 1986. Antarctica and the new Law of the Sea. Cornell Int. L. J. 19:211.

Pardo, A. 1983. The Convention on the Law of the Sea: A preliminary appraisal. San Diego L. Rev. 20:489.

Pinto, M. 1985. Emerging concepts of the law of the sea: Some social and cultural impacts. *In* Managing the oceans: Resources, research, law. pp. 297–309. Ed. by J. Richardson. Lomond Publications, Inc., Mt. Airy, MD. 407 pp.

Restatement, Foreign Relations Law of the United States (Revised). Tentative Draft No. 6. 1985. Vols. 1 & 2. American Law Institute, Philadelphia.

Roe, R. 1987. Some thoughts on the management of interjurisdictional fisheries. *In* Proceedings, coastal states are ocean states. pp. 33–40. Center for the Study of Marine Policy, Newark, DE. 137 pp.

Schneider, J. 1985. State responsibility for environmental protection and preservation. *In* International law: A contemporary perspective. pp. 602–623. Ed. by R. Falk, F. Kratochwil, and S. Mendlovitz. Westview Press, Boulder, CO. 702 pp.

Sherman, K. 1986. Measuring strategies for monitoring and forecasting variability in large marine ecosystems. *In* Variability and management of large marine ecosystems. pp. 203–236. Ed. by K. Sherman and L. M. Alexander. AAAS Selected Symposium 99. Westview Press, Boulder, CO. 319 pp.

Smith, G. 1984. The United Nations and the environment: Sometimes a great notion? Texas Int. L. J. 19:335.

Sohn, L. B. 1984. Implications of the Law of the Sea Convention regarding the protection and preservation of the marine environment. *In* Law of the Sea Institute, The developing order of the oceans. pp. 103–116. Ed. by R. Krueger and S. Riesenfeld. Law of the Sea Institute, Honolulu. 749 pp.

Speranskaya, L. 1986. Marine environmental protection and freedom of navigation in international saw. *In* Ocean yearbook 6. pp. 197–202. Ed. by E. Borgese, N. Ginsberg, J. Baylson. N. Dunning, and D. Dzurek. Univ. Chicago Press, Chicago.

Steele, J. 1982. Strategies for marine pollution research. *In* Center for Ocean Management Studies, Impact of marine pollution on society. pp. 279–283. J. F. Bergin Publishers, Inc., South Hadley, MA. 313 pp.

Stockholm Declaration. 1972. Report of the U.N. Conference on the Human Environment. U.N. Doc. A/Conf. 48/14. *Reprinted in* Int. Legal Materials 11:1416.

United Nations Convention on the Law of the Sea [UNCLOS]. 1982. U.N. Doc. A/Conf. 62/121. *Reprinted in* Int. Legal Materials 21:1245.

United Nations. 1980. Draft World Charter for Nature. G.A. Res. 35/7, 35 U.N. GAOR Supp. (No. 48) at 14, U.N. Doc. A/35/48. *Reprinted in* Int. Legal Materials 20:462.

United Nations Environment Programme [UNEP]. 1987. Report of the Working Group of Experts on Environmental Law on its Second Session on Environmental Impact Assessment, UNEP.WG.152/4.

VanderZwaag, D. 1983. The fish feud. Lexington Books, Lexington, MA. 135 pp.

Vienna Convention on the Law of Treaties, U.N. 1969. Doc. A/Conf. 39/27. *Reprinted in* Int. Legal Materials 8:679.

Walsh, A. 1982. The political context. *In* Center for Ocean Management Studies, Impact of marine pollution on society. pp. 3–23. J. F. Bergin Publishers, Inc., South Hadley, MA. 313 pp.

Wilkinson, C., and Connor, D. 1983. The law of the Pacific salmon fishery: Conservation and allocation of a transboundary common property resource. Kansas L. Rev. 32:17.

Yuru, L. 1985. Amassing scientic knowledge to preserve the marine environment. *In* Managing the ocean: Resources, research, law. pp. 125–129. Ed. by J. Richardson. Lomond Publications, Inc., Mt. Airy, MD. 407 pp.

Index

Acoustic backscatter sensor, 89
Acoustic Doppler system, 88, 90
Acoustic samplers, multispectral, 122
Acoustic surveys, 55, 62–63, 195–196
Acoustic techniques, 88–90, 105
Acoustic tomography, 88
Adriatic fisheries, 179–180
Advection, 140, 148, 153, 156–157
Aggregation, 183
Agricultural contaminant inputs, 215
Airborne Oceanographic Lidar (AOL), 92
Air pressure, ocean current patterns and, 40
Algal assemblages, 60
Aliasing, 151, 153, 183
Alkaline phosphatase, 111, 113
Alleles, 107–108
Allozymes, 106–110
Amphipods, 13, 124
Amundsen Sea, 6
Anchoveta, 198
Anchovy
   annual catch, 214
   cormorant abundance and, 198
   mortality, 136, 138
   Peruvian catch, 222–223
   spawning, 92
   spawning biomass off California, 213
   upward trends, 207
Antarctic
   characteristic features, 5–6
   marine biomass variability, 215
   recruitment-related hypotheses, 139, 141
Antarctic Circumpolar Current, 5–6, 91
Antarctic Convergence, 6
Antarctic Peninsula, 3, 6, 9–10, 13–14
Antigens, population-specific, 113
Arafura Sea, 54–57, 59–61, 63
Arctic, 190–191
   Belgica Bank, 13
   Canadian, 221
   characteristic features, 5–6
   Northeast, recruitment-related hypotheses, 140
Arctic Basin, 15
Arctic Front, 6
Arctic Mediterranean, 5
Arctic Ocean, 6
Artedidraconids, 13
Aru Basin, 54–58, 60–62
Asteroids, 13
Atlantic Ocean, 190–191
Atlantic Shelf, primary production, 59, 63
Atmosphere-controlled systems, temporal scales, 172
Atmospheric-oceanic coupling, 40–41
Atolls, 47

Bahamas, 45
Balsfjord, 189, 196, 198–201
Baltic Sea, 140, 192–193, 210, 215
Banda Sea, 4, 54–63
Barents Sea, 20, 29–30, 32, 39, 140, 202, 215
Barrier reefs, 47
Batfish system, 90
Bathydraconids, 13
Bellingshausen Sea, 6
Benguela Current ecosystem, 141, 210, 213, 215
Benthos, 12–13, 16, 189
Bering Sea, Eastern, 189, 192, 197–199, 202–203
Bermuda, 45
Biodynamics, 69, 71–80
Biomass potential, 188–204
Biomass yields, 206–216
BIONESS, 88
Bio-optical profiling system (BOPS), 90–91
Birds, 14–15, 198, 203
Biscayne Bay, 136
Block Island, 125
Boobies, masked, 171
Boundaries, maritime, 222
Box models, 196–197
Bristol Bay, 189
Browns Bank, 221
Brown water, 8
Bryozoans, Weddell Sea, 5, 12–13
Buoyancy, 23, 31, 75

Calanoides acutus, 10–11
Calanoids, Banda Sea, 61
Calanus finmarchicus, 24, 26, 28, 32, 126, 128
Calanus pacificus, 114
Calanus propinquus, 10
California Current, 92–93, 140, 210, 213, 215
Cannibalism, 167, 189, 197–200
Cape Cod, 221
Cape Hatteras, 162
Capelin, 156–157, 159–161
Carbon dioxide, atmospheric, 167, 216
Caribbean Sea, 4, 44–51, 175
Causation, correlations and, 183–184
Cayman Trough, 44
Chaetognaths, 10–11, 124–125, 159
Channel catfish, 117
Channichthyids, 13
Chesapeake Bay, 110
Chlorophyll, 7–9, 170, 182
Chlorophyll a, 57–61, 88–90, 92
Chlorophyll b, 60
Ciguatera poisoning, 47
Cladocerans, 126, 175
Cloning, 111–112
Clupea harengus, 107, 162, 213

Clupeids, 106, 140
Clupeoids, population increases, 215
Coastal Zone Color Scanner (CZCS), 92
Cod
    Arcto-Norwegian, 3–4, 19–32
    Atlantic, 214–216
    biomass flow from, 200
    biomass fluctuations, 202
    directed fishery, 200
    egg mortality, 32
    fishing pressure, 202–203
    Georges Bank stocks, 132–133, 136–137, 139–149
    icelandic, 39–40
    keystone predator, 198–199
    Newfoundland fishery, 160
    Northeast Arctic, 20
    O-group, 29–30, 39–40
    postlarval, 29–30
    recruitment-related hypotheses, 140–141
    size at age, 38, 42
    warm-core ring activity and, 162
    West Greenland fishery, 36–42
    young, 29–30
Coherence spectra, 181–184
Colombian Basin, 44
Commission for the Conservation of Antarctic Marine Living Resources (CCAMLR), 3, 209, 227
Common Fisheries Policy, 222
Community studies, 180
Comparative ecosystem approach, 207
Complexity, biological, 184–185
Component models, 183–184
Conch, 47
Conductivity, temperature, depth (CTD) systems, 88–89
Conservation, 222
Continental shelf
    finfish biomasses, 192
    U.S. Northeast, 4, 140, 210, 213–216, 221
    U.S. Southeast, 141
Cooperative Investigations in the Caribbean Sea and Adjacent Areas (CICAR), 45, 51
Copepods
    cod and, 3
    feeding, 114, 123
    Georges Bank, 142
    image analyzer studies, 123–128
    nauplii, 24–29
    Reynolds numbers for, 86
    transport, 107
    Weddell Sea, 9–11, 13
Coral reefs, 46–47, 50–51
Corals, 4, 44–45, 50
Cormorants, Cape, 198
Correlations, causation and, 183–184
Counter, electronic gate (Coulter), 123
Crab, decorator, 189
Croaker, 214
Crown-of-thorns starfish, 210, 215

Ctenophores, Newfoundland, 159
Current meters, vector-measuring, 88
Current transport, 137, 140
Cyanobacteria, 60
Cyclones, 161, 172

Davis Strait, 39–40
Dawson-Lambton Glacier, 11
Denmark Strait, 40–41
Density dependence, 79–80, 133–137, 173, 180
Diadema, 4, 44, 50, 175
Diatoms, 8, 60
Diet, 193–194, 196
Diffusion, 153, 156, 182
Dimensionality, reduction of, 69, 184–185
Discriminant analysis, 122, 126, 128
Dissipative structures, 185
Dissolved organic carbon (DOC), 60
DNA, 110–111
Doubling times, 83–84
Drifters, 88, 90
Drogues, 88, 90
Dynamic analyses, 188
Dynamic simulation, 201–203

East Bering Sea, recruitment-related hypotheses, 141
Eastern Boundary Current, 56
East Greenland Current, 39
East Greenland Sea ecosystem, biomass variability, 215
East Greenland shelf, 13
East Wind Drift, 6, 10
Economics, 222–223
Ecosystems, 188–189
    boundaries, transport across, 189
    management unit, 188–189, 203–204
    models, 188, 199
    nontropical, diet variability, 193–194
    simulation, 188–189
    See also Large marine ecosystems
Eddies, 69
Eels, North American and European, 109
Eggs. *See* Fish eggs
Elephant Island, 14
El Niño, 50, 83, 199, 222–223
El Niño-Southern Oscillation (ENSO), 170
Emigration, 153, 156–157
Encounter models, 183
Encounter rate theory, 216
Energy
    dissipation, 172
    flow, 167, 199–200
    storage, 174–175
    transfer in food chain, 207–208
    turnover times, 176
Environment
    ecosystem yields and, 210–211, 213, 215
    genetics and, 108, 137
Environmental impact reviews, 228
Enzymes, digestive, 114

Epizootic disease, 4, 47–51
Error analysis, 199
Euphausia crystallorophias, 9, 11, 14
Euphausia superba, 9, 11–12, 14–15, 179
Euphausiids, 5–6, 9–15, 124–125
European Economic Community, 222
Evolutionary processes, scale, 171, 176–177
Exclusive Economic Zones, 206, 209, 222, 225, 227, 231
Exponential decay model, 154–155, 157–158
Extinction rates, periodicity, 170, 177

Fat reserves, 175
Fatty acids, cod larvae, 24
Feeding currents, 87
Filchner/Ronne Ice Shelf, 12–13
Filchner Trough, 11
Finfish, 47, 190–193, 197–199
Fish
    coral reef, 46–47
    critical period in early life stages, 134–135, 137–138
    demersal, 3, 13–14
    growth, 132–134
    growth hormone levels, 114–115
    hybrids, identification, 106–107
    juvenile, mortality of, 137, 147–148
    matriarchal lineages, 108
    migration, 209
    pelagic, 3, 14
    population, year-class strength and, 19
    salmonid, 114–116
    starvation, 115–116
    traps, 47
    variation, scale of, 171
Fish eggs
    abundance, 151, 153
    image analyzer studies, 124–125
    mortality, 32, 135, 137–138, 208
    mortality estimates, 151, 153–157
Fisheries
    Adriatic, 179–180
    artisanal, 44, 46–47
    Banda Sea, 54–63
    Caribbean, 44–51
    coralline reef, 46–47
    directed, 201
    Gulf of Mexico, 44, 46
    hierarchical classification, 221
    Jamaica, 47
    law, 225
    management, 44, 47, 51, 107
    mixed, geographical origin of, 107–109
    models, 179, 181–182, 214
    Newfoundland, 160
    Puerto Rico, 47
    reproductive potential and, 213, 215
    resources, jurisdiction, 222
    seals and, 199, 203
    spiny lobster, 46
    trawl, 201
    U.S., ecosystem strategy, 209–210
    West Greenland, 36–42
Fishing, selective, 214
Fish larvae
    density-dependent growth, 136–137
    ecology and recruitment processes, 151–162
    image analyzer studies, 124–125
    minute, identification of, 113
    mortality, 19, 137, 139–149, 208
    mortality estimates, 151, 153–157
    predators of, 158–159
    settling behavior models, 183
    theory of survival, 215–216
    Weddell Sea, 9
Fish stocks
    cycles of abundance, 176
    exclusive economic zones and, 225
    geographical distribution, 189
    Georges Bank, 132, 139–148
    identification, 107, 110
    periodic recruitment, 170
    size estimates, 179
    transient, perturbations by, 189
Flatfish, 200
Flips, biomass dominance, 212–213
Fluid particle trajectories, 99–101
Fluorescence, 181–182
Fluorescent age pigment, 127–130
Fluorometers, in situ, 88
Food chain, energy transfer, 207–208
Food signals, 74, 76–80
Food webs, 6, 15, 114, 180
Foraging, optimal, 183
Fringing reefs, 47
Fronts, 172
Fucoxanthin, 60
Fundulus heteroclitus, 108–109, 117
Fylla Bank, 37, 39

Galpagos Archipelago, 170
Gametogenic capacity, 116
Gene flow, 107–109, 175
Gene frequencies, environmental variables and, 108
Gene probes, signature, 110
Genetic drift, 108
Genetic exchange, 107
Genetic markers, 107
Genetics, 69, 137
Genetic variation, 73, 105–111
Geography, 220–223
Georges Bank, 167
    cod and haddock stocks, 132–133, 136–137, 139–149
    ecosystem model, 199
    herring, 223
    keystone predator, 198–199
    overfishing effects, 199–200
    U.S./Canadian boundary, 222
Global warming, 215
^{14}C-Glycine, incorporation in fish scales, 115

Gonadosomatic index, 116
Grand Bank, 161–162, 221
Grazers, Weddell Sea, 13
Great Barrier Reef, 140, 210, 215
Great South Channel, 221
Greenhouse effect, 167
Greenland, climate, 36–37
Gridded models, 196–197
Growth, 115–116, 132–134, 172
Growth hormone levels, 115
Growth-mortality interaction, 134
Growth-predation interaction, 136
Gulf of Alaska, recruitment-related hypotheses, 141
Gulf of Carpentaria, 61
Gulf of Maine, 167
Gulf of Maine case, 221
Gulf of Mexico, 44–46, 93, 141, 161
Gulf Stream, 91–92, 162
Gulf of Thailand, 63, 140, 210, 213–215, 222
Gyres, 5–6, 69, 142

Habitat selection, 183
Haddock
    Georges Bank stocks, 132–133, 136–137, 139–149
    recruitment, 215
    recruitment-related hypotheses, 140–141
    West Greenland, 40
Hairtail, annual catch, 214
Hake, 162, 198–199, 213–214
Halibut/turbot complex, 202
Halley Bay, 6
Hardy Continuous Plankton Recorder, 89
Hawaiian bonefish, cryptic species, 106
Heat-shock proteins, 116–117
Herring
    Atlantic, 213–214
    biomass flows to, 200
    cod egg mortality and, 32
    drift pattern and spawning origin, 107
    Georges Bank, 223
    indigenous and migratory stocks, 189
    mortality estimates, 156
    mtDNA analysis, 110
    Pacific, 162, 214
    recruitment-related hypotheses, 140
19′-Hexanoylfucoxanthin, 60
High dimensionality, 71–80
Hokkaido, 211–212
Holographic systems, 89
Holothurians, 5
Horse-mackerel, 213
Horseshoe crab, mtDNA analysis, 110
hsp70 gene complex, evolutionary conservation, 117
Humboldt Current ecosystem, 210, 212, 215
Humboldt/Peru Current, recruitment-related hypotheses, 140
Hurricane Allen, fish mortalities and, 50
Hurricanes, 50–51, 161, 172

Hybridization
    fish, 106
    in situ, 111
    nucleic acid, 110–111
    probes, 113
Hydrocarbons, offshore, 225
Hydrography changes, biological effects, 54–57
Hypotheses, scale of, 151

Iberian Coastal ecosystem, 210, 213, 215
Iberian Peninsula, recruitment-related hypotheses, 140
Ice algae, 5–6, 8–9, 11–12, 15
Iceland cod, 39–42
Iceland Sea, 41
Ichthyofaunal regions, 190–192
Image analysis, 105, 122–130
Image identification, 69
Immigration, 153, 155–157, 192
Immunochemical methods, 113
Indian Ocean, 63, 190–191
Input-output analysis, 188, 200
Insect populations, 172
In situ sampling, 87–91
International Council for the Exploration of the Sea (ICES), 207
Intertidal studies, 180
Inverse problems, 72–73, 80, 184
Irminger Current, 39
Irminger Sea, 4, 36, 40
Irminger Water, 40–41
Irradiance, predator-prey relations and, 74–75
Isozymes, 104–108, 110

Jamaica, fishery, 47
Japan, fish catch, 211–212
Jellyfish, Newfoundland, 159
Jurisdiction, 167, 222, 226
Juveniles, fish, mortality, 137, 147–148

Kiel Sea Rover, 90
Kolmogorov scales, 99–100
Krill
    aggregation, 183
    Antarctic, 6, 127–129, 167, 179, 215
    distribution, 181
    triglycerides in, 9
    Weddell Sea, 5, 9–12, 16
Kuroshio Current ecosystem, 210–211, 215
Kuroshio/Oyashio Current, recruitment-related hypotheses, 140
Lactate dehydrogenase locus Ldh-B, 108
Lagrangian models, 93–94, 99–100, 183
Langmuir circulation, 31, 69
Large marine ecosystems
    biomass potential, 188–204
    biomass yields, 206–216
    comparative studies, 223
    demographic studies, 122–130
    early theoretical approaches, 180
    growth, survival, and recruitment, 132–149

hierarchical classification, 167
hypothesis testing, 151
interactions within, 221–223
law in management of, 224–231
locational component, 220–221
overlap, 167, 221
yield controls, 209–216
Larvae. *See* Fish larvae; Planktonic larvae
Law, 167, 222–231
Law of the Sea, 208
Leptocephalus larvae, 109
Leucine amino peptidase (*Lap*) locus, 108
Light-imaging systems, 89
Light intensity, larval feeding and, 25
Limit cycles, 173
Linkages, tight, 185
Lipids, 9, 114, 175
Lipovitellin, 116
Lizardfish, cryptic species, 106
Lofoten spawning grounds, 3, 20–21, 23, 25–28, 32
Logistic growth, 172
Lotka-Volterra interaction, 173, 184

Mackerel, 106, 213–214
Malaysia, 222
Mammals, 14–15, 198, 203
Management, 167
  cooperative strategies, 222
  fisheries, 44, 47, 51, 107
  ocean, 167, 188
  total ecosystem, 207, 224
  units, 122, 130, 206
Markers, chemical, 111
McMurdo Sound, 8
Menhaden, 161–162
Mesocosm, 3, 142, 145
Mesoscale, 69
Micronekton, 61–63, 88–89
Mid-Atlantic Bight, 94, 162
Migration, 182, 209
Minisatellites, 110
Mitochondrial DNA (mtDNA), 108–110
MOCNESS, 88
Modeling, 93–95, 99–100
Models
  box, 196–197
  component, 183–184
  diagnostic, 93
  ecosystem, 188, 199
  encounter, 183
  Eulerian, 93, 99–100
  exponential decay, 154–155, 157–158
  gridded, 196–197
  krill fishery, 179, 181–182
  Lagrangian, 93–94, 99–100, 183
  large-scale, 93–95
  larval settling, 183
  mesoscale, 94
  multispecies, 214
  North Sea, 184

ocean circulation, 93, 95
patchiness, 182, 185
predator-prey, 179–185
predictive, 93, 201
predictive capabilities, 151
random walk, 100
reaction-transport, 183
recruitment, 160–161
small-scale, 94–95
stochastic growth, 183
turbulence, 99–103
Mollusks, genetic variation and growth rate, 105
Monoclonal antibodies, 113
Monsoons, 4, 54–57, 63
Monte Carlo analysis, 199
Mortality
  age and, 136
  anchovy, 136, 138
  between life stages, 104
  density-dependent, 134–138
  egg, 32, 135, 137–138, 208
  estimates, 151, 153–157
  fishing, 4, 193–195, 197–198, 213, 215
  growth and, 134
  harvest-induced, 107, 135, 139, 198
  juvenile, 137, 147–148
  larval, 19, 137, 139–149, 156–157, 208
  major sources, knowledge of, 189
  pollock, 197–198
  predation and, 146–149, 157–158
  prerecruit, 132–134
  sampling intervals and, 154–157
  sea urchin, 50
  severe storms and, 145–146
  starvation and, 147–148, 157
  total, 188
Multifrequency Acoustic Profiling System (MAPS), 89–90
Multispecies analysis, 167, 188, 192–197, 199–202
Multispecies model, North Sea, 214
Multivariable moored system (MVMS), 90
Multivariable profiler (MVP), 90
Mussel, *Lap* locus, 108

Namibia, 198, 213
Narragansett Bay, 125–126, 221
National sovereignty, 224
Nauplii, 24–29, 31
Net systems, 88–89
Newfoundland, 159
Nitrate concentration, Banda Sea, 57, 59
Nitrate enrichment, 215
Noosphere, 189
North Atlantic, circulation patterns, 40–41
North Atlantic Current, 39
Northeast Channel, 221
North Pacific, turbulence scales, 85
North Sea, 4, 107
  fishing mortality, 198
  fishing patterns, long-term effects, 202

jurisdiction, 222
keystone predator, 199
models, 184
multispecies analysis, 192–193
multispecies fisheries model, 214
population declines, 222
primary production, 59, 63
recruitment-related hypotheses, 140
sand eel population, 213
Norwegian coastal current, 20
Norwegian Sea, 3, 19–32, 215–216
Nototheniids, 13–14
Nucleic acid hybridization, 110–111
Nutrients, 56–57, 182
Nutritional status, 113–115

Ocean-controlled systems, temporal scales, 172
Ocean, natural regions of, 190–191
Oceans commons, 226–227
Ocean Topography Experiment (TOPEX), 91
Ophiurids, 13
Ostracods, Weddell Sea, 11
Overfishing, 199–200, 210, 213–215
Oyashio Current ecosystem, 210–211, 215
Oyashio-Kuroshio confluence, 210–212
Oysters, cryptic species, 106
Ozone depletion, 215–216

Pacific, Insular, recruitment-related hypotheses, 139, 141
Pacific Ocean, 190–191
Pacific rim, biomass decline, 213–214
Pack ice, 6–9, 11–12, 15
Paracel Islands, 222
Particle counters, 89
Particulate organic carbon (POC), Banda Sea, 60
Patchiness
   larval, 155–156
   models, 182, 185
   nutrient, 182
   plankton, 84, 94
Pelagic fish, 3, 6, 14
Penguins, 5, 14–16
Permanent ice zone, 6–7, 9, 16
Petroleum pollution, Caribbean, 50
Phase statistics, 90
Phosvitine, 116
Photosynthetically available radiation (PAR) systems, 88
Phylogenetic analysis, 110, 112–113
Physical-oceanographic variability, 73
Physiological stress assessment, 116–117
Phytoplankton
   abundance measurement, 88
   distribution, 182
   fraction used directly, 192
   size-frequency spectra, 123
   time scales, 83–84
   variation, scale of, 171–172, 175
   Weddell Sea, 5, 7–9
Pigment concentrations, 92

Pilchard, biomass estimates, 213
Plankton
   Banda Sea, 57–61
   buoyancy and, 31
   contact rates, turbulence and, 99–103
   patches, 84, 94
   polar sea, 5–6
Planktonic larvae, settling, 170
Platforms, observational, 87, 91
Pleuragramma antarcticum, 5, 11, 14–16
Poikilotherms, 116
Pollock
   cannibalism, 189, 197–199
   fishing pressure, 202–203
   predation, 189
   walleye, 141, 197–199
   warm-core ring activity and, 162
Pollution, 140
   coastal and marine, 167, 225
   ecosystem yield and, 210–211
   resistance to, 105
   transnational, 225
Polychaetes, 13
Polyclonal antibodies, 113
Polymerase chain reaction, 110–113
Polynyas, 7, 15
Population
   dynamics, 71–74, 80
   fluctuation, 201
   oscillations, 175–176
   patterns, scale of, 171
Prawn, 200, 214
Predation
   density dependent, 215
   dominant role, 198–202
   ecosystem yield and, 210–211, 213–216
   mortality and, 146–149, 157–158
   patterns from stomach content analysis, 193–195, 203
   recruitment regulation by, 136–137
Predator, 192, 198–199
Predator-prey contact rate, 78, 99, 101–103
Predator-prey relations, 71–80, 136
   field example, 158–160
   modeling, 179–185
   simulation, 157–158
   turbulence and, 85–87, 94
Predictive models, 201
Prey density, cod larval feeding and, 26
Prey-switching algorithm, 193
Primary production
   Banda Sea, 58–59, 63
   correlation with physical factors, 181
   estimates, 208
   finfish biomass and, 190
   ocean color data and, 92
Protozooplankton, Weddell Sea, 8
Providence River, 125
Prydz Bay, 127
Prymnesiophycea, 60
Pteropods, 13, 124

Puerto Rican Trench, 44
Puerto Rico, fishery, 47
Pump-based systems, 89
Punta Cevallos, Española, 171

Queuing theory, 215

Radar, side-looking, 92
Radiometer sensors, 92
Radio tracking, 170–171
Rainbow trout, mtDNA analysis, 110
Random processes, 174
Random walk, 100–102
Reaction-transport models, 183
Recruit, 104
Recruitment, 82, 104, 132
    Arcto-Norwegian cod, 19–32
    Atlantic cod and haddock, 215
    forecasting, 216
    genetic variation and, 105
    hypotheses, 139–142
    large marine ecosystem, 132–149
    larval, 151–162
    model, 160–161
    molecular techniques applied to, 104–117
    Norwegian Sea, 19–32
    overfishing and, 213–214
    periodicities, 170
    physical-optical-biological scales, 82–95
    predation and, 136–137
    processes, 133–137
    starvation and, 136
    temperature and, 32
    variability, 132, 151–152
Red drum, 161
Redfish, warm-core ring activity and, 162
Regressions, multiple, 183
Release/recapture studies, 107
Remote sensing, 87–88, 91–93
Reporter systems, 113
Reproductive potential, fisheries-imposed loss, 213, 215
Reproductive status, 113, 116
Resource surveys, information from, 195–196
Restriction length polymorphism, mtDNA, 109–111
Restriction maps, mtDNA, 108
Reynolds number, 86, 94
Rhode Island Sound, 125
Ricker-Foerster hypothesis, 134, 136
Rings, warm core, 145, 162
RNA/DNA ratio, 104, 115–116, 145
RNA sequencing, 112–113
Ross Sea, 12–13
R/P FLIP, 91
RV *Polarstern*, 3, 5, 7, 12
RV *Samudera*, 55
RV *Tenggiri*, 55, 62
RV *Tyro*, 55, 62–63

Salinity, growth and, 139, 142–144
Salmon, 105, 110, 189
Salmonid fishes, 114, 116
Salps, 10
Sampling, 152–157, 207
Sand eel, population explosion, 213–214
Sand lance, 223
Sardine, 140, 207, 211–214
Sargasso Sea, 109
Satellite infrared imagery, 92–93
Satellite oceanographic observations, 90–91
Scale
    generation, 172–174
    patterns and processes, 169–177
    physical and biological, 179–185
    problem of, 69
    relationships, 151–162
    time and space, 82–87, 91–92, 95, 151–153, 169–172, 181, 185, 196, 207
    variation, statistical analysis of, 183
Scatterometers, 92
Schlieren video systems, 89
Scotia Sea, fish community, 13
Scotia Shelf, 85, 141, 221
Seabirds, breeding productivity, 170
Sea of Japan, 222
Seals, 5, 14–16, 199, 203
Searching behavior, 183
Search theory, 183
Seas, polar, 5–6
SEASAT, 91
SEASAT-A Satellite Scatterometer (SASS), 93
Seasonal ice zone, 6–7, 12, 16
Sea urchin, 4, 44–45, 50
Seawater, kinematic viscosity, 99
Sensitivity analysis, 188, 199
Seymour Narrows, turbulence scales, 85
Shrimp, 202
SIBEX II, 127
Silhouette photography, 130
Single-species analysis, 188–189, 197–201, 204
Siphonophores, 10
Size-frequency distributions, 123, 125–127
Snellius II Expedition, 54–55, 63
Sole, yellowfin, fishing pressure, 202–203
South China Sea, 63, 222
Southern Ocean, 5, 8, 10, 12, 15
    krill fisheries model, 179, 181–182
South Georgia, 3, 14
South Orkneys, 3
South Shetlands, 3
Spatial variability, 196–197
Spawning, storms and, 161–162
Species
    abundance estimation, 189–190
    cryptic, 106
    identification, 105–113, 126
    interactions in ecosystem, 199–203
Spectral analysis, 82–83
Spectroradiometers, multiwavelength, 88
Spiny lobster, fishery, 46
Spitsbergen, 20
Sponges, Weddell Sea, 5, 12–13
Spratly Islands, 222

Squid, 15, 106
Starvation, 136, 147–148, 157
Static analyses, 188, 199–201
Stochastic growth models, 183
Stochastic variation, 174
Stockholm Conference on the Human Environment, 228
Store Hellefiske bank, 40
Storms, 145–146, 161–162
Straits of Florida, 45
Stratification, 75–76, 140–141, 144–146
Striped bass, 106, 109–110
Suspension feeders, Weddell Sea, 5, 13

Taxonomic identification, 122, 124–125
Temperature
    abundance changes and, 213, 215
    Caribbean surface water, 44
    C. finmarchicus spawning and, 28
    cod recruitment and, 32
    growth and, 137, 139, 142–144, 147
    predator-prey relations and, 77
    productivity correlations, 181–182
    time series, 4
    West Greenland cod fishery and, 36, 38, 40–42
Test, scale of, 151
Thailand, 222
Thermocline, 22, 146, 148
Time series, 82, 91
Time-space relationship, 171–172
Tornadoes, 172
Tourism, Caribbean, 4, 44, 51
Trade winds, 44
Transmissometers, beam, 88
Traps, fish, 47
Trawl survey, biomass estimates from, 195–196
Triglycerides, in polar marine animals, 9
Triotrophic principle, 134–135
Trophic scales, 83
Trophodynamics, 69, 139, 142
Tropical storm Charley, 161
Tuna, 63, 92–93
Turbulence
    cod eggs and, 22–23
    encounter rates and, 184
    fluorescence distribution and, 182
    model, 99–103
    plankton distribution and, 31–32, 99–103
    predator-prey relations and, 74–75, 77–80, 85–87
    scales, 99
Turbulent energy dissipation, 75–76, 78, 85, 99–102
Turbulent velocity field, 99–100
Turtles, sea, 47

Undulating oceanographic recorder, 90
United Nations
    Charter of Economic Rights and Duties of States, 226
    Conference on the Environment (1972), 226
    Convention on the Law of the Sea, 206, 227, 230–231
    Draft World Charter for Nature, 226–227
Upwelling, 4, 190
    larval survival and, 159
    monsoons and, 54–57, 63
    productivity and, 54, 215
    recruitment-related hypothesis, 140–141

Venezuelan Basin, 44
Vesterølen spawning grounds, 20–21
Vestfjorden, 24, 31
Vestkapp, 11
Viscosity, kinematic, 99–101
Vitellogenin, 116
Volterra equations, 180

Water flea, 169–170, 175
Water masses, 151, 160, 170
Waves, 22, 173
Wax esters, in polar marine animals, 9
Weddell-Scotia Confluence, 6
Weddell Sea, 3, 5–16
Westerlies, 40
West Greenland Sea ecosystem, 4, 36–42
Westwind Drift Region, 190–191
Whales, 6, 14, 160, 189
White perch, identification, 106
Wind, 21–22, 24, 75, 159–160
Witch flounder, isozyme analysis, 107
World Ocean Circulation Experiment (WOCE), 41
Wright's F_{st}, 107–108

Year-class strength, 152
    first feeding larval condition and, 25
    fish population size and, 19
    O-group indices and, 30
    prediction, 203
    regulators, 19–20, 161
    spawning biomass and, 208
    temperature and, 30, 32
    variation, 173–174
    West Greenland cod, 38–39
Yellow Sea ecosystem, 140, 210, 214–215
Yellowtail flounder, warm-core ring activity and, 162
Yucatan Basin, 44
Yucatan Channel, 44

Zeaxanthin, 60
Zooplankton
    abundance measurement, 88–89
    abundance peaks, 170
    Banda Sea, 60
    demographic studies, 122–130
    time scales, 83–84
    variation, 171, 175
    Weddell Sea, 9–12